THE WOMAN WHO SMASHED CODES

A True Story of Love, Spies,
and the Unlikely Heroine Who Outwitted
America's Enemies

JASON FAGONE

Grateful acknowledgment is made to the following for the use of the images that appear throughout the text: George C. Marshall Research Foundation (pages ix, 21, 37, 93, 110, 127, 168, 177, and 327); New York Public Library, Manuscripts and Archives Division (pages 33 and 45); Hathi Trust Digital Library (page 88); National Archives of the United Kingdom (page 223); and U.S. National Archives and Records Administration (page 249).

HarperCollins books may be purchased for educational, business, or sales promotional use. For information please e-mail the Special Markets Department at SPsales@harpercollins.com.

FIRST EDITION

Designed by Paula Russell Szafranski

Frontispiece © Jonathan Weiss/Shutterstock, Inc.

Library of Congress Cataloging-in-Publication Data has been applied for.

ISBN 978-0-06-243048-9

17 18 19 20 21   DIX/LSC   10 9 8 7 6 5 4 3 2 1

The king hath note of all that they intend,
By interception which they dream not of.
—SHAKESPEARE, *HENRY V*, 1599

Knowledge itself is power.
—FRANCIS BACON, *SACRED MEDITATIONS*, 1597

CONTENTS

Umol-huun   tah-tiyal

William   Frederick

yetel

Elizebeth   Smith   Friedman

Lay   ca-huunil   kubenbil   tech   same.
This   our   book   we   entrusted   you   a   while-ago.

Ti   manaan   apaclam-tz'a   lo   toon
It   not-being   you-return-give   it   us,

Epahal   ca-baat   tumen   ah-men.
Is-being-sharpened   our-axe   by   the   expert.

Prying Eyes

This is a love story.

In 1916, during the First World War, two young Americans met by chance on a mysterious and now-forgotten estate near Chicago. At first they seemed to have little in common. She was Elizebeth Smith, a Quaker schoolteacher who found joy in poetry. He was William Friedman, a Jewish plant biologist from a poor family. But they fell for each other. Within a year they were married. They went on to change history together, in ways that still mark our lives today. They taught themselves to be spies—of a new and vital kind.

What they learned to do, better than anyone in the world, was reveal the written secrets of others. They were codebreakers, people who solve secret messages without knowing the key. Puzzle solvers. In a time when there were only a handful of experienced codebreakers in the entire country, the two lovers became a sort of family codebreaking bureau, a husband-and-wife duo unlike any that existed before or has since. Computers didn't exist, so they used pencil, paper, and their brains.

Over the course of thirty years, while raising two children, Elizebeth and William Friedman unscrambled thousands of messages spanning two world wars, prying loose secrets about smuggling networks, gangsters, organized crime, foreign armies, and fascism. They also invented new techniques that transformed the science of secret writing, known as cryptology. Today the insights of this one couple lurk at the base of everything from huge government agencies to the smallest fluctuations of our online lives. And the Friedmans did it all despite having little to no training in mathematics. The basic unit of their life was not the equation but the word. At heart they were people who loved words—words kneaded and pulled and torn, words flipped and arranged in grids and squares and strips and in lines marching down the pale sheet of scratch paper.

In the decades since the Second World War, the husband, William Friedman, has become a revered figure to intelligence historians. He is called "the world's greatest cryptologist" by the eminent chronicler of secret writing David Kahn: "Singlehandedly," Kahn writes, "he made his country preeminent in his field." William Friedman is also widely considered to be the father of the National Security Agency, the part of the U.S. government that intercepts foreign communications and sifts them for information—"signals intelligence." He wrote the definitive textbooks that trained generations of NSA analysts who are still working today. In 1975 the agency named its main auditorium after William Friedman, at its headquarters in Fort Meade, Maryland, and a bronze bust of William's head still stands guard there, above a plaque that reads CRYPTOLOGIC PIONEER AND INVENTOR, FOUNDER OF THE SCIENCE OF MODERN AMERICAN CRYPTOLOGY.

Today Elizebeth isn't nearly as famous, despite her talent and contributions. Early on she worked side by side with William and collaborated on several of their groundbreaking scientific papers; she was considered by some of their friends to be the more brilliant of the pair; she ultimately carved out a spectacular career of her own; and by 1945 the government considered *both* Friedmans to be pioneers of their field. A then-secret document said of Elizebeth, "She and her husband are among the founders of American military cryptanalysis"—cryptanalysis is another word for codebreaking— and a federal prosecutor told the FBI that "Mrs. Friedman and her husband . . . are recognized as the leading authorities in the country." Yet in the canonical books about twentieth-century code-breaking, Elizebeth is treated as the dutiful, slightly colorful wife of a great man, a digression from the main narrative, if not a foot-note. Her victories are all but forgotten.

I started reading about the Friedmans in 2014, after Edward Snowden shocked the world by revealing that the NSA was gathering the phone records of millions of ordinary Americans. Curious to know more about Elizebeth, I found a brief bio on the website of a Virginia library, along with a set of pictures. There she was, a petite woman in a white dress, standing on a patch of grass almost one hundred years ago, skin porcelain, head cocked at the photographer, smiling and squinting slightly in what must have been a blinding sun.

The library held the Friedmans' personal papers. One morning I drove down to Virginia and asked the chief archivist to show me what Elizebeth had left. In the back of an office, he unlocked a solid gray metal door and an inner door of silver metal bars, led me into a darkened, humidity-controlled vault, and pointed to multiple shelves of gray archival boxes, twenty-two boxes in all. "We try to tell people that Elizebeth's stuff is amazing," the archivist said, but usually researchers want to see William's papers.

You get these moments sometimes as a journalist, if you're lucky. You hear a voice that bursts from a body or a page with beauty or urgency or insight. Elizebeth's boxes contained hundreds of her letters. Love letters. Letters to her kids *written in code.* Handwritten diaries. A partial, unpublished autobiography. I'm not a mathematician, and I'll never be an expert on codes and ciphers, but Elizebeth's descriptions of her work gave me a sense of what it must have felt like to be her—the excitement of solving the kind of puzzle that could save a life or nudge a war. She liked to say that codes are all around us: in children's report cards, in slang, in headlines and movies and songs. Codebreaking is about noticing and manipulating patterns. Humans do this without thinking. We're wired to see patterns. Codebreakers train them-selves to see more deeply.

As rich as Elizebeth's papers were, they struck me as incom-plete. The records trailed off around 1940. What was she doing in the Second World War? No one seemed to know.

It took me almost two years to find the answer. She spent the war catching Nazi spies, among other little-known feats. Working with an elite codebreaking unit that she founded in 1931 and collaborating closely with both British and U.S. intelligence, Elizebeth became a secret detective, a Sherlock Holmes on the trail of fascist agents infiltrating the Western Hemisphere. She tracked and exposed them, smashing the spy rings, ruining Nazi dreams.

In a broader sense, she filled gaps in agencies that weren't prepared for the battle of wits that now faced them, a pattern that repeated throughout her entire career. The FBI, the CIA, the NSA—to different degrees Elizebeth pressed her thumb into the clay of all these agencies when the clay was still wet. She helped to shape them and she battled them, too, a woman hammering herself into the history of what we now call the "intelligence community." But when powerful men started telling the story, they left her out of it. In 1945, Elizebeth's spy files were stamped with classification tags and entombed in government archives, and officials made her swear an oath of secrecy about her work in the war. So she had to sit silent and watch others seize credit for her accomplishments, particularly J. Edgar Hoover. A gifted salesman, Hoover successfully portrayed the FBI as the major hero in the Nazi spy hunt. Public gratitude flowed to Hoover, increasing his already considerable power, making him an American icon, virtually untouchable until his death in 1972.

It's not quite true that history is written by the winners. It's written by the best publicists on the winning team.

What follows is my attempt to put back together a puzzle that was fragmented by secrecy, sexism, and time. I relied on the Friedmans' letters and papers, declassified U.S. and British government files, Freedom of Information Act requests, and my own interviews. Anything between quotation marks in this book is from a letter or other primary source document.

In these files I found a story of a true American adventure. A young woman with no money or connections is hired during the First World War by a millionaire to probe an odd theory about the works of Shakespeare. Through the millionaire's sleight of hand and the urgencies of war, this eccentric literary project turns into a life-or-death hunt for actual enemy secrets, one that spawns a completely new science of codebreaking. The woman goes on in the 1930s to become one of the world's most famous codebreakers, a front-page celebrity, before the government recruits her for one of the most closely guarded missions of the Second World War. And through it all she serves as muse and colleague to her husband, a troubled genius who lays the foundation of modern surveillance.

All democracies ride the line between security and transparency, secrecy and disclosure. What do people have a right to know? What must stay secret and why? The Friedmans lived these tensions more deeply than most. Their journey took them to great heights in the service of their country—and also to the depths of paranoia, poverty, and madness.

Jason Fagone
PHILADELPHIA

Terminology: A Cheat Sheet

You don't need to know math to enjoy this book, just a bit of lingo.

A *code* is a fixed relationship between one set of symbols or ideas and another. It can be a very ordinary and everyday thing. Slang is code, emoji are code. Think of Paul Revere hanging lamps in the Boston steeple to signal the route of the British invasion: one if by land, two if by sea. That's a code.

A *cipher* is a rule for altering the letters in a message. Usually it involves a one-to-one exchange: one letter gets replaced with one other letter, or a digit. For instance, if A=B, B=C, and so on, SMASH becomes TNBTI.

A *cryptogram* is a catchall term for a string of garbled text, solution unknown. It can be generated by a code or a cipher.

You can think of codes and ciphers as different sorts of locks that protect words, like a padlock protects money in a safe. In this analogy, the security professionals who make the locks and the keys are called *cryptographers,* and the thieves who try to pick the locks without having the keys are called either *codebreakers* or *cryptanalysts,* two terms that mean exactly the same thing.

The broad science of codes and ciphers—making them, breaking them, studying them, writing about them—is *cryptology.*

At different points in their careers, Elizebeth and William were asked to make codes, and they were good at this task, but their most significant feats involved codebreaking. They snuck into vaults of text, sometimes alone, sometimes together, feeling for the click of the bolt. Their lives became a series of increasingly spectacular and improbable heists. They used science to steal truth.

PART I

RIVERBANK

1916–1920

CHAPTER I

Fabyan

Sixty years after she got her first job in codebreaking, when Elizebeth was an old woman, the National Security Agency sent a female representative to her apartment in Washington, D.C. The NSA woman had a tape recorder and a list of questions. Elizebeth suddenly craved a cigarette.

It had been several days since she smoked.

"Do you want a cigarette, by the way?" Elizebeth asked her guest, then realized she was all out.

"No, do you smoke?"

Elizebeth was embarrassed. "No, no!" Then she admitted that she did smoke and just didn't want a cigarette badly enough to leave the apartment.

The woman offered to go get some.

Oh, don't worry, Elizebeth said, the liquor store was two blocks away, it wasn't worth the trouble.

They started. The date was November 11, 1976, nine days after the election of Jimmy Carter. The wheels of the tape recorder spun. The agency was documenting Elizebeth's responses for its classified history files. The interviewer, an NSA linguist named Virginia Valaki, wanted to know about certain events in the development of American codebreaking and intelligence, particularly in the early days, before the NSA and the CIA existed, and the FBI was a mere embryo—these mighty empires that grew to shocking size from nothing at all, like planets from grains of dust, and not so long ago.

Elizebeth had never given an interview to the NSA. She had always been wary of the agency, for reasons the agency knew well—reasons woven into her story and into theirs. But the interviewer was kind and respectful, and Elizebeth was eighty-four years old, and what did anything matter anymore? So she got to talking.

Her recall was impressive. Only one or two questions gave her trouble. Other things she remembered perfectly but couldn't explain because the events remained mysterious in her own mind. "Nobody would believe it unless you had been there," she said, and laughed.

The interviewer returned again and again to the topic of Riverbank Laboratories, a bizarre institution now abandoned, a place that helped create the modern NSA but which the NSA knew little about. Elizebeth and her future husband, William Friedman, had lived there when they were young, between 1916 and 1920, when they discovered a series of techniques and patterns that changed cryptology forever. Valaki wanted to know: What in the world happened at Riverbank? And how did two know-nothings in their early twenties turn into the best codebreakers the United States had ever seen—seemingly overnight? "I'd be grateful for any information you can give on Riverbank," Valaki said. "You see, I don't know enough to . . . even to ask the first questions."

Over the course of several hours, Valaki kept pushing Elizebeth to peel back the layers of various Riverbank discoveries, to describe how the solution to puzzle A became new method B that pointed to the dawn of C, but Elizebeth lingered instead on descriptions of people and places. History had smoothed out all the weird edges. She figured she was the last person alive who might remember the crags of things, the moments of uncertainty and luck, the wild accelerations. The analyst asked about one particular scientific leap six different times; the old woman gave six slightly different answers, some meandering, some brief, including one that is written in the NSA transcript as "Hah! ((Laughs.))"

Toward the end of the conversation, Elizebeth asked if she had thought to tell the story of how she ended up at Riverbank in the first place, working for the man who built it, a man named George Fabyan. It was a story she had told a few times over the years, a memory outlined in black. Valaki said no, Elizebeth hadn't already told this part. "Well, I better give you that," Elizebeth said. "It's not only very, very amusing, but it's actually true syllable by syllable."

"Alright."

"You want me to do that now?" Elizebeth said.

"Absolutely."

The first time she saw George Fabyan, in June 1916, he was climb-ing out of a chauffeured limousine in front of the Newberry Library in Chicago, a tall stout man being expelled from the vehicle like a clog from a pipe.

She had gone to the library alone to look at a rare volume of Shakespeare and to ask if the librarians knew of any jobs in the literature or research fields. Within minutes, to her confusion and mystification, a limousine was pulling up to the curb.

Elizebeth Smith was twenty-three years old, five foot three, and between 110 and 120 pounds, with short dark-brown curls and hazel eyes. Her clothes gave her away as a country girl on an adventure. She wore a crisp gray dress of ribbed fabric, its white cuffs and high pilgrim collar imparting a severe appearance to her small body as she stood in the lobby and watched Fabyan through the library's glass front doors.

He entered and stormed toward her, a huge man with blazing blue eyes. His clothes were more haggard than Elizebeth would have expected for a person of his apparent wealth. He wore an enormous and slightly tattered cutaway coat and striped trousers. His mustache and beard were iron gray, and his uncombed hair was the same shade. His breath shook the hairs of his beard.

Fabyan approached. The height differential between them

was more than a foot; he dwarfed her across every dimension. With an abrupt motion he stepped closer, frowning. She had the impression of a windmill or a pyramid being tipped down over her.

"Will you come to Riverbank and spend the night with me?" Fabyan said.

Elizabeth didn't understand any part of this sentence. She didn't know what he meant by spending the night or what Riverbank was. She struggled for a response, finally stammering a few words. "Oh, sir, I don't have anything with me to spend the night away from my room."

"That's all right," Fabyan said. "We'll furnish you anything you want. Anything you need, we have it. Come on!"

Then, to her surprise, Fabyan grabbed Elizabeth under one elbow, practically lifting her by the arm. Her body stiffened in response. He marched her out of the library and swept her into the waiting limousine.

People often guessed that she was meek because she was small. She hated this, the assumption that she was harmless, ordinary. She despised her own last name for the same reason; it seemed to give people an excuse to forget her.

"The odious name of Smith," she called it once, in a diary she began keeping at age twenty. "It seems that when I am introduced to a stranger by this most meaningless of phrases, plain 'Miss Smith,' that I shall be forever in that stranger's estimation, eliminated from any category even approaching anything interesting or at all uncommon." There was nothing to be done: changing her name would cause horrendous insult to blood relations, and complaining provided no satisfaction, because whenever she did, people asked why she didn't just change her name, a response so "inanely disgusting" that it made her feel violent. "I feel like snipping out the tongues of any and all who indulge in such common, senseless, and inane pleasantries."

Her family members had never shared this fear of being ordinary. They were midwestern people of modest means, Quakers from Huntington, Indiana, a rural town known for its rock quarries. Her father, John Marion Smith, traced his lineage to an English Quaker who sailed to America in 1682 on the same boat as William Penn. In Huntington he worked as a farmer and served in local government as a Republican. ("My Indiana family," Elizabeth later wrote, "were hide-bound Republicans who had never under any instances voted for any other ticket.") Her mother, Sopha Strock, a housewife, delivered ten children to John, the first when she was only seventeen. One died in infancy; nine survived. Elizabeth was the last of the nine, and by the time she was born, on August 26, 1892, most of her brothers and sisters had already grown up and scattered. She got along with only two or three, particularly a sister named Edna, two years her elder, a practical girl who later married a dentist and moved to Detroit.

Sopha had decided to spell "Elizebeth" in a nonstandard way, with *ze* instead of the usual *za*, perhaps sensing that her ninth child named Smith would want something to set her apart in the world. But Elizebeth didn't need the hitch in her first name to know she was different. Prone to recurring fits of nausea that began in adolescence and plagued her for years, she had trouble sitting still and keeping her tongue. She clashed with her father, a pragmatic, stubborn man who ordered his children around and believed women should marry young. She questioned her parents' faith. John and Sopha, though not devout, were part of a Quaker community and believed what Quakers do: that war is wrong, silence concentrates goodness, and direct contact with God is possible. Elizebeth's God was more diffuse: "We call a lot of things luck that are but the outcome of our own bad endeavor," she wrote in the diary, "but there is undoubtedly something outside ourselves that sometimes wins for us, or loses, irrespective of ourselves. What is it? Is it God?"

Her father didn't want Elizebeth to go to college. She defied him and sent applications to multiple schools, vowing to pay her own tuition; a friend later recalled that she was full of "determination and energy to get a college education with no help or encouragement from her father." (John Smith did end up loaning her some money—at 4 percent interest.) After being rejected from

Swarthmore College in Pennsylvania, a top Quaker school, she settled on Wooster College in Ohio, studying Greek and English literature there between 1911 and 1913. Then her mother fell ill with cancer and Elizebeth transferred to another small liberal arts school, Hillsdale College in Michigan, to be closer to home. At both schools she earned tuition money as a seamstress for hire. Her dorm rooms were always cluttered with dresses in progress and stray ribbons of chiffon.

College took Elizebeth's innate tendency to doubt and gave it a structure, a justification. At Wooster and Hillsdale she discovered poetry and philosophy, two methods of exploring the unknown, two scalpels for carving up fact and thought. She studied the works of Shakespeare and Alfred, Lord Tennyson, carrying books of their poems and plays around campus, annotating and underlining the pages until the leaves separated from the bindings. A course on philosophy introduced her to a new hero, the Renaissance scholar Erasmus, who "believed in one aristocracy— the aristocracy of intellect," she wrote in a paper. "He had one faith—faith in the power of thought, in the supremacy of ideas." Elizebeth, a smart person from a working-class family, found this concept liberating: the measure of a person was her ideas, not her wealth or her command of religious texts. She wrote a poem about this epiphany:

> I sit stunned, nerveless, amid the ruins
> Of my fallen idols. The iconoclast Philosophy
> Has shattered for me
> My God . . .
> But through the confusing ruins, Faith, still hoping,
> Somehow raises her hands and bids me—
> Yearn on! Finally
> Through the mazes of error and doubt and mistrust
> You will come, weary heart
> To the final conclusion upon which you will build anew.
> You will find triumphant
> The Working Hypothesis,
> The Solid Rock.

In addition to the well-worn volumes of Shakespeare and Tennyson, she lugged her own diary from place to place, a book with a soft black binding that said "Record" on the cover in silver script. The round-cornered pages were lined. She wrote in wet black ink with a quill pen, in a slanted cursive hand that was not too beautiful, about the importance of choosing the right words for things, even if those words offended people. She didn't like it when she heard a friend say that a person who had died had "passed away" or that a staggering drunk at a party was "a bit indisposed." It was more important to be honest. "We glide over the offensiveness of names and calm down our consciences by eulogistic mellifluous terms, until our very moral senses are dulled," she wrote. "Let things be shown, let them come forth in their real colors, and humanity will not be so prone to a sin which is glossed over by a dainty public!"

Sometimes Elizebeth had trouble channeling these energies and frustrations into cogent work. Her professors found her intensely bright, yet unfocused and argumentative. More than one told her, she said, that "I have marvelous abilities, yet do not use them." A philosophy professor wrote on the back of her Erasmus paper, "Very suggestive, with lots of good ideas and phrases. Also novel. But the style is choppy and the ideas are not in proper sequence." Next to these words Elizebeth scribbled a defiant note, dismissing the criticism on the grounds that she had recently won second place in a state oratorical contest.

She found herself attracted to male artists. Attending a choral concert one night, "my musical heart was carried completely away by a baritone," she wrote. "He loved the very act of singing—it could be seen in his eyes, in his mouth, in his very hands, as they irrepressibly moved in half-gestures. It made me want to be able to sing well myself, so badly that—well, I just couldn't sit still with the desire of it." At Hillsdale she dated a poet named Harold Van Kirk, called Van by his friends, handsome and athletic. He typed French sonnets for her and later joined the army and moved to New York. Van's roommate, Carleton Brooks Miller, wooed her when the relationship with Van fell apart, urging Elizebeth to read James Branch Cabell's erotic science fiction novel *Jurgen* because "it reveals the naked man-soul as it is." Carleton joined the army, too, then became the minister of a Congregational church near the college, writing to Elizabeth a few years later that he was still looking for a mate.

When graduation came around in the spring of 1915, Elizabeth still felt like "a quivering, keenly alive, restless, mental question mark." She had no sense of where to go or what she wanted to do with her life. That fall she accepted a position as the substitute principal of a county high school twenty miles west of her childhood home. The landscape of small-town Indiana was depressingly familiar to Elizabeth, and while she enjoyed parts of her job—she taught classes in addition to running the school—she also felt trapped. For an educated American woman in 1915, teaching at the high school level or below was what you did. Almost 90 percent of professors at public universities were male; only 939 women in the country received master's degrees in 1915, and 62 women earned Ph.D.s. Elizabeth had arrived at the last stop on a dreary train. There was no path from teaching that led anywhere else she might want to go. A woman taught, had kids, retired, died.

All her life, Elizabeth assumed that her restlessness was a defect that adulthood would somehow remove. She had called it "this little, elusive, buried splinter" and hoped for it to be "pricked from my mind." But she was learning to see the splinter as a permanent piece of her, impossible to remove. "I am never quite so gleeful as when I am doing something labeled as an 'ought not.' Why is it? Am I abnormal? Why should something with a risk in it give me an exuberant feeling inside me? I don't know what it is unless it is that characteristic which makes so many people remark that I should have been born a man."

Wanting something more, and ready to take a risk, Elizabeth quit her job at the Indiana high school in the spring of 1916 and moved back in with her parents to think about what was next. She soon remembered how unpleasant it was to live with her father. She reached her limit and packed a suitcase in early June. Nervous, but forcing herself to be brave, she boarded a train for

Chicago, hoping to find a new job there, or at least a new direction.

That month, the war in Europe—the First World War, then called the "Great War"—was two years old. America had not yet joined the battle. Woodrow Wilson was finishing his first term as president and campaigning for reelection in November on a platform of peace. More than a thousand Republican delegates had just kicked off their national convention to nominate a challenger to Wilson. They were gathered in the same city that now lured Elizebeth: Chicago, the young capital of the Midwest, an upstart empire of stockyards and skyscrapers.

The scale of the city jangled her. Pedestrians brushed past each other on sidewalks that cut mazes through the downtown office buildings, banks, apartments, hotels. It rained most every day, a cold, miserable rain, sheets of fat, icy drops that saturated the wool coats of the political delegates and swamped the grass at the baseball parks, canceling Cubs and White Sox games. Elizebeth stayed in the apartment of a friend on the South Side and ventured out each morning in search of work, visiting job agencies and presenting her qualifications. She told the receptionists she would like to work in literature or research. She pictured herself at a desk in a room of desks, taking notes with a sharp pencil. Not clerical work but something that required the brain. The people at the job agencies said they were sorry, but they didn't have anything like that.

She had no other cards to play. No money or connections, no means of bending Chicago to her will. She felt small and anonymous. After a week she decided to return home.

Before boarding the train, though, she wanted to make one more stop in the city, at a place she had heard about, the Newberry Library, which owned a rare copy of the First Folio of William Shakespeare, a book whose backstory had intrigued her when she learned it in college. The Bard's plays were never collected and printed in one place during his lifetime, because the culture in which he worked, Queen Elizabeth I's England, revered the spoken word above the written. It wasn't until 1623, seven years after his death, when a group of admirers gathered thirty-six of Shake-

speare's comedies and tragedies in a single hefty volume that came
to be known as the First Folio. Simply publishing the book was a
radical act, a statement that the phrases of a playwright deserved
to be documented with the same care as the Gospels. A team of
London artisans produced about a thousand copies, each typeset
and bound by hand. Five men memorized portions of the plays to
help them set type faster, stacking metal letters one by one into
words and sentences.

Over the centuries, most of the copies were lost or destroyed.
The Newberry had one of the few on display in America. So, on
what she thought would be her final day in Chicago, Elizabeth
made her way to the library.

The library was an odd institution, created by a dead man's will and
a quirk of fate. A rich merchant named Walter Newberry died on
a steamship in 1868. The crew preserved his body for the remainder of the voyage in an empty rum cask before returning it to his
beloved city, where lawyers discovered that Newberry had left
behind almost $2,150,000 for the construction of a public library.

According to his will, the library had to be free to use, and it
had to be located in North Chicago. These were the only conditions. The library didn't even have books to start with, because
three years after Newberry's death, his own hoard of rare volumes was destroyed in the Great Chicago Fire of 1871.

The slate of the library was blank. Now the library's trust-
ees wrote their status anxieties upon it. These were wealthy
Chicago businessmen who felt they lived in one of the finest
cities in the world and were painfully aware that the world
did not agree. For all of Chicago's sudden material success, its
skyscrapers and factories and department store empires and
slaughterhouses, it lacked the institutions of art and music and
science that elevated New York and Boston and Paris in traditional measures of civic greatness but omitted Chicago and
made its large men feel small.

They wished to prove that they were men of culture and re-
finement, and they were willing to spend whatever it took.
This was the same insecurity that drove the fathers of Chi-

cago to raise the dreamlike White City, the temporary pavilions of the Chicago World's Fair that soared along the southern edge of the lake in the summer of 1893. The White City exhibited the future in prototype, pieces of an unfinished puzzle. On August 26, 1893, a day of demonstrations at the Palace of Mechanical Arts, a building twice as large as the U.S. Capitol, the palace rumbled and whirred with machines that turned raw sugar into candy, made sausages and horseshoe nails and bricks, and sewed ten thousand button holes per hour. All day long one hundred thousand people wandered the sprawling aisles, eardrums split by machine roar, drinking lemonade that spurted from a fountain. An entire newspaper was printed in exactly sixty-three minutes starting with raw planks of wood that were pulped as people watched. "Everywhere was a demonstration of the almost irresistible power of mind when matter is set to do its bidding," the *Tribune* reported. The world's tallest man, Colonel H. C. Thurston of Texas, eight foot one and a half in boots, mingled with the throngs, and in the afternoon fifty thousand gathered outdoors to watch a fat man dive for a bologna sausage that dangled from a pole above the lake's lagoon.

This was the day Elizabeth turned one year old in Indiana. And as crowds of Americans roamed the White City in awe, builders completed construction of the Newberry Library, ten miles north of the noisy fairgrounds, and the first patrons entered the library in reverent silence.

Unlike the White City, a spectacle for the masses, the library was designed as "a select affair" for "the better and cleaner classes," the *Chicago Times* wrote with approval when the Newberry first opened. It was an imposing five-story building of tan granite blocks. All visitors had to fill out a slip stating the purpose of their research and they were turned away if they could not specify a topic. The books, available for reference only, were shelved in reading rooms modeled after the home libraries of wealthy gentlemen, cozy and intimate spaces containing the rarest and most sophisticated books that vulgar Chicago money could buy. During the library's first decades, the masters of the Newberry acquired books with the single-mindedness of hog merchants. They bought hundreds of incunabula, printed volumes from before 1501, written by monks. They bought fragile, faded books written by hand on unusual materials, on leather and wood and parchment and vellum. They bought mysterious books of disputed patrimony, books whose past lives they did not know and could not explain. One book on the Newberry's shelves featured Arabic script and a supple, leathery binding. Inside were two inscriptions. The first said that the book had been found "in the palace of the king of Delhi, September 21st, 1857," seven days after a mutiny. The second inscription said, "Bound in human skin."

In one especially significant transaction, the library acquired six thousand books from a Cincinnati hardware merchandiser, a haul that included a Fourth Folio of Shakespeare from 1685, a Second Folio from 1632, and most exceptional of all, the First Folio of 1623, the original printing of Shakespeare's plays.

This is the book that Elizebeth Smith was determined to see in June 1916, when she was twenty-three.

Opening the glass front door of the Newberry, she walked through a small vestibule into a magnificent Romanesque lobby. A librarian at a desk stopped her and sized her up. Normally Elizebeth would have been required to fill out the form with her research topic, but she had gotten lucky. The year 1916 happened to be the three hundredth anniversary of Shakespeare's death, and libraries around the country, including the Newberry, were mounting exhibitions in celebration.

Elizebeth said she was here to see the First Folio. The librarian said it was part of the exhibition and pointed to a room on the first floor, to the left. Elizebeth approached. The Folio was on display under glass.

The book was large and dense, about 13 inches tall and 8 inches wide, and almost dictionary-thick, running to nine hundred pages. The binding was red and made of highly polished goatskin, with a large grain. The pages had gilded edges. It was opened to a pair of pages in the front, the light gray paper tinged with yellow due to age. She saw an engraving of a man in an Elizabethan-era collar and jacket, his head mostly bald except for two neatly combed hanks of hair that ended at his ears. The text said:

MR. WILLIAM SHAKESPEARES
COMEDIES,
HISTORIES, &
TRAGEDIES.
Publifhed according to the True Originall Copies.
LONDON
Printed by Ifaac Iaggard, and Ed. Blount. 1623.

Elizebeth later wrote that seeing the Folio gave her the same feeling "that an archaeologist has, when he suddenly realizes that he has discovered a tomb of a great pharaoh."

One of the librarians, a young woman, must have noticed the expression of entrancement on her face, because now she walked over to Elizebeth and asked if she was interested in Shakespeare. They got to talking and realized they had a lot in common. The librarian had grown up in Richmond, Indiana, not far from Elizebeth's hometown, and they were both from Quaker families.

Elizebeth felt comfortable enough to mention that she was looking for a job in literature or research. "I would like something unusual," she said.

The librarian thought for a second. Yes, that reminded her of Mr. Fabyan. She pronounced the name with a long *a*, like "Faybe-yin."

Elizebeth had never heard the name, so the librarian explained. George Fabyan was a wealthy Chicago businessman who often visited the library to examine the First Folio. He said he believed the book contained secret messages written in cipher, and he had made it known that he wished to hire an assistant, preferably a "young, personable, attractive college graduate who knew English literature," to further this research. Would Elizebeth be interested in a position like that?

Elizebeth was too startled to know what to say.

"Shall I call him up?" the librarian asked.

"Well, yes, I wish you would, please," Elizebeth said.

The librarian went off for a few moments, then signaled to Elizebeth. Mr. Fabyan would be right over, she said.

Elizebeth thought: What?

Yes, Mr. Fabyan happened to be in Chicago today. He would be here any minute.

Sure enough, Fabyan soon arrived in his limousine. He burst into the library, asked Elizebeth the question that so bewildered and stunned her—"Will you come to Riverbank and spend the night with me?"—and led her by the arm to the waiting vehicle.

"This is Bert," he growled, nodding at his chauffeur, Bert Williams. Fabyan climbed in with Elizebeth in the back.

From the Newberry, the chauffeur drove them south and west for twenty blocks until they arrived at the soaring Roman columns of the Chicago & North Western Terminal, one of the busiest of the city's five railway stations. Fabyan hurried her out of the limo, up the steps, between the columns, and into the nine-hundred-foot-long train shed, a vast, darkened shaft of platforms and train cars and people rushing every which way. She asked Fabyan if she could send a message to her family at the telegraph office in the station, letting them know her whereabouts. Fabyan said no, that wasn't necessary, and there wasn't any time.

She followed him toward a Union Pacific car. Fabyan and Elizebeth climbed aboard at the back end. Fabyan walked her all the way to the front of the car and told her to sit in the frontmost seat, by the window. Then he went galumphing back through the car saying hello to the other passengers, seeming to recognize several, gossiping with them about this and that, and joking with the conductor in a matey voice while Elizebeth waited in her window seat and the train did not move. It sat there, and sat there, and sat there, and a bubble of panic suddenly popped in her stomach, the hot acid rising to her throat.

"Where am I?" she thought to herself. "*Who* am I? Where am I going? I may be on the other side of the world tonight." She wondered if she should get up, right that second, while Fabyan had his back turned, and run.

But she remained still until Fabyan had finished talking to the other passengers and came tramping back to the front of the car. He packed his big body into the seat opposite hers. She smiled at him, trying to be proper and polite, like she had been taught, and not wanting to offend a millionaire; she had grown up in modest enough circumstances to be wary of the rich and their power.

Then Fabyan did something she would remember all her life. He rocked forward, jabbed his reddened face to within inches of hers, fixed his blue eyes on her hazel ones, and thundered, loud enough for everyone in the car to hear, "Well, WHAT IN HELL DO YOU KNOW?"

Elizebeth leaned away from Fabyan and his question. It inflamed something stubborn in her. She turned her head away in a gesture of disrespect, resting her cheek against the window to create some distance. The pilgrim collar of her dress touched the cold glass. From that position she shot Fabyan a sphinxy, sidelong gaze.

"That remains, sir, for you to find out," she said.

It occurred to her afterward that this was the most immoral remark she had ever made in her life. Fabyan loved it. He leaned way back, making the seat squeak with his weight, and unloosed a great roaring laugh that slammed through the train car and caromed off the thin steel walls.

Then his facial muscles slackened into an expression clearly meant to convey deep thought, and as the train lurched forward, finally leaving the station, he began to talk of Shakespeare, the reason he had sought her out.

Hamlet, he said. *Julius Caesar, Romeo and Juliet, The Tempest,* the sonnets—the most famous written works in the world. Countless millions had read them, quoted them, memorized them, performed them, used pieces of them in everyday speech without even knowing. Yet all those readers had missed something. A hidden order, a secret of indescribable magnitude.

Out the train window, the grid of Chicago gave way to the silos and pale yellow vistas of the prairie. Each second she was getting pulled more deeply into the scheme of this stranger, destination unknown.

The First Folio, he continued. The Shakespeare book at the Newberry Library. It wasn't what it seemed. The words on the page, which appeared to be describing the wounds and treacheries of lovers and kings, in fact told a completely different story, a secret story, using an ingenious system of secret writing. The messages revealed that the author of the plays was not William Shakespeare. The true author, and the man who had concealed the messages, was in fact Francis Bacon, the pioneering scientist and philosopher-king of Elizabethan England.

Elizebeth looked at the rich man. She could tell he believed what he was saying.

Fabyan went on. He said that a brilliant female scholar who worked for him, Mrs. Elizabeth Wells Gallup, had already succeeded in unweaving the plays and isolating Bacon's hidden threads. But for reasons that would become clear, Mrs. Gallup needed an assistant with youthful energy and sharp eyes. This is why Fabyan wanted Elizebeth to join him and Mrs. Gallup at Riverbank—his private home, his 350-acre estate, but also so much more.

Genius scientists lived there, on his payroll, working in laboratories unlike any on earth. Celebrities made pilgrimages to get a glimpse of projects under way. Teddy Roosevelt, his personal friend. P. T. Barnum. Famous actresses. Riverbank was a place of wonders. She would see.

After they'd been riding west for ninety minutes or so, traveling thirty-five miles across the plain, the train began to slow, hissing as it came to a full stop. Fabyan opened the door and he and Elizebeth walked down the length of a platform and emerged into a handsome waiting room of dark enameled brick with terra-cotta flourishes. They continued out the front door, into the main street of Geneva, Illinois, a village of two thousand. Originally settled by a Pennsylvania whiskey distiller, Geneva had swelled with foreign immigrants in recent years, Irish and Italians and Swedes leaving crowded Chicago for the open spaces of the prairie. Whiskey still accounted for a good portion of Geneva's commerce, grain from the fields mixing with the sweet water of the Fox River, which bisected the town north to south.

To Elizebeth's amazement, a limousine was waiting for her at Geneva Station—not the one she'd ridden in an hour ago in Chicago but a second limousine with a second chauffeur. She climbed in with Fabyan and was carried south along a local road known as the Lincoln Highway for a bit more than a mile, until a long, high stone wall appeared to the left. Then a gate.

The limousine slowed. It pulled off the highway, to the right, across from the wall and the gate, and came to a stop in front of a two-story farmhouse with a wide front porch.

The Lodge, Fabyan announced. Elizebeth would be staying here tonight.

CHAPTER 2

Unbelievable, Yet It Was There

Elizebeth Smith and George Fabyan
at Riverbank, summer 1916.

A naked woman was living in a cottage at Riverbank. This was the story going around town in Geneva. The woman was said to be young, in her late teens or early twenties. Above the entrance of the cottage hung a sign that read *Fabyan*.

The story mutated as it passed from teller to teller. The cottage at Riverbank was stocked with attractive women, kept by Fabyan to satisfy his lust. They had been seen disrobing. Five women, ten.

Rumors about Fabyan and his strange laboratory were always spreading through the small farm towns surrounding the estate. The grounds were private and only open to the public at particular times. A stone wall protected part of the 350 acres, patrolled by Fabyan's guards, and at night the lighthouse on the island in the river broadcast a continual warning to intruders in code: two white flashes followed by three red ones, signifying "23-skidoo," meaning "keep out." Sometimes, on Sundays, he opened Riverbank to local residents as a gesture of goodwill, a benevolent king allowing his people to roam the castle grounds. The electric trolley operated by the Aurora-Elgin and Fox River Electric Company, which usually ran past the estate without stopping, was permitted to stop by the river. People tumbled out and wandered in awe through an elaborate Japanese Garden. And then Monday came, and the trolley did not stop at Riverbank anymore, and people once again had to guess from afar what might be happening there.

They heard loud noises from the direction of the estate, things that sounded like bombs exploding. They saw what looked like warplanes buzzing around the buildings and making an incredible racket. The press often called him "Colonel Fabyan" or simply "The Colonel." It seemed obvious that Fabyan was performing military research, but the townspeople did not know exactly what. They gleaned clues from newspapers and magazines. Fabyan was always inviting journalists and professors to tour the laboratories, under controlled conditions, and their reports spoke of Riverbank as a wonderland, a place almost beyond earthly reckoning. Visitors called Riverbank, variously:

A Garden of Eden on Fox River
Fabyan's colony
a wonder-working laboratory near Chicago
one of the strangest and, at the same time, most beautiful
    country estates in America

As for George Fabyan himself, visitors described him as:

one of the greatest cipher experts of the world
one who has achieved triple success in three distinct fields of
    activity, those of business, letters, and science
the man of a thousand interests
the lord and master
Chicago inventor
multi-millionaire country gentleman
the seer of Riverbank
the caliph on a grand scale

Guests of Riverbank went away telling two main types of stories. On the one hand, the visitors spread bizarre rumors and anecdotes of Fabyan's personal behaviors, portraying him as a mad king: "Credible persons," one newspaper reported, "say that a pair of sprightly, highly groomed zebras dash down with a station wagon to the Geneva station . . . to meet him mornings and evenings." These tales of bacchanalia were mingled with incredible stories of scientific experimentation at the laboratories, hints of anatomical investigations, and tales of secret cipher messages divined from old books.

Before he built the laboratories, Fabyan had often appeared in the Chicago newspapers in connection with more conventional tycoon activities: donations to political figures, board meetings of the stock exchange. People thought they knew his story. The black sheep of a prosperous New England family, he had dropped out of boarding school at age sixteen after repeated clashes with his father. He ran away from home and wandered the West for several years in the 1880s, making a living by selling lumber and railroad ties. Later, moving to Chicago, he reconciled with his father, and when the old man died, George inherited his $3 million fortune—equal to almost $100 million today—along with the reins of the family business, Bliss Fabyan & Company, one of the largest fabric companies in America. George used his gift for salesmanship to grow the company. After a Bliss Fabyan textile mill in Maine started making a type of striped seersucker cloth, he christened it

"Ripplette," a wonder fabric that required no ironing and resisted stain, undyed white bedspreads staying white after repeated washings, "white and clear as the driven snow . . . the name 'Ripplette' on a bedspread is the only sure indication of Ripplette quality. . . ."

Fabyan never claimed to be an altruist. "I ain't no angel," he said once, "and there are no angels in the New England cotton textile business, and if there are, they will all be broke." But in one part of his life he did strive toward some kind of greater good, and he wanted people to know it. In his free time, for his own amusement, he had made himself into a man of science. The steel magnates of Pittsburgh collected paintings, old and contemporary masterpieces. Newspaper tycoon William Randolph Hearst would soon build a 165-room castle in California full of marble statues. Fabyan was thinking bigger. "Some rich men go in for art collections, gay times on the Riviera, or extravagant living, but they all get satiated," he said. "That's why I stick to scientific experiments, spending money to discover valuable things that universities can't afford. You can never get sick of too much knowledge."

The atom had not been split in 1916. The structure of DNA was undescribed. There were no antibiotics. Aspirin, vitamins, blood types, and the medical uses of X-rays had all been discovered in the last twenty-one years. Einstein's theory of general relativity was only a year old. According to Einstein, space and time were one and the same, related by the universal force of gravity, and people did not know what to make of it. They came to Riverbank knowing that major scientific discoveries had emerged from the private laboratories of Thomas Edison and Nikola Tesla, so they were primed to believe that a new age of wonders was just over the next hill. And Fabyan gave them a peek. He paraded them from lab to lab, marvel to marvel, a former teen runaway and dropout showing off his Eden of science.

There seemed to be experiments happening everywhere, even inside his own house, known as the Villa. A man from the *Chicago Herald* was shocked to notice a swarm of bees flying through the Villa's open window. Fabyan laughed. "Those bees are just going into the music room to deposit their honey," he said. "You see I didn't trust that particular bunch of bees, so I had their hive placed inside the [Villa] and had it glassed in so we could watch them and see that they didn't cheat. . . . It's made honest bees out of them—this constant supervision."

A correspondent from the *Chicago Daily News* visited on a clear spring morning. Fabyan asked him, "Do you ever think? No, I don't think you do. Ninety-nine percent of the people don't, so why should you? I can make you think. We're all thinkers out here. Yes-siree, every one of the 150 souls in the Riverbank community." Fabyan, wearing a bowler hat, a lavender scarf, a tailored vest, and a frock coat with his French Légion d'Honneur rosette pinned to the lapel, added that he himself was "just a worker" like all the others, that there were no bosses at Riverbank, no time clocks, no iron regulations. Then he removed a gold-tipped cigarette from a cigarette case, snapped it in two, and lowered the halves toward a nearby monkey enclosure. The monkeys took the cigarette halves from Fabyan's outstretched hand, peeled off the paper, and jammed the tobacco into their mouths.

"Yes," Fabyan continued, "a community of thinkers." He took the reporter to the farm and explained how his scientists were taking cows and pigs and sheep and freezing them with rocks of ice and then slicing them thin as salami, to study their anatomy; he showed off the statues of the duck and the Egyptian thrones next to the Villa and pointed out with glee that they weren't made of marble or stone but of concrete, which lasts longer than stone and can be carved like stone; he pointed to the Dutch windmill and bragged that it was fully functional, that he used the mill to grind flour and bake fresh loaves of bread for the workers. He invited the reporter into the Acoustics Laboratory, built around an ultraquiet test chamber where the buzz of a stray mosquito seemed as loud as an air siren, and a pencil writing on paper sounded like a dozen people coughing. Fabyan said that experiments here would someday make cities more livable by eliminating the "racket ogre" of machines and crowds.

"Look through this telescope thing," boomed Colonel Fabyan proudly. He struck a tuning fork. The visitor squinted, and saw a flickering light, like a gas flame in a wind. "That's the sound made by this tuning fork! Sure, you're seeing it!"

And all through the tour, Fabyan kept circling back to the primary mission of the laboratories, the glimmering idea at the bottom of it all: immortality. Extending human life. Each person could live to be one hundred or more, he said. The thinkers of Riverbank had sequestered themselves in this lush, remote location to learn how not to die.

"Over there in that hothouse, they're trying genetics on nasturtiums, orchids, roses, and tulips," Fabyan told the Chicago correspondent, jabbing a finger in the direction of Riverbank's greenhouse. "What for? Why, look at the average human being. A mighty pitiful contraption of flesh and bones. If we of the Riverbank community can improve the human race by experimenting first with flowers and plants—say, won't that be a wonderful thing?"

Some experiments veered into ethically dubious territory. A journalist visiting from Philadelphia stopped and asked for directions from "a pretty girl, clad in blue overalls," with a "slim young figure—one of Colonel Fabyan's colony crowned with a head of bobbed blonde hair." Fabyan told the reporter that he had enrolled a number of young women in a series of studies to correct their defective posture. He recruited these human subjects from a boarding school adjacent to the Riverbank property, the Illinois State Training School for Delinquent and Dependent Girls at Geneva, really a low-security juvenile prison in the countryside, a place where judges across Illinois sent "wayward girls" deemed mentally deficient or sexually promiscuous. The founder of the Training School ordered the girls beaten with rawhide whips and thought society should force them to be sterilized: "When they begin to grow and attain some size the blood that runs in their veins will begin to tell and the incorrigible girl is the result." The school housed a rotating population of five hundred women ages ten to eighteen, and some lived in a cottage built with a donation from George and Nelle Fabyan. This was the cottage that townspeople gossiped about. The donation explained why the Fabyan sign hung above the door, and the posture experiments explained the rumor of nudity; the girls were required to undress for physical examinations. "The results of our experiments on the girls at Geneva have been marvelous," George Fabyan boasted. "Their so-called 'debutante slouch' has disappeared. They are learning to stand erect and not like anthropoid apes just learning to walk. I am trying to improve the human race, to discover what's wrong with the female figure. What will the next generation be like if all the women have hollow chests?"

The Philadelphia reporter also revealed that "in his effort to impress on the young women the terrors of crooked spines," Fabyan maintained a laboratory at Riverbank that he called "The Chamber of Horrors," containing actual human skeletons with grotesquely deformed spines, procured through methods that Fabyan never explained. Multiple Riverbank employees later told an Illinois historian that Fabyan "collected from hospitals and cemeteries numerous unclaimed cadavers that his scientists would radiate, cut, probe, and dissect, and then bury the remains in secret graves around the estate." At night in the laboratories "the beams would creak and the chairs would seem to move," and several staffers "recalled looking out the windows into the dark yard and seeing running girls with flowing white trains." Opinions about these visions differed: some staffers thought they were seeing "wayward girls" from the Training School escaping momentarily, and others believed in ghosts.

One summer a science journalist visited. Austin Lescarboura was a professional debunker, a man who had once partnered with Houdini to prove that fortune-tellers were liars and frauds. George Fabyan led Lescarboura into a darkened room in one of the laboratories. "The staff in charge moved about like so many Egyptian priests of old guarding the darkest secrets," Lescarboura later reported in *Scientific American*.

To deepen the mystery still further, a pretty girl was brought in. We were ushered into a small booth with dull black curtains for walls. It reminded us strongly of our psychic experiments back in New York, when we exposed one of the leading mediums after three sittings. At the command of the Colonel, the demonstration got under way. In a few minutes, we were astounded by what we were witnessing. It seemed unbelievable, yet it was there, in plain black and white. We had been brought face to face with certain facts regarding the human mechanism which we would hardly dared to have surmised in the absence of such a convincing demonstration. We were shown how—well, at this point we can go no further. Colonel Fabyan made us promise that nothing would be said about the nature of this investigation until some later date, when the experiments have progressed further.

What he was seeing was a woman standing behind an X-ray screen, the structure of her bones illuminated by the penetrating energies of $750,000 worth of radium. X-rays had been discovered in 1895, so they were hardly new technology by the time Lescarboura arrived at Riverbank, but the aura of mystique at Riverbank was so thick, the range of scientific experiments so wide, that even a trained skeptic like Lescarboura could not necessarily distinguish between the real and the fantastic. "Every so often the world reaches a point bordering on stagnation, because everything seems to be fully developed," he wrote. "But the scientist, pegging away at the secrets of nature, sooner or later breaks down existing barriers, opens the way to a new field, and we are soon confronted with brand new opportunities for exploration."

Twenty-three-year-old Elizebeth Smith climbed the stairs of the Lodge to the porch, opened the front door, and found herself in a warm, spacious drawing room. The walls were lined with double-paned casement windows that looked out across a grassy field on one side and back toward the road. There were people milling about.

In a brusque, hurried way, Fabyan introduced Elizebeth to a pair of magnificently dressed women, then disappeared, leaving her with these strangers.

The aristocratic appearance of the women was so incongruous that Elizebeth looked them over a few times to be sure they were real. They were sisters. The first, an older woman, wore a dark dress and a necklace that glittered with jewels. Her gray hair was tied in a bun and escaped in wisps that framed a delicate face. It seemed as if a French duchess had been teleported to the prairie, and her voice dripped with learning. Her name, she said, was Mrs. Elizabeth Wells Gallup. She ran the Riverbank Cipher School. The other woman was her younger, darker-haired sister, Miss Kate Wells.

Young Elizebeth gathered from this brief conversation that Mrs. Gallup and Miss Wells lived and worked here at Riverbank in this very building, which also contained quarters for the cooks and servants who fed and catered to the sisters and to the other scholars and scientists who worked on the estate.

The sisters informed Elizebeth that dinner would soon be served here at the Lodge, and that she would be dining with the two of them and some of the scientists as well. Mrs. Gallup and Miss Wells suggested that she head upstairs and freshen up in a certain spare bedroom where she would be sleeping. Elizebeth did as asked, and some minutes later, when she descended the stairs, she saw that Fabyan had returned, in striking new clothes: riding pants, a billowing shirt with a riding collar, and a big, broad cowboy hat. He looked ready to jump on a horse and gallop away. It didn't make sense to her at the time, given that he was about to eat dinner; later she would realize that Fabyan simply enjoyed dressing up as the ideal of a country squire. She would never once see him wearing a traditional business suit at Riverbank.

People began streaming into the Lodge in ones and twos, walking up the steps to the porch in the fading prairie light. Elizebeth sat on the bannister of the staircase, looking out across the Lincoln Highway, listening to crickets chirp and cicadas sing, watching the guests arrive. They all wore semiformal clothes with a country feel, except for a slim man in a pinstriped shirt and pants, a neat bow tie, and sparkling white buck shoes. He had short, dark hair parted in the exact middle of his head and pomaded to each side, and his ears were pointy; he seemed like the youngest of the arriving guests, and by far the best dressed, as if he were attending a society dinner at some mansion in the city. The young man reminded Elizebeth of Beau Brummell, the eighteenth- and nineteenth-century British fashion icon who polished his boots with champagne and peach marmalade.

They sat down to dinner at a long communal table covered with fine china and linens. Swedish and Danish servants appeared in crisp white uniforms and brought heaping plates of meat and vegetables from Riverbank's working farm. To keep food costs low and to satisfy his own taste for meat, Fabyan kept the farm stocked with chickens, ducks, sheep, and turkeys, and his wife, Nelle, bred prize-winning livestock here. Mrs. Gallup sat at the head of the table, flanked by the other guests, all of them lured here by Fabyan to investigate different pieces of the world. Elizebeth spoke little and tried to get a sense of who these people were and what they were doing here. A sweet-seeming man in his fifties introduced himself as J. A. Powell, president of the University of Chicago Press and the top public relations man at that university; his job there was to "cause the University of Chicago to be known as well in Peking as in Peoria," the *Tribune* once put it. Another dinner guest was Bert Eisenhour, Riverbank's chief engineer and builder of structures, a short man with a ruddy complexion who struck Elizebeth as a country bumpkin.

Then there was the well-dressed man with the white buck shoes. He smiled shyly at Elizebeth and introduced himself as William Friedman, head of the Genetics Department at Riverbank. He worked here studying seeds and plants, breeding new strains of corn, wheat, and other crops, trying to infuse them with desirable properties.

Altogether it was a curious bunch of characters. Elizebeth couldn't see an obvious thread that connected them. Literary scholars, an engineer, a geneticist. Perhaps Fabyan was the kind of rich person who collected people in addition to banknotes and stocks.

The dominant personality that night was Mrs. Gallup. As the smell of meat and the noise of clinking silverware filled the room, and the servants whisked away the empty plates, Mrs. Gallup told stories of her travels while researching Francis Bacon and Shakespeare, staying in the homes of wealthy patrons around the world, in France and in England, who believed in her theories and had sponsored her work. When she spoke about the details of her investigations and findings, no one interrupted her to ask skeptical questions. It was obvious to Elizebeth that people here were used to treating Mrs. Gallup with great deference, that she was an important person at Riverbank, and that dinner conversations like this had probably happened many times before, with Mrs. Gallup holding court and the others nodding and smiling. Elizebeth got the sense that "Mrs. Gallup had dwelt only among those who agreed with her premise and that she had little personal contact with the viewpoint of those who did not believe."

After dinner, the guests went their separate ways. Fabyan gave Elizebeth a set of men's pajamas to wear to bed, telling her they would have more to discuss in the morning, and wishing her good night. She went upstairs to her room and found that a pitcher of ice water had been left on her bedside table along with an enormous bowl of fresh fruit, plus knives to carve it up.

On Elizebeth's second day at Riverbank, after she woke in the Lodge and got dressed, Fabyan found her and said she ought to see the rest of the estate. He assigned an employee to give her a short tour.

Fifty or sixty yards along the highway from the Lodge was a smattering of buildings known collectively as Riverbank Laboratories, where many of the scientists worked. Elizebeth was told that a new laboratory was under construction for the study of sound waves, designed by the top acoustics expert in the country, Professor Wallace Sabine of Harvard University, who would move to Riverbank when the new lab was complete. Adjacent to Riverbank Laboratories was the ordnance building, a low concrete hut where Fabyan and several scientists tested bombs and mortars for potential use by the U.S. military.

Elizebeth wasn't shown inside these buildings but instead was led across the highway to an iron gate she had seen yesterday while riding in the limo. She walked through it. A short, curving driveway led down a gentle slope to Fabyan's personal residence, known as the Villa, a long, low two-story house in a cruciform shape with a heavy roof and thin clapboard siding that seemed to press the house downward into the hill. Originally a far smaller farmhouse, it had been expanded in 1907 by the famous architect Frank Lloyd Wright, who produced a mansion for Fabyan that looked like a peaceful part of the countryside. Strange objects dotted the lawn: a concrete wading pool, a concrete table and concrete semicircular bench carved with elaborate Egyptian hieroglyphs, a concrete chair whose front legs were sphinxes, and a concrete duck the size of a human eight-year-old.

Inside the Villa the walls were paneled with squares of dark walnut, and the sun shone through a series of thin slats that Wright's builders had carved in the hill-facing wall and that decorated the opposite wall with rhombuses of light. Elizabeth was amazed to see that the chairs and divans in the living room and drawing room and even the beds upstairs were suspended from the ceiling on chains—no chair or bed legs anywhere to be seen. She had no idea what to make of this. Fabyan and Nelle each had their own private bedchamber. It was unclear whether they slept in the same room. On a wide veranda that looked down to the river, another piece of furniture swung on chains, a wicker chair with arms woven from thick reeds.

Taxidermized animals stared out from walls and glass display cases inside the house, beasts that Fabyan or his wealthy friends had killed and stuffed: a buck, an alligator, a Gila monster, a shark, birds of all kinds (grouses, owls, hawks), and hundreds of bird eggs, speckled with blue, yellow, and pink. There was also a valuable work of art in the Villa that had once been displayed to millions at the White City: a life-size marble statue of a naked woman petting a lion, her right hand falling across the lion's mane, the lion looking calmly to the side. The statue was called *Diana and the Lion, or Intellect Dominating Brute Force.*

The mastery of Nature. This appeared to be Fabyan's preoccupation.

Back outside, Elizebeth walked down the steepening hill toward the Fox River until it leveled out a hundred yards from the water. A curving path took her through a Shinto arch of wood and into a pristine garden ringed by buildings, benches, and lanterns of Japanese design. She was told it had all been devised by one of the emperor's own personal gardeners. Flowering trees were aflame with pink and red and blue and orange blossoms that breathed their reflections onto a circular pool at the garden's center, the surface of the water like a painter's palette smeared with color. A half-moon footbridge spanned the pool. Every leaf and flower, every plank of wood and drop of water seemed designed for maximum tranquility, except for the low concrete structure to the right of the pool, shaped like a hexagon, protected by heavy black iron bars. It was a bear cage.

Fabyan kept two pet grizzly bears inside. Their names were Tom and Jerry.

The river lay beyond, a placid silver width flowing southward, from left to right, away from the center of Geneva. Elizebeth could see a small island in the middle of the river, connected to the near shore by two bridges, and on the far bank, an impressive Dutch windmill, a giant X spinning against the sky. It was explained to her that Fabyan had bought the windmill in Holland and transported it here, piece by piece.

Later that day, after the tour, Elizebeth sat down with Mrs. Gallup in the Lodge to discuss the work they might do together if Elizebeth were to accept the research position. They talked for two or three hours, not long enough for Elizebeth to grasp the full nature of the project but sufficient to get a sense of Mrs. Gallup's immediate needs and her personality.

Unlike Fabyan, Mrs. Gallup spoke in a restrained, careful manner, the tones of a scholar. There was nothing hucksterish about her at all. She illustrated her explanations with oversize sheets of paper that were curled up like scrolls. She rolled them out to their full length to show Elizebeth, and placed weights on the ends to prevent them from curling up again. The sheets were beautiful and full of hand-drawn letters of the alphabet in subtle variations, lowercase and uppercase, roman and italic:

Mrs. Gallup said she had drawn these letterforms from photographic enlargements of the Newberry Library's First Folio of Shakespeare, and the drawings had helped reveal the secrets that Francis Bacon had woven into Shakespeare's plays. In some way Elizebeth didn't understand yet, the hidden messages were embedded in the shapes of the letters themselves, in small variations between an *f* on one page of the Folio and an *f* on another.

According to Mrs. Gallup, she had already discovered these messages. She knew what they said; she was certain they existed. The problem, as she saw it, was that some literary experts disputed her method and doubted that the messages were really there. So Elizabeth's job at Riverbank would be twofold. First she would use Mrs. Gallup's method to reproduce her existing results, providing scientific confirmation and silencing the critics. Then Elizabeth would assist Mrs. Gallup with new investigations. Mrs. Gallup believed that in addition to authoring Shakespeare, Bacon also secretly wrote works commonly attributed to Christopher Marlowe, Ben Jonson, and other major figures of the age. Together Elizabeth and Mrs. Gallup would rewrite the history of seventeenth-century England—and by extension, the history of all English literature.

George Fabyan popped in briefly, to see how the women were getting on, rolling out one of Mrs. Gallup's scrolls and regarding it with evident satisfaction.

It was a lot for Elizabeth to process. Over the last twenty-four hours she had been accosted by an eccentric tycoon, dragged to his hall-of-wonders laboratory in the countryside, introduced to a merry team of scientific experts, told about a secret cipher embedded in the heart of the First Folio, and invited to assist them in turning history upside down.

That evening Elizabeth returned to her room in the Lodge to find a vase of flowers and another bowl of fresh fruit by the bed, filled to abundance. She lay awake for a time, thinking about all she had seen and heard, stunned and a bit jangled by the weirdness of Fabyan's kingdom, yet impressed with Mrs. Gallup's erudition and quiet confidence. To be sure, their theory was unconventional, but what if it was correct? If there was even a chance that Fabyan and Mrs. Gallup were on to something here, how could Elizabeth pass up a chance to join an effort of such magnitude? The next day, when she rode the Union Pacific back to Chicago, she was buzzing with "a mixture of astonishment, incredulity, and curiosity."

On the morning of June 7, around the time Elizabeth Smith had first arrived in the city to look for a job, five thousand women marched toward the Republican National Convention, being held at the Coliseum, to demand the right to vote. The wind and rain shoved the women this way and that by the handles of their increasingly useless umbrellas. Dye from their yellow sashes streamed down their legs. Reaching the convention hall, the women surged through the entryway. Water poured from straw hats, hems, and sleeves, pooling at their feet in a spreading puddle. Many held rain-blurred signs. "We want to be citizens. Do we look desirable?" The protesters demanded that the GOP support a constitutional amendment granting women the vote, but after debating the issue, the delegates decided that an amendment would violate "the right of each state to settle this question for itself."

Elizebeth Smith wasn't involved in the suffrage movement or any other. Her views on women's rights at age twenty-three were complicated. She idolized the suffrage pioneers but doubted that men would give up their power without a vicious fight. Earlier that year she had found herself riding a crowded bus when a heavyset woman looked straight at her and then used her rump to shove her out of the way. Elizebeth fumed in her diary, "No woman's rights was adequate to the situation then! I wanted genuine masculine title; if I had been a man she'd never have dared do it."

There in the city, she reviewed her options. Should she take the job that George Fabyan had offered, or go back home to Indiana? She couldn't decide. Fabyan scared her. But she had wanted an unusual job, and in all her life she had never seen a place so unusual as Fabyan's estate.

She was running out of clean clothes in her suitcase. She was out of time.

Elizebeth made her way to the Chicago & North Western train station. At the ticket counter she asked, in her firm, polite voice, for a fare to Geneva, Illinois.

When she arrived once again at Riverbank, Fabyan and Mrs. Gallup were glad to see her. They wasted no time, and began teaching her how to dive for what Francis Bacon had left behind: a sunken treasure of words, a ship of gold at the bottom of the sea.

CHAPTER 3

Bacon's Ghost

The Bacon-Shakespeare investigators at Riverbank, 1916.
Mrs. Gallup is seated in the front row on the left, and
Elizebeth Smith is third from the left standing in the middle row.

Mrs. Gallup had to know if her new assistant, this Elizebeth Smith, could be trained to see. So this is where they began—with a deciphering test.

In the Lodge, Mrs. Gallup placed several pages in front of Elizebeth. One was a worksheet of white paper with some typing

on it, eight lines of text from the Shakespeare Folio. The text was
broken into five-letter blocks:

```
TheWo rkeso fWill iamSh akesp earec ontai ninga llhis Comed
iesHi stori esand Trage diesT ruely setfo...
```

When Elizebeth read the text and skipped over the spaces, she
could make sense of it as English:

> The workes of William Shakespeare containing all his
> Comedies Histories and Tragedies Truely set...

She recognized these words from one of the early pages of the
First Folio, "The Names of the Principall Actors."

According to Mrs. Gallup, Francis Bacon had concealed a
message on this page. She already knew the secret, but needed to
know if Elizebeth could find it, too.

Mrs. Gallup always said that as a devout Christian she was appalled
when she first discovered the secret messages of Francis Bacon. She
did not traffic in such matters as Bacon discussed: deception, black-
mail, adultery, the insatiable lusts of queens and earls. "Surprise followed surprise," Mrs. Gallup wrote, "as the hidden messages were
disclosed, and disappointment as well was not infrequently encountered. Some of the disclosures are of a nature repugnant, in many
respects, to my very soul." However, her own moral beliefs were
irrelevant. "The sole question is—what are the facts? These cannot
be determined by slight and imperfect examinations, preconceived
ideas, abstract contemplation, or vigor of denunciation."

She was not the first person who claimed that Shakespeare was
really Francis Bacon in disguise. This idea, known as the "Baconian" theory, enjoyed broad appeal and made a certain sense. Francis Bacon and Shakespeare had lived in the same country in the same
era, the England of Queen Elizabeth I, and of the two men, Bacon
was by far the more distinguished, a child prodigy who graduated
from Trinity College at age fifteen, studied law, served in Parlia-
ment, became lord chancellor, won the lofty titles of Baron Verulam
and Viscount St. Albans, and wrote manifestoes that heralded the
dawn of the scientific age and inspired generations of inventors and

revolutionaries. Charles Darwin idolized Francis Bacon; Thomas Jefferson thought Bacon was among the two or three greatest men who ever lived; Teddy Roosevelt's love of Bacon's writings encouraged him to create America's system of national parks.

The radical idea that made Bacon a legend is one of the epigraphs of this book: "Knowledge itself is power" (his admirers often shortened it to "knowledge is power"). What people called science in Bacon's day was more like philosophy or logic: the thinking of beautiful thoughts. Bacon said no, science is about *physical evidence.* Knowledge is found not in the skull but in contact with Nature. And Bacon made it his mission to collect and classify all forms of knowledge, arguing that if enough knowledge was gathered and sorted and pinned to the page, there was nothing men could not achieve. In an unfinished utopian novel, *The New Atlantis,* Bacon imagined a lush, remote island ruled by superintelligent scientists. The people spend their days studying the native beasts and plants, running experiments in towers, caves, artificial lakes, and specially constructed laboratories. The island is a like a cross between a research university and a nature preserve, a place devoted to the investigation of light, acoustics, perfumes, engines, furnaces, mammals, fishes, flowers, seeds, geometry, illusions, deceptions, and, above all, methods of extending human life, of achieving immortality. He thought humans might learn to live forever, be immortal, become like gods.

All in all, Bacon was such an impressive person that it seemed perfectly plausible to writers and scholars in the late nineteenth and early twentieth centuries that Bacon might have written Shakespeare's plays under a pseudonym. Mark Twain believed it. So did Nathaniel Hawthorne. Was there proof? Had Bacon left a hidden signature? Men and women romped through the grassy fields of the texts and scraped at the individual letters with every kind of tool imaginable. Anagrams: whereby existing letters are rearranged to create new words and phrases. (The phrase "Maister William Shakespeare" in the 1623 Folio can be anagrammed into "I maske as a writer I spelle Ham.") Numerology: whereby letters are converted into numbers that seem significant. (If A=1 and B=2, the name "Bacon" is 2+1+3+14+13, equaling 33; the appearance of 33 in any count of Shakespeare's words is a signature of Bacon.) One man, Orville Ward Owen, a physician from Detroit, invented a machine he called a "wheel," two large wooden spools stretched with one thousand feet of canvas on which he had printed thousands of pages of different Elizabethan texts. Dr. Owen and his team of assistants would spin the wheel, look for instances of four "code words" they believed were important (FORTUNE, HONOUR, NATURE, and REPUTATION), write down words that appeared next to those four words, and arrange them into sentences.

What made Mrs. Gallup different from these other investigators was that she presented herself first and foremost as a scientist, and her system for finding the messages was the most scientific and plausible yet. It wasn't something that came to her in a dream. The method had been demonstrated by Francis Bacon himself, in his book *De Augmentis Scientarium,* published the same year as Shakespeare's First Folio, 1623.

Bacon revealed that year that he had invented a new type of cipher, a method to signify "omnia per omnia": anything by means of anything. It possessed what he said were the three virtues of a good cipher: it was "easy and not laborious to write," it was "safe," and it did not raise suspicion—that is, an enciphered message would not appear, at first glance, to be in cipher at all. These are still sound principles today. His insight was that all letters of the alphabet can be represented with only two letters, if the two letters are combined in different permutations of five-letter blocks. The letters *i* and *j*, and *u* and *v*, were interchangeable in Bacon's time, so, choosing *a* and *b* for the two letters that represent all the rest, the new alphabet looks like this:

A	B	C	D	E	F
aaaaa	aaaab	aaaba	aaabb	aabaa	aabab

G	H	I, J	K	L	M
aabba	aabbb	abaaa	abaab	ababa	ababb

N	O	P	Q	R	S
abbaa	abbab	abbba	abbbb	baaaa	baaab

T	U, V	W	X	Y	Z
baaba	baabb	babaa	babab	babba	babbb

Each letter becomes five, so a word like *Riverbank*, written in this cipher, grows five times as long: *baaaa abaaa baabb aabaa baaaa aaaab aaaaa abbaa abaab.*

This is exactly like binary code, the language at the root of computers, and Morse code as well. In all of these systems, just two symbols, arranged in different combinations, can stand for many others. Binary code uses 0s and 1s, Morse code dots and dashes. Francis Bacon discovered the basic principle in 1623.

Crucially to Mrs. Gallup, he also showed the flexibility and power of his cipher *by example*. Bacon pointed out that the two letters that represent the others in his cipher don't have to be *a* and *b*. They can be *c* and *d*, or *x* and *y*. They can be physical objects, like apples and oranges arranged on a table; they can be sounds, like the alternating and audibly distinct shots of a musket and a cannon. In Bacon's cipher, the plaintext for ♫♪♫♪♫♪♪♫♪♫♪♪♫♪♫♫ ♫♪♫ is "deaf." ♥♥♥♡♡♥♡♥♥♥♥♥♡♥♥ means "die." All that's required is a "biliteral alphabet," an alphabet made of any two forms that are recognizably different. Write a manifesto with candies, send a love letter with bullets. As long as you specify an *a*-form and *b*-form, you can make anything stand for anything else. *Omnia per omnia.*

You can even camouflage a secret message in plain sight.

A message that reads *aaaba abbab aaabb aabaa* is obviously written in cipher, and anyone who intercepts it will know it contains a secret. Bacon suggested creating a "bi-formed alphabet" to overcome this problem—an alphabet with two slightly different versions of each letter, an *a*-form and a *b*-form. For example, an italic letter might be the *a*-form, and a normal, "roman" letter the *b*-form. A string of text like

k*nowle*dge i*s* power

might translate to

run

This was the heart of Mrs. Gallup's method. She scoured photo enlargements of pages from Shakespeare's First Folio and other Elizabethan books, looking for minute differences in the shapes of letterforms to discover the "biformed alphabet" she believed Bacon had planted in the text—the two alphabets with letters of different shapes. Then she drew charts of the *a*-form letters and the *b*-form letters. Then she went back through the original texts of the old books and compared each letter to the drawings of the letters on her charts, deciding if a letter was an *a*-form or a *b*-form. After classifying five letters in this manner, she was able to check Bacon's key (*aaaaa*=A, *aaaab*=B, *aaaba*=C) and write one letter of the final message. And that was when she found the secrets that troubled her Christian conscience.

Queene Elizabeth is my true mother and I am the lawfull heire to the throne. Finde the cypher storie my bookes containe. It tells great secrets, every one of which, if imparted openly, would forfeit my life. —F. Bacon.

Francis of Verulam is author of all the plays heretofore published by Marlowe, Greene, Peele, Shakespeare, and of the two and twenty now put out for the first time. Some are alter'd to continue this history.

Francis St. Alban, descended from the mighty heroes of Troy, loving and revering these noble ancestors, hid in his writings Homer's *Iliad* and *Odyssey* (in cipher), with the *Aeneid* of the noble Virgil, prince of Latin poets, inscribing the letters to Elizabeth. . . . He in this way, and in his Cypher workes, gives full directions, in a great many places, for finding and unfolding of severall weightie secrets, hidden from those who would persecute the betrayer.

You will either finde the guides or be lost in the labyrinth.

—Fr. St. Alban.

First published in her 1899 book *The Biliteral Cypher of Sir Francis Bacon Discovered in His Works and Deciphered by Mrs. Elizabeth Wells Gallup,* the messages told an alternate history of Elizabethan England that riveted journalists and divided scholars. According to Mrs. Gallup's decipherments, Francis Bacon wasn't just the great genius of his age. He was a secret king: the bastard son of Queen Elizabeth, known for her indiscretions, and the Earl of Leicester. During his own lifetime, Bacon was afraid that if he claimed his royal blood, he would be killed to suppress a scandal, so he had found a way to smuggle the truth into history, using his cipher to conceal messages in "great dramaticall works" he wrote under pseudonyms (Shakespeare, Marlowe) and also in "workes of science" published under his own name. He conspired with printers to sneak the cipher into books without anyone catching on, and he taught the cipher to a clandestine society of engineers, the Rosicrucian Society of England, who conducted scientific experiments in secret, fearing accusations of witchcraft. Using the cipher, Bacon and the Rosicrucians were able to exchange dangerous knowledge without fear of discovery and design technologically advanced machines.

Her work unleashed a furor. Mrs. Gallup seemed to come out of nowhere with an impressive scientific procedure and reams of proof. "Here are 360 pages of deciphered matter," one journalist wrote in a typical review, "with sufficient means of proof to satisfy any investigator." Skeptics questioned the veracity of the messages and said Mrs. Gallup must be imagining them; she savaged her critics in icy pamphlets and letters to the editor, writing that her style of analysis was "impossible to those who are not possessed of an eyesight of the keenest and most perfect accuracy of vision in distinguishing minute differences in form, lines, angles and curves in the printed letters. Other things absolutely essential are unlimited time and patience, and aptitude, love for overcoming puzzling difficulties, and, I sometimes think inspiration." She argued that if other people could not replicate her findings, it was their own fault—they had poor eyesight, they were lazy, they were uninspired.

She traveled to Oxford, England, and won converts in the literary community there. She produced testimony from researchers in England and America who swore that they had been able to replicate her decipherments: Mrs. Gertrude Horsford Fiske, Mrs. Henry Pott, Mr. Henry Seymour, Mrs. D. J. Kindersley, Mr. James Phinney Baxter. And of all her supporters, no one had more faith than George Fabyan. He invited Mrs. Gallup and her sister to Riverbank in 1912 and gave them carte blanche to pursue their investigation to its ultimate end. There was nothing he would not buy or build to support her work, no mode of investigation too outlandish or expensive. After establishing herself at Riverbank, Mrs. Gallup reported that she had deciphered a message from Bacon describing an "acoustical levitation device," an antigravity machine he apparently invented in the seventeenth century. It used the vibrations of musical strings to lift a rapidly rotating cylinder off the ground. Fabyan ordered his chief engineer, Bert Eisenhour, to build the machine out of wood. The result looked like a water wheel. Eisenhour couldn't get it to work. Something about the tuning of the strings. Fabyan was undaunted, saying, "The inheritance which the world received from Mrs. Gallup's work is the greatest that has ever been given to posterity."

He wrote that in 1916, the same year Elizebeth Smith arrived at Riverbank and was handed her first deciphering test by Mrs. Gallup.

Elizebeth looked at the test worksheet:

TheWo rkeso fWill iamSh akesp earec ontai ninga llhis Comed
iesHi stori esand Trage diesT ruely setfo...

Along with the typed worksheet, Mrs. Gallup had provided a photo enlargement of the actual page from the First Folio on which these words appeared. There was a copy of the biformed alphabet that Mrs. Gallup had already extracted from this part of the Folio—a list of all the *a*- and *b*-forms apparently inserted by Bacon. Mrs. Gallup also gave Elizebeth a looking glass of her own, and the key to the biliteral cipher: *aaaaa* means *A, aaaab* means *B,* and so on. To find the secret message, Elizebeth would need to squint at the Folio page through the glass, decide if each letter was an *a*-form or a *b*-form, and write a dash or a slash on the typed worksheet above the corresponding letter: a dash for *a*-form, a slash for *b*-form. Once Elizabeth made five dashes or slashes, she should check the key and write one letter of the final message. For instance, if her pattern of dashes and slashes looked like

--//-

then she would write the letter *G*.

Elizabeth knew nothing about secret writing at this point. She had never studied codes and ciphers. She had never even been particularly fond of puzzles. She was as fresh to the whole subject as any person off the street. But Mrs. Gallup had given her the rules of the game, and now she tried to follow them.

TheWo rkeso fWill iamSh akesp earec ontai ninga llhis Comed iesHi stori esand Trage diesT ruely setfo...

She started looking back and forth between the Folio page and the biformed alphabet, trying to tell if the letters were *a*- or *b*-forms. It was slow going. She got stuck on the first couple of letters, staring and staring through the glass, unable to decide if a letter was an *a*- or *b*-form. The variations were subtle: a slight wobble in the stem of an *H,* a tilt in the ovals of a *g.* It was like trying to sort blueberries by color, or beach pebbles by smoothness. Ultimately she needed Mrs. Gallup's help to get the answer, and it still took her eight hours to produce the twenty-four-word plaintext translation: "As I sometimes place rules and directions in other ciphers, you must seeke for the others, soone to aide in writing. Fr. of Ve (Francis of Verona)." She signed it on the bottom:

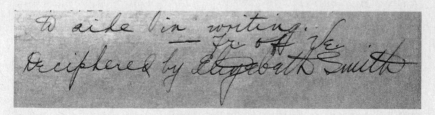

It went like this with the tests that followed. Mrs. Gallup handed Elizabeth a new Folio passage to decipher, and Elizabeth struggled for hours, solving it only with her boss's intervention. From time to time Elizabeth carried her materials over to Gallup's desk and set them down; Gallup pressed her eye to her looking glass, made some sharp pencil marks on Elizebeth's sheet, and handed it back. Impressed, Elizebeth always asked Mrs. Gallup how she succeeded when Elizebeth failed—had she modified the list of *a*- and *b*-forms, tweaking the alphabet to get the "right" answer? No, she hadn't: Elizebeth, being a novice, had failed to see the subtleties in the letters, had overlooked a little angle or an accent or a tiny shift in the position of the dot above an *i*.

At first she didn't worry that she struggled with Mrs. Gallup's system. Elizebeth awoke each morning in a dreamland. She had arrived to begin her new job during Riverbank's sweetest season, the moment of peak summer pleasure, the colors and smells dialed all the way up, the food most plentiful: breakfasts of fresh eggs from the chickens on the farm, dinners of meats and fruits prepared by the nimble Danish and Swedish cooks. She took walks along the river. Wild orchids grew on the banks. Shapes of sunlight twitched on the water like spinning coins. Ragtime music played from somewhere. She turned her head searching for the source. Fabyan had installed a series of loudspeakers across the property, operable from a single control panel in the Villa, so that he and Nelle might listen to music at any spot on the estate, and the songs changed throughout the day, switching directions, coming now from the garden and now from the veranda, ragtime shifting to jazz, then to a Beethoven symphony, the boss, intense and restless, wanting to hear every kind of music all at once and never getting his fill.

Fabyan assigned her a bedroom in a two-story building called Engledew Cottage, named after a local florist and friend of the Fabyans', one of the larger of the many cottages spread across the estate where "brain workers" lived. Engledew stood down the road from the Lodge a few hundred yards to the south, next to the farm with its big barn and Nelle Fabyan's prize cows. There were shared work areas in Engledew Cottage as well as the Lodge, and during the day men and women walked back and forth between the cottages along the highway, carrying papers and books, as horse-drawn carriages and automobiles drove past.

Elizebeth wasn't the only young woman assigned to cipher research at Riverbank. When she arrived there were at least two others, sisters from Chicago, barely out of high school. Fabyan tended to hire women out of clerical pools because it was convenient, but he had come to believe that in many ways they were better than men at analyzing ciphers. Women had the stamina and patience to look at text all day, and complained less. "Our experience at Riverbank," Fabyan wrote, "has demonstrated that women are particularly adapted for this kind of work."

After a few weeks Elizabeth fell into a routine, adjusting to the rhythm of her new job. Mrs. Gallup often worked with her assistants in the Lodge's spacious living room, with its tall casement windows that looked east across the highway, toward the Villa and the river. The work atmosphere in the Lodge was a bit like how she imagined a museum of natural history or a lepidopterist's lab to be, a place where people analyzed delicate objects, pinning dead butterflies to pages, drawing pictures. Mrs. Gallup sat at a handsome wooden desk, peering through an oblong looking glass at photo enlargements of pages from old books. The enlargements were made by William Friedman, the geneticist with the white buck shoes who had caught Elizebeth's eye at her first dinner in the Lodge. Because William happened to be handy with a camera and a darkroom, Fabyan had roped him into the cipher project, even though it wasn't his job, and he often visited the Lodge to drop off new prints for Mrs. Gallup.

The woman would raise the looking glass, lower it, write a few words in a notebook, raise it and lower it again, and write some more, hour after hour. When Elizabeth asked the other girls what Mrs. Gallup was doing, they said she was attempting to complete Bacon's unfinished science fiction novel *The New Atlantis*, to recover the remainder of the text, which Mrs. Gallup believed was woven throughout Bacon's works.

The cursive line of her pen was exceedingly fine. Each page of her notebook resembled a piece of art. She kept images of Bacon close at hand, for inspiration: an engraving of Bacon in his prime, a handsome youth with curls and a ruff; a picture of Gorhambury House, Bacon's mansion outside London. She filled small wooden boxes with news clippings about her own research and that of her competitors, ultimately pasting the clips into scrapbooks.

The women worked long hours, into the evening, sun dipping

low, flies swarming on the porch. "We lived hard and fast," Elizebeth later recalled to the NSA's Valaki, then paused, embarrassed. No, she did not mean to imply anything salacious. "I mean, there was absolutely no carousing, no parties, no nothing. Fabyan had use for only one kind of worker, and that was one that knew his business and worked at it damned hard." He paid the codebreakers and scientists tiny salaries but promised to take care of them in all other ways. Food, lodging, recreation: they would live like the "minor idle rich" as long as they stayed under his wing at Riverbank.

And who would ever want to leave? On weekends Elizebeth put on her bathing suit, skidded down the hill to the river's edge, and walked across the bridge to the island, which Fabyan called "Isle of View" because he liked that it sounded similar to "I love you." The lighthouse rose above the northern bank of the island, and the southern bank was crowned with tall trees and a magnificent swimming pool built by Fabyan for the enjoyment of his staff, lit at night by floodlamps and lined with soaring Roman columns. Swimming there made Elizebeth feel like an Italian princess or an actress in a movie. The cool water licked off the sweat and she dried herself in the sun, talking and laughing with Mrs. Gallup's other female assistants.

In August 1916 she turned twenty-four.

The men of Riverbank noticed Elizebeth early on, particularly the brain workers. They smoked pipes stuffed with cheap tobacco. They asked around about her story, tried to figure out if she was single, looked for ways to get her attention. Bert Eisenhour, the carpenter and engineer, pulled strings to borrow Fabyan's Stutz Bearcat, a roadster with a four-stroke engine, and invited Elizebeth to climb in; seconds later she was ripping along the Lincoln Highway at 60 or 70 miles per hour with the top off. In an era when most roads were dirt and gravel, the highway was paved, a result of cooperation between Fabyan and other local business owners; as a result, "no billiard ball on the smoothest billiard table ever made could have more pleasure in motion than that enjoyed by the ordinary Illinoisan skimming in a powerful car over that gleaming, winding stretch of concrete," a visitor once wrote. In the passenger seat of the Bearcat, Elizebeth raced past barns and silos, her body inches from the ground, wind plastering her curls to her head, engine roaring in her ears. She didn't know if her head would blow off.

Another man who crossed her path during free hours was William Friedman, the geneticist and Mrs. Gallup's photo assistant. He didn't like to swim, because he was afraid he'd catch cold, but he enjoyed bicycling, and he and Elizebeth took leisurely rides together around the estate, stopping to picnic on the grass, sand-hill cranes and red hawks circling above.

At twenty-five he was one of the younger male scientists, closest to her own age, so it felt natural to spend time with him. She appreciated his shy, precise way of speaking, his soft, halting voice that seemed to encode its own refutation, as if he were constantly checking a mental ticker tape of his words for correctness. One day he showed her where he lived on the estate. It was a working windmill—not the big showy Dutch contraption on the other side of the river but a smaller windmill on the same side of the road as the Lodge and the ordnance lab. It was two stories tall. William opened the door and she walked into an old, creaky structure, damp and warm, with a powerful smell of soil. She saw that the ground floor contained some microscopes and work shelves. An interior door led to the greenhouse that William managed, which is where Fabyan had him breeding new strains of crops and flowers, violets and wheat, and a type of corn with no cob.

Upstairs, he said, was his sleeping quarters, and down here was a little laboratory where he ran genetics experiments with living fruit flies, *Drosophila melanogaster*. Elizebeth could see his bottles full of the teensy-weensy flies. Each bottle was about the size of a coffee mug, only thinner, and was smeared with some over-ripe banana that the flies ate. William explained that geneticists like to use fruit flies in experiments because they reproduce very quickly, then die. If you marry a normal fruit fly with a fly that has yellow eyes, say—a genetic mutation, an alteration in the biological code—they will produce children in three weeks, and you can look at the children to see if they have yellow eyes, showing they inherited the yellow-eye gene. There was something incongruous and surreal about seeing this good-looking young man in a crisply pressed white shirt and bow tie working in a rustic prairie windmill that smelled of banana and the sweetish decay of plant matter. Elizebeth used to watch him there, mating the flies, carefully pouring one bottle of flies into another, getting them to exchange their codes.

The size and scope of Riverbank was dawning on her. What had appeared on her first visit to be a sparsely populated stretch of land now revealed itself to be a small self-contained village, a community of 150 workers, some of whom had been with Fabyan for more than decade: the caretaker of the Japanese garden, Susumu Kobayashi; the boathouse manager, Jack "the Sailor" Wilhelmson, a happy and well-built Norwegian; Fabyan's personal secretary, Belle Cumming, originally from Scotland, who kept Riverbank's financial records in black folders and hurled torrents of profanity at guests she felt were overstepping their bounds; Silvio Silvestri, Fabyan's personal sculptor. Fabyan hired them on whims. He trusted his own impressions of people instead of their accomplishments or educations. He brought people to Riverbank if he decided they were spectacular. He was always saying that to Elizebeth and everyone else: "Achieve success! Be spectacular! Then things break your way."

To entice spectacular individuals to stay, he welcomed their spouses and children. Every child born on the estate received a sum of money from Fabyan, placed in a bank account to grow and pay for future schooling. This was another aspect of the place that made Elizebeth marvel: there were families here, boys and girls growing up at Riverbank. Fabyan seemed to genuinely love children. He handed out shiny dimes he kept in his coat pocket. He stopped whatever he was doing to answer their questions about Riverbank's zoo creatures and explain the curious behaviors of the animals, to remove a snake from a cage with his own hands and demonstrate how a snake was able to disengage its jaw in order to swallow an egg.

Elizebeth realized that everything that appeared so hallucinatory to her about Riverbank must seem perfectly normal to these children. It was normal for them to live where two monkeys roamed outdoors wearing red diapers, one a kleptomaniac with a habit of stealing men's keys. It was normal for Jack the Sailor to sing sea shanties to the children, dance the jig on their command, and teach them how to tie knots. It was normal to be outside playing and see a famous actress walk by, or Teddy Roosevelt, who liked to stroll the grounds with Fabyan and talk about crops, genetics, and Francis Bacon. In summer Jack the Sailor always wove a gigantic spiderweb out of rope that spanned two elm trees; squirrels climbed it, and children tried, and so did Lillie Langtry, a stage and vaudeville actress and an adventurous horsewoman. Other celebrities vacationed at Riverbank: the curly-haired actress and pilot Billie Dove; the aviator and polar explorer Richard Byrd; Broadway producer Flo Ziegfeld and his actress wife, the elegant Billie Burke, who would later play the role of Glenda the Good Witch in *The Wizard of Oz;* the titans of the Chicago Stock Exchange, of which Fabyan was a member. They ate dinner with the Fabyans, George and Nelle, then drank and smoked around a campfire.

Elizebeth wasn't impressed by the celebrities. She met and talked with Lillie Langtry, one of the most famous women in America at the time, and only mentioned it later in passing. Elizebeth was proud not to be "afflicted with the star-complex and hero-worship," she would write later. "Whatever quality it is which is possessed by those who love the adulation and star worship seems to be, in my case, supplanted by an intense reach for freedom from observation—and for privacy."

This is one reason Elizebeth became wary of George Fabyan almost immediately after starting her job: He seemed interested in prying into every part of her life. Soon after she arrived and settled in, Fabyan told her that the modest white and gray dresses she liked to wear were inadequate and she needed to buy a new wardrobe at Marshall Field's in Chicago. Frugal by nature, Elizebeth resisted paying a premium for a name brand, but when she raised her voice to complain, Fabyan told her to hush. "That's so typically Fabyan," Elizebeth recalled: if you told him he was wrong, "[t]he next thing you know there'll be a gun rammed down your throat." He sent her into the city with one of his secretaries, who accompanied her to the department store and made sure she bought Fabyan-approved items.

She figured out within a week or two that she was dealing with a half-crazy individual of unlimited funds and a split personality.

There was a side of him that was authoritarian, that craved order and ceremony, which explained the bugler who played reveille in the morning and taps at night, and the American flag that was raised each morning and lowered in the evening, folded into a triangle as a cannon fired a ceremonial ball into the prairie dusk. The staff called him "Colonel" or "the Colonel." Elizabeth was told to address Fabyan as "Colonel Fabyan," and she did, assuming he must have served in the military; it was only later when she learned the truth, that the title of Colonel was an honorary one, bestowed by the governor out of gratitude after Fabyan allowed the Illinois National Guard to use the estate as a training ground. The governor even named a group of cavalry scouts the Fabyan Scouts. Chest leaping with pride, Fabyan recruited some local farmhands to join a militia he called the Fox Valley Guards, as if he aimed to build a personal army.

He liked to sit in the wicker chair that hung on chains from the tree next to his villa. Elizabeth heard people call it the "hell chair" and soon understood why. If Fabyan was angry at an employee or a guest, he brought that person over to the hell chair and sat there, screaming at the offender, giving them hell, while he swung to and fro, chains creaking. He sat there at night sometimes, in the hell chair, stoking the coals of a campfire in the dark.

The other side of Fabyan loved chaos and ripped through the days under power of impulse and inspiration. He had a habit of buying supplies sight unseen from train boxcars—a skyscraper's worth of steel I-beams; seventy-five plows—and storing them in a warehouse next to the Dutch windmill that he called the Temple de Junk. He seemed to glory in randomness to the point of mocking the foundations of his world. He published a book, *What I Know About the Future of Cotton and Domestic Goods,* by George Fabyan, and kept copies in his office. Visitors grabbed the book with sweaty hands and flipped through, hoping for a stock tip from a wealthy cotton magnate. Inside were one hundred blank pages. It was Fabyan's joke about the riddle of commerce, the arbitrary American system that kinged him with enough money that he didn't have to care about money anymore.

He liked to dress up as a horseman or hunter, in riding coats and knee-high boots, but no one ever saw him riding a horse or shooting wild game, and he liked to dress up as a yachtsman, in a white sweater and jaunty blue cap, but no one ever saw him sailing a boat. He spilled across his kingdom on foot, thundering from place to place, dropping heavy ideas and moving on, letting others do the lifting. One day he walked past the swimming pool and saw a little girl, Sumiko Kobayashi, the daughter of his head gardener, resisting her first swimming lesson, crying because she was afraid of the water. Fabyan commanded an adult to throw her in the pool and let her *learn by doing*. He watched the terrified girl thrash for her life in the water, then walked away, satisfied with his solution, while the adults dove in and saved poor Sumiko from drowning.

He may have been a monster. But he was no idiot. To underestimate his intelligence was dangerous, Elizebeth sensed. She considered him to be, despite his lack of formal education, "a very bright man" with a cunning mind and a proven ability to predict how people and institutions would react to moments of stress and crisis. He could get anyone to listen to him. He didn't read scientific papers; Elizebeth never saw him read anything longer than a newspaper headline. But he had been blessed with a near-photographic memory, and whatever his scientists told him, he could repeat back verbatim. This skill for mimicry, combined with his innate abilities as a salesman, made Fabyan seem like a credible prophet of science even when he was talking about things that science said were impossible. He pursued schemes for perpetual motion, infinite energy from nothing, and once showed Elizebeth a prototype of a perpetual-motion machine that he kept behind one of the labs. She was unimpressed: "I remember going, looking at it for quite a while, and it just seemed to me like a great, huge, metal something-or-other." He argued that common human ailments could be traced to the fact that our primate ancestors crawled on their stomachs and humans have never properly learned to walk. And he wasn't selling these ideas cynically; he really believed them. He was good at blurring the line between fantasy and reality because he didn't believe any such line existed. As he once told William Friedman, *I have seen impractical and improbable things accomplished.* All it took to achieve improbable things was an optimistic attitude and a refusal to give up.

"We play the game day-to-day as best we can," he was fond of saying.

Of all the investigations at Riverbank, Fabyan sold the Bacon cipher project the hardest. Though he gave every visitor at least a taste of the cipher work, presenting it as one element of the general package of wonders, he organized separate junkets to persuade hesitant or openly hostile academics that Riverbank had found the answer.

In the late summer of 1916 he began to lean on Elizebeth for help. He had already realized that when she spoke, even though she was only twenty-four, people listened to her—her good looks caught the eye of men and her precision and earnest intelligence held attention. He started to let her know that Professor So-and-So from Such-and-Such College was coming to learn about the cipher discoveries and Elizebeth needed to persuade this person that Riverbank's approach was correct. "We'll get along fine," Fabyan told Elizebeth. "We'll see if we can induce him to stay."

The academic would come, all expenses paid. There would be a lot of food, a lot of wine. Fabyan usually delivered a presentation on the ciphers using lantern slides, square photographic negatives printed by William Friedman and projected onto the wall of a darkened room through a curved piece of glass. He had contempt for what he saw as the timid and conformist mind-set of literary intellectuals while at the same time wanting to win them over, and took pains to present himself as the sort of careful and factual man he felt they would be likely to respect. While displaying slides of the Folio and Mrs. Gallup's lovely drawings of biformed alphabets she claimed to find within, Fabyan explained that he did not care about the "useless Bacon-Shakespeare controversy" of who wrote Shakespeare; that he only cared about getting to the bottom of the cipher contained in the plays; that he and his Riverbank colleagues had no use for anything but "hard, cold facts"; that the existence of the cipher was such a fact; that it had passed careful tests; that no one at Riverbank was making any money from these investigations; and that they were doing it for the benefit of humanity, committed to sharing their discoveries with the world. Who could object? The combination of his gravelly voice in the shadows and the delicate letters on the wall tended to disorient the guest and lull him into a state of increased charity toward the Riverbank view. Heads, ever so slightly, began to nod. Jaws to relax. And Elizebeth played her part. If a visitor grew sick of listening to Fabyan and turned to Elizebeth, asking what she thought, she said she was convinced that the work was solid, that the messages were really there.

Privately, though, she was beginning to doubt. Skeptical visitors made arguments difficult to refute. The head of the English department at the University of Chicago, John Matthews Manly, an authority on Chaucer and an amateur cryptologist, stayed at Riverbank for a time and concluded that it was all bunk. Manly was already famous in his field, didn't need money or anything else Fabyan could offer, and he took delight in pointing out the holes in Mrs. Gallup's method, like a boy in a roomful of red balloons, stomping them flat one by one. Fabyan asked Elizebeth to "wrassle" with Manly for a weekend, and she found him a pompous ass. At one point during an argument, Manly's voice rising sharply, Elizebeth staying calm but arguing back, Manly pushed her on the shoulder, baffled and upset that anybody might challenge the great John M. Manly—she never forgot it. "Oh, my! That was too much to take. Ahhh!"

But there was substance to what he and other skeptics were saying, a stubborn logic that tugged at the hem. Mrs. Gallup's technique depended on discerning small yet consistent fluctuations in letterforms in books made long ago, with the technology of a more primitive era. It strained credulity to think that the printers, setting the type by hand in 1623, could have duplicated these minute fluctuations across hundreds of copies of the First Folio, and in fact the variations between different Folio copies were sometimes larger than the variations Mrs. Gallup thought she saw in a single book.

Another skeptical argument moved Elizebeth. It was the literary case against Bacon's secret messages. There was no kind way to put it: the messages were badly written. *Francis St. Alban, descended from the mighty heroes of Troy, loving and revering these noble ancestors*—was this tedious author the same one who gave such light and supple voice to Romeo's desire for Juliet? *See how she leans her cheek upon her hand. O that I were a glove upon that hand, that I might touch that cheek.* To believe Mrs. Gallup's theory, you had to believe that the plays, these warm-blooded, ravishing beasts, had been conceived almost as an afterthought, as mere envelopes for a stilted memoir about a guy whose mom was the queen. It made no sense. It would be like God creating a galaxy simply to tell a knock-knock joke to some distant deity, enciphered in the shapes of stars.

The big question then became: If the secret messages discovered by Mrs. Gallup weren't really there, what was she seeing?

Elizabeth never once suspected that Mrs. Gallup was a fraud. Deception was not in her. The only possibility was that she had been somehow deceiving *herself*. Humans are so good at seeing patterns that we are often able to see patterns even when they aren't really there. Mrs. Gallup must have been altering the rules of her method to fit the desired result, changing the all-important assignment of letters to the two baskets (*a*-form and *b*-form) until she saw words that made sense. Decades later Elizabeth and William would describe what they thought was happening with Mrs. Gallup, in a book they wrote as coauthors:

> She could go through the texts extracting from them what she unconsciously wished to see in them. . . . With each successive letter deciphered she had a choice—limited but definite—of possibilities; and so, as she went on, there would be a kind of collaboration between the decipherer and the text, each influencing the other. Hence perhaps the curious maundering wordy character of the extracted messages, very like the communications of the spirit world: with some sense but no real mind behind them, just a sort of drifting intention, taking occasional sudden whimsical turns when the text momentarily mastered the decipherer.

This was the clarity of hindsight. In the moment, at Riverbank, Elizabeth didn't know what to do with her doubt. She saw how her bosses responded to criticism. Mrs. Gallup restated her conclusions in combative letters and articles, denying that it was all a figment of her imagination and comparing herself to Galileo: "The idea that the earth moves, was once thought an illusion."

And Fabyan doubled down on publicity. He released a picture book for children, *Ciphers for the Little Ones*, that taught the story of Bacon and his biliteral cipher. He printed business cards alleging that Bacon was the bastard son of Queen Elizebeth and added at the bottom:

ALL INQUIRIES REGARDING THE SOURCE OF, AND AUTHORITY FOR, THESE HISTORICAL, BUT HITHERTO UNKNOWN FACTS, WILL BE PROMPTLY ANSWERED FROM

RIVERBANK LABORATORIES
Geneva, Illinois.

When people did inquire, Fabyan replied with a form letter describing the cipher project:

> *Riverbank Laboratories are a group of serious, earnest researchers, digging for facts. It is supported by Colonel Fabyan at his country home in Geneva, for his own information and amusement....*

A pressure was building in Elizebeth's chest. It was the old scalding sensation she remembered from college when she realized people valued politeness more than truth. For now she kept her doubts to herself. She doubted her doubt. Who was she to declare that she was right and everyone else was wrong? Was it her vanity telling her that? How would she prove her case if she did speak up? Would she lose her job? Would anyone stick up for her? She was twenty-four. She was a nobody here. She was a nobody anywhere.

During conversations in the Lodge she looked around the room at the faces of her colleagues, trying to tell if they really believed or if they were just pretending. She sometimes met the eye of William Friedman. She wondered what he was thinking.

Lately they'd been talking more and more. William carried his camera everywhere, a black box that hung from his neck. Elizebeth was becoming his favorite photo subject. He would ask her to stand in a garden or on a square of grass, and he would hold the camera at chest level and look down at the image of her face in the glass.

She was learning more about him, where he came from and how he got here. His family was Jewish, originally from a town in Russia called Kishinev, where he was born with a different name, Wolfe. His parents changed it to William when they sailed to America a year after his birth, escaping a famine in Russia and the anti-Jewish laws of the Czar. They settled in Pittsburgh.

William said his father was a serious, bookish man, fluent in eight languages, a student of the Talmud. In Russia he had been a postal clerk but had trouble finding a good job in America and resorted to selling Singer sewing machines door-to-door. His mother worked as a peddler for a clothing company. So William and his four siblings grew up poor in Pittsburgh, poorer than Elizebeth's family in Indiana.

He went to Cornell on a scholarship and chose to study genetics because it was a young field that "seemed to offer great possibilities for research and ingenuity." He earned his degree and stayed on to teach a few courses as an untenured lecturer. That was when the unsolicited letter arrived from Fabyan in the general mailbox of the biology department. William didn't know who Fabyan was. He said he ran a private research facility in Illinois and needed an expert in heredity to launch a genetics department and supervise experiments in the breeding of crops and fruit flies. William wrote back, introducing himself, and over the next three months, Fabyan courted him by mail, promising a life of intellectual freedom and adventure: "I am not looking for an agricultural expert, the woods are full of them; and I am not looking for a man to duplicate work that is being done at every agricultural station in the country, and at every advanced school and university. . . . If I should hear of something anywhere this side of Hell that I thought would do us any good, I might want you to go there and find out about it; in other words, I don't want to go backwards."

William replied with deference, formality, and gratitude: "I re-alize the value of the opportunity you are giving me to make good and I hope that our future relations will be mutually agreeable and profitable."

There were hints in this exchange that Fabyan would be difficult to work with. William, cautious by nature, asked about salary. Fabyan responded with a vague, long-winded riff: "I want to get some practical level-headed fellows that will carry themselves, and a community which is asking no favors and yet having the best there is, where people will have to come for what we have." William asked what he raised on the farm at Riverbank. Fabyan said he raised hell. His analogies were beautiful and bizarre. Writing of his desire to breed a new strain of wheat that would thrive in dry climates and help feed the hungry, Fabyan told William, "Here is a problem that has come up in my mind, that I want you to work on. I want the father of wheat, and I want a wife for him, so that the child will grow in an arid country." He added that "one of my wealthy Jewish friends" was also working on the problem, but "if I can beat him to it, he will foot the bills, and be damned glad to. . . . This may seem impractical and improbable, but I have seen impractical and improbable things accomplished."

Eventually Fabyan offered to pay William a hundred dollars a month, on top of free lodging, and William accepted. It happened to be an old yearning of his to live on a farm, a dream wrapped up in his Jewish identity. As a kid he'd heard stories from his parents of the pogroms in Russia, mob violence against Jews that they barely escaped, and by the time he got to high school he realized that anti-Semitism was spreading in the United States, too. Popular American magazines portrayed Jewish immigration from Eastern Europe as a "Jewish Invasion," a threat to the jobs of whites, with the Russian Jew said to be especially conniving thanks to his "nervous, restless ambition." Concerned and wanting to protect themselves, William and some of his high school classmates fell under the spell of the "back-to-the-soil" movement, a homespun brand of Zionism that encouraged Jewish kids in America to resist anti-Semitism by tilling the land, making themselves strong and self-sufficient. William took this idea seriously enough to enroll in some courses at a Michigan agricultural college. When he actually tried farming, he realized that everything about it, from the physical labor to the grit in his clothes, made him miserable. He went to Cornell instead.

Now, at Riverbank, he found himself on a farm again. Of sorts.

Elizebeth had never gone for shy men. But she liked William, and so did her elder sister, Edna, who visited Riverbank to see how Elizebeth was getting along. Edna's dentist husband had recently died, leaving her a widow, and the dapper geneticist left an impression. She wrote William two flirty letters. Edna informed him that her sister was growing fond—"I think E[lizebeth] cares a very great deal more for you than she lets herself or anybody else believe"—but also implied that perhaps she, Edna, the mature and responsible sister, might make a better mate for William than younger, flakier Elizebeth. Edna wrote to William, "My idea of real love-making is sort of the Lochinvar kind"—Lochinvar, the hero of an old Scottish poem: *So faithful in love, and so dauntless in war / There never was knight like the young Lochinvar.* "Him riding up furiously, sweeping his bride up before him with one hand and riding away."

Although Elizebeth found William attractive, she was drawn at first to his way of carrying himself, his scrupulous precision about words and facts and clothes, his modesty—qualities that made him George Fabyan's opposite. She liked checking in with William after spending hours in the blast zone of Fabyan's hype cannon. It felt healing, like drinking a glass of cold lemonade after a long walk. And what a mind he had! Talking to most people, Elizebeth felt like she could see the rough carpentry of their thoughts, the joints and tenons that never quite fit, but with William, ideas emerged smooth and whole, as if from a workshop. And he was so *playful* about it all, unlike Fabyan and Mrs. Gallup. Science to them was about results: defeating gravity, rewriting literary history, finding the secret to eternal life. Huge, epic, shocking, revolutionary ends. William never used such words. He didn't care about the answers so much as the questions. He enjoyed science because it was an interesting way of being alive.

He had a feel for ciphers thanks to his work with Mrs. Gallup, and also a youthful fascination with Edgar Allan Poe's short story "The Gold-Bug." The plot of the story revolves around a cryptogram whose solution points to a buried treasure chest full of diamonds, rubies, emeralds, sapphires, and gold coins, placed there by a murderer. Poe wrote articles about codes and ciphers, bragging that he could solve any cryptogram and daring readers to stump him, and for decades to come, Americans associated codebreakers with the sunken-cheeked, disreputable figure of Poe. William took a more whimsical approach to ciphers.• He liked to blend them with his knowledge of botany to make jokes and works of art. At Riverbank he drew a sketch of a long-stemmed plant with many fine veins in its leaves; although from a distance it looked like an ordinary botanical illustration, closer examination revealed patterns of notches in the roots and leaves and petals that spelled out the words "Bacon" and "Shakespeare" in the biliteral cipher. He captioned the drawing, "CIPHER BACONIS GALLUP," "A MOST INTERESTING AND PECULIAR PLANT, PROPAGATED AT RIVERBANK RESEARCH LABORATORIES."

The autumn weeks burned away. The temperature dropped and Elizebeth experienced her first Riverbank winter, a gray duration of pitiless wind that scraped across the plains unbroken and slammed into the estate. The sky threw down a tarp of pale blue light. Your breath crystallized in the air like clouds of cigar smoke, and the cold groped into your lungs. The cottages and labs burned coal all day and the dark gray smoke rose from the chimneys. Elizebeth and William were growing closer. She didn't know what to call it—more than friends, less than lovers. William would perch in a rocking chair sometimes and she would sit on his lap as he pushed the chair forward and back, his thin arms around her thin waist, the chair creaking in a steady rhythm, neither of them saying much at all.

It took a while for her to get up the courage to share her doubts about Mrs. Gallup's work; she worried that William would look at her strangely, would think she was wrong and think less of her. But Elizebeth's mind wouldn't let it rest, and eventually, she asked what he thought. Wasn't it strange how Mrs. Gallup could see these things that no one else could see?

To her enormous relief, William said he had been wondering the same. Sometimes a thought floated to the front of his mind, the deepest heresy at Riverbank: *There are no hidden messages in Shakespeare.*

The idea rang in the air between them like a broken chime. Ugly, dissonant notes. Elizebeth and William exchanged a look. For the first time, but not for the last, each gathered strength from the other, and the notes resolved into a chord: *There are no hidden messages in Shakespeare.*

- What if everyone involved in the Bacon work was crazy, except for the two of them?

CHAPTER 4

He Who Fears Is Half Dead

and then begins *step step leap*
she continues these leaps
scramble the code scramble uphill scramble eggs
and without premeditation but in full arc if possible
have a good time.
—ANNE CARSON

The intercepted and decoded telegram burned its way from hand to hand, from junior diplomat to senior diplomat, first in London and then in Washington, producing involuntary noises of surprise and bulging eyes. It was obvious that the president himself needed to see it. At 11 A.M. on February 27, 1917, the U.S. secretary of state, Robert Lansing, carried a copy of the intercepted telegram to the White House and showed it to Woodrow Wilson. The president read it and grew uncharacteristically angry: "Good Lord!" he said. "Good Lord!"

The telegram had been sent from Germany to Mexico on January 16, traveling by three separate telegraph routes and encoded as a series of number blocks: 130 13042 13401 8501 115 3528 416 17214. The British had intercepted the message, and a small team of civilian codebreakers toiled for a month in a secret office inside Whitehall to scrub away the grime of code and make the plaintext visible. What they saw, to their shock, was nothing less than a conspiracy plot against the United States.

Written by Germany's foreign minister, Arthur Zimmermann, the telegram proposed an alliance between Germany and Mexico: "We intend to begin unrestricted submarine warfare on the first of February. We shall endeavor in spite of this to keep the United States neutral. In the event of this not succeeding, we make Mexico a proposal of alliance on the following basis: make war together, make peace together, generous financial support, and an understanding on our part that Mexico is to reconquer the lost territory in Texas, New Mexico, and Arizona. The settlement in detail is left to you."

The Zimmermann Telegram, as it came to be known, was indisputable proof of a German plot against America, "clear as a knife in the back and near as next door," as the historian Barbara Tuchman has put it. Residents of Texas were particularly displeased to learn that the Kaiser was trying to give them to Mexico, but outrage against Germany was general across the States. The telegram sped up history. It pushed America into war with Germany, whether America was ready for war or not.

It was not.

And this is how the telegram changed the destinies of Elizebeth Smith and William Friedman: as American codebreakers, they happened to possess an extraordinarily rare and suddenly indispensable set of skills.

Elizebeth got word in January 1917 that her mother, Sopha, long ill from cancer, was on the verge of death, and Elizebeth should come to Indiana to say goodbye. She packed a bag and rode the train back to Huntington and her childhood home. Her father was there, and her sister, Edna. The two sisters consoled each other as physicians prowled through the old house, murmuring about a growth. Sopha was in a lot of pain and vomited violently. A doctor turned her on her belly and spread iodine across her back. He used cocaine to numb a particular spot and tapped a metal rod into the skin, removing what Elizebeth felt was a horrifying quantity of pinkish fluid.

She had brought some cipher materials with her, hoping to get work done. "My book-bag lies here unopened," she wrote to William at Riverbank. "I try to make myself work, but I cannot.

I sit a moment, then spend the hours pacing back and forth from Mother's bed, in the vain hope that there is something I can do. It is so awful—Billy Boy—to look on the face of death like that—the beckoning face—Do you know it makes me think a lot about posterity, and responsibility, and all that?" She wasn't sure what to call William in these letters, or to call herself in relation, so she mostly kept things platonic, signing her letters "yours, Elsbeth," and thanking William for being "one of the truest friends I've ever had," although she did admit that she missed William's "rocking," his comforting way with a rocking chair, and in one letter she slipped in something stronger: "I love you / Elsbeth."

When Sopha died, in February 1917, Edna stayed behind to arrange the funeral, while Elizebeth returned to Riverbank, seized by a new impatience. She had no desire to spend any more time on the Bacon ciphers. Life was too short to waste on fruitless quests. When she reunited with William, he said he felt the same. They both agreed they had to remove themselves from the project. The question was how.

Confronting Mrs. Gallup seemed a little cruel. She had worked too many years in a single direction to admit her compass was broken. She had treated both of them with kindness. They tried talking to Fabyan instead. On a few occasions the two youngsters buttonholed him and tried to get him to listen. Mrs. Gallup's theory was unsound, they said. Fabyan's money might be better spent on other projects. He shouted them down, as they expected. Fabyan said he wasn't paying them to question the theory, only to persuade the academy that it was correct.

But by now, even if he didn't want to admit it, a new scheme was diverting Fabyan's attention from the literary ciphers. Shakespeare, Bacon, Mrs. Gallup, old books, dead men—it paled in urgency to the world of the living.

For months now Fabyan had been advertising his patriotism and his willingness to place Riverbank at the disposal of the flag. He had ordered his groundskeepers to expand the network of model trenches next to the Lodge, and after months of digging by a team of mud-spattered workers the trenches reached a total length of three miles, enough to be useful for the Fox Valley Guards to conduct infantry drills complete with live mortar rounds. And Fabyan had told officials in Washington that if they needed help with codebreaking, Riverbank stood ready to serve.

"Gentlemen," he wrote to Washington on March 15, 1917, "I offer anything I have to the government, and if you care to have any of your local men call on me, and see the work that is being done, I should be very glad to show it to them." He described his interest in old ciphers, especially the biliteral cipher of Francis Bacon, and added: "To avoid any possibility of being considered a crank, or a theorist, I respectfully call your attention to the fact that I was the business partner of the late Cornelius N. Bliss, formerly Secretary of the Interior, whom most of the older men in Washington remember with a great deal of respect and admiration."

Military officials were of course reluctant to give any power or responsibility to a fake colonel in Illinois, but they had little choice but to accept Fabyan's offer. They were desperate for codebreakers because of the way radio and wireless technology was changing the art of war.

In earlier conflicts, codebreakers had mattered less; fewer military and diplomatic messages were encrypted because the messages were harder to intercept. If you wanted to steal an enemy message, you had to capture a messenger on horseback, or open an envelope at a post office, or install a tap on a telegraph wire. But with radio, all it took to intercept a message was an antenna. The air was suddenly full of messages in Morse code, dots and dashes that registered as audible pings and whines. You could pluck them out of the sky. So to protect their secrets, armies had begun encrypting their wireless messages before sending them over the wireless in Morse.

This simple fact transformed codebreakers from disreputable freaks into potential superheroes, wizards with power over life itself. Now the air was full of encrypted information of enormous tactical significance and the utmost stakes. The routes of ships at sea. Troop movements on the ground. Airplane sightings. Diplomatic negotiations and gossip. Reports of spies. Thousands upon thousands of puzzles zipping through the atmosphere, any one of which, if decrypted, might win or lose a battle, wipe out a regiment, sink a ship. In this new world, a competent codebreaker was suddenly a person of the highest military value—a savior, a warrior, a destroyer of worlds.

And yet, as Elizebeth would later write, "There were possibly three or at most four persons" in the whole United States who knew the slightest thing about codes and ciphers. She was one of them, William another.

The government lacked the capacity to reliably intercept foreign messages, much less break the codes and read them. The CIA didn't exist in 1917. There was no NSA, and the FBI was a crumb of its future self, a nine-year-old organization known as the Bureau of Investigation, which fielded only three hundred agents, on a total budget of less than half a million dollars. There simply was no intelligence community as we think of it today. The Department of Defense was called the War Department then, which operated the army, and though the War Department did contain an intelligence-gathering section, the Military Intelligence Division (MID), it was tiny and underfunded. On the day Congress declared war, April 6, 1917, the MID employed just seventeen officers. The officer in charge of the MID, Major Ralph van Deman, considered the government's ignorance of codes and ciphers an "emergency."

So, in the second week of April, the War Department dispatched an emissary to Riverbank, an army colonel named Joseph Mauborgne, to check out the place and report back on its suitability.

Mauborgne was one of the three or four people in America who knew something about codebreaking. In 1912, while stationed at the Army Signal Corps School in Kansas, a bare-bones airfield and laboratory to probe radio technology, Mauborgne had made history by figuring out how to send a radio signal from a plane to the ground for the first time, and in 1914 he became the first American to break the British army's field cipher, known as the Playfair Cipher, based on a table of letters arranged in a five-by-five grid.

When Mauborgne arrived at Riverbank, Fabyan greeted him with the usual overwhelming gusto and brought him to the second floor of the laboratory building, declaring with a flourish that the Riverbank Department of Ciphers was open for business. The office appeared busier and more crowded than it had ever been. In anticipation of the army man's visit, Fabyan had gone out and hired a dozen clerks, stenographers, and translators fluent in German and Spanish, to provide support for Elizebeth and William. Fabyan hoped the two young people would be able to lead the effort, to break codes for the government, while Mrs. Gallup continued her long labors on the Bacon ciphers. Superficially, the office looked like the picture in Fabyan's imagination, the pitch he had sold to Washington. It looked like a codebreaking agency on the prairie.

There in the new Department of Ciphers, Elizebeth and William introduced themselves to Mauborgne. They clicked with him immediately. He was thirty-six and big—big body, big voice, big brain, with perfectly round, black glasses. He was the only man they had ever seen stand eye to eye with Fabyan and not seem intimidated. Mauborgne liked Elizebeth and William, too. He saw a spark in the pair of young codebreakers. (He would later call them "the two greatest people I have ever known.") They had little formal training but were bright and eager. Fabyan made him wary—a mess of a man, lunging wildly from promise to promise—but it was undeniable that Riverbank had excellent security from a military standpoint. Aside from the virtue of being in the middle of nowhere, safe from enemy attack, it was protected by the lighthouse, and Fabyan also had the Fox Valley Guards nearby—his own private army. If all else failed and the Germans invaded, Fabyan said he would open the bear and wolf cages in the garden and sic the beasts on the intruders.

On April 11, Mauborgne informed his commanders that Riverbank was ready. He urged the army and also the Justice Department "to take immediate advantage of Colonel Fabyan's offer to decipher captured messages," owing to "the mass of data" in his private library of cipher books, the security of his compound, and the quality of his employees, "a force of eight or ten cipher experts who spend their time delving into the works of antiquity, discovering historical facts hidden away."

After reading Mauborgne's enthusiastic report, Van Deman of the MID wrote to Fabyan with gratitude, thanking him for "your

exceedingly kind and patriotic offer of assistance," and soon encrypted messages started arriving at Riverbank from Washington.
They came in the mail and by telegram, sent by different parts of
the government: the War Department, the navy, the Department
of State, the Department of Justice. The messages had been intercepted by covert means, mostly from various telegraph and cable
offices across the country.

Fabyan had gotten his wish: for the foreseeable future, Riverbank would become ground zero for military codebreaking in
America, a de facto government agency. He had drafted Elizebeth and William into the war, assuming they would be able to
handle what was coming. But when they looked at the messages,
the fresh piles of gobbledygook spilling from the mail sacks onto
their desks, they weren't sure that he was right.

A woman and a man are sitting side by side in a busy room. People
come and go and the door opens and closes and there is the sound
of typewriter keys smacking ink into pages. Outside the window,
hawks fly and cows moo and a bear scratches himself in an iron
cage and a parrot sings and a river runs and there are also monkeys in diapers for some reason.

The two people, Elizebeth and William, notice nothing except
what is in front of them on a slab of desk. They are looking down at
a sheet of paper. All of their intensity is shining down at the paper,
a bright beam of desire to understand the text that is typed there.

It looks like nothing. It is clearly not written in the biliteral cipher of Francis Bacon that they are familiar with. It is something
else, a new level of mystery. They must understand it. But they
don't know what they are looking at.

BGVKX	TLXWB	SHSFW	KWGRI	KZTZG
RKZFE	YDIWT	KOFOB	GUHGD	SFVRE
UIUQX	HSLDS	OHSRM	HTWKY	VHUIK
BJDUH	VSART	BGVNG	VBAFO	AZOXG
PQPMJ	DRODW	RCNML	MTMXL	SSVAR

A hiss of symbols, a raw block of babble. A cryptogram. Someone wrote and sent it for a purpose, and someone else intercepted

it, and now it is here on your desk. These letters contain meaning. How to unlock it without knowing the key?

The basic task of codebreaking might seem impossible if you think about how many different ways a message might be encrypted. Each human language has its own quirks and curiosities. Then, within each language, a cryptographer can choose from among dozens of varieties of locks—codes and ciphers. And each lock will accommodate only one of a vast array of possible keys.

For instance, one of the simplest kinds of ciphers, called a mono-alphabetic substitution cipher, or MASC, swaps out one set of letters for another. Perhaps A=B, B=C, and C=D, or perhaps A=X, B=G, and C=K—or any other map between the 26 letters of the plaintext alphabet and 26 different letters in the ciphertext. A MASC is a very basic method of encrypting a message. But even here, there are 403,291,461,126,605,635,584,000,000 potential alphabets: 403 *septillion*. A thousand computers, each testing one million alphabets per second, would take more than a billion years to exhaust the possibilities.

And yet anyone who has ever solved a cryptogram on a newspaper puzzle page has conquered the 403 septillion possibilities, because, of course, there are shortcuts, ways of taming the task by grabbing on to certain patterns in the text.

This is the essence of codebreaking, finding patterns, and because it's such a basic human function, codebreakers have always emerged from unexpected places. They pop up from strange corners. Codebreakers tend to be oddballs, outsiders. The most important trait is not pure math skill but a deeper ability to pay attention. Monks, librarians, linguists, pianists and flutists, diplomats, scribes, postal clerks, astrologers, alchemists, players of games, lotharios, revolutionaries in coffee shops, kings and queens: these are the ones who built the field across the centuries and pushed the boundaries forward, stubborn individuals with a lot of time to sit and think and not give up.

Most were men who did not believe women intellectually or morally capable of breaking codes; some were women who took advantage of this prejudice to steal secrets in the shadows. One of the more cunning and effective codebreakers of the seventeenth century was a Belgian countess named Alexandrine, who upon the death of her husband in 1628 took over the management of an influential post office, the Chamber of the Thurn and Taxis, which routed mail all throughout Europe. The countess had a taste for espionage and transformed the Chamber into a brazen spy organization, employing a team of agents, scribes, forgers, and codebreakers who melted the wax seals of letters, copied their contents, broke any codes, and resealed the letters. This was an early example of what the French would later call a *cabinet noir*, or black chamber, a secret spy room in a post office. The countess's male contemporaries were slow to discover her true occupation because they couldn't imagine that a woman was capable of such deceptions. "What if this countess does not merely open our letters but is also capable of deciphering their contents?" one diplomat wrote in panic to another. "God knows what she is capable of doing to us!"

The two most prominent codebreakers in America when Elizabeth and William started were a married couple, Parker Hitt and Genevieve Hitt. Parker was a tall, dashing Texan with weathered skin, an army infantry commander in his thirties who had gotten interested in cryptology after volunteering to fight in the Spanish-American War and trying his hand with messages intercepted from the Mexican army. His wife, Genevieve, a proper southern girl, had scandalized her family by falling in love with a man they saw as a cowboy. She also studied cryptology, eventually becoming chief of the code operation for the War Department's Southern Division, based in San Antonio. "This is a man's size job," she wrote to her mother-in-law, "but I seem to be getting away with it, and I am going to see it through. . . . I am getting a great deal out of it, discipline, concentration (for it takes concentration, and a lot of it, to do this work, with machines pounding away on every side of you and two or three men talking at once)." Parker supported Genevieve and was proud of her: "Good work, old girl," he wrote to her in one letter.

Parker was the only American to have written a serious book about cryptology. Aimed at army units with no prior training, *Manual for the Solution of Military Ciphers* showed how to set up a quick-and-dirty deciphering office in the field with five or six soldiers, some radio equipment to intercept enemy signals, and a day or two of study. Hitt went over the basics of military cryptography and explained, accurately, that the methods of the world's armies had not changed much in hundreds of years. Just like there are millions of chicken recipes in the world but only several basic methods to cook the bird (roasting, frying, poaching, boiling), there are countless ciphers but only a handful of common types. Then he laid out some basic steps for solving a cryptogram written in cipher. Today a computer could do any of these steps in picoseconds, but in 1917 it all had to be done by hand, with a pencil and paper.

The first step was usually very simple: count the letters in the cryptogram. In English, the most frequently used letter is E, the most frequent two-letter group is TH, and the most frequent three-letter group is THE. So if you count the letters in the ciphertext and the most common letter is B, it might stand for E, and if the most common three-letter group is NXB, it might stand for THE.

You can count other things in a cryptogram, like the total number of vowels and consonants, and how often particular letters or groups of letters appear before or after other letters. All of these counts give hints to the hidden structure. A frequency count can also reveal if the plaintext was written in English, German, French, Spanish, or some other language, because the frequency of letters in a language is like a unique signature. The most common six letters in German, starting with the most common, are E, N, I, R, T, S. In French, E, A, N, R, S, I. In Spanish, E, A, O, R, S, N.

It's best to do the counting in a systematic way. You might start by drawing a thing called a frequency table. You chop the cryptogram into its component parts and sort them according to the letters by which they're surrounded. It looks like this:

A	B	C	D	E	F	G	H	I	J	K	L	M	N	O	P	Q	R	S	T	U	V	W	X	Y	Z
IP	HF		NR	PX	BA	JN	DJ	UJ	IG	NR	RJ		GD	JW	RE	AD	DP	FQ	AR	–I	OR	EY	XJ		
OQ	FR		QH		WS	PR	UN	RA	YO	PR	AI		HH	RA	AG	PJ	WI	JF		JH	JF	RP	JQ		
FT	FN		HR		JB	JJ	NB	WQ	HU				BG	WR	RQ	SW	GO			RI	QI	QP			
GL	PR				SB	NA	RD	QR	QW				QQ	RJ	XQ	II	TP				RO				
IP	HP						NB	LQ	LF				PH	IJ	XQ	PJ	IL								
							JR	PN	QG				IK		RB	IX	BH								
								QR	GS						RN	PN	DX								
								UA	OY						BI	NR	QP								
								QR	OH						AK	RI	BW								
								QO								YI	OR								
																RI	RP								
																	KQ								
																	IU								
																	KO								
																	IQ								
																	H–								

Though it may look like gibberish, it's a powerful tool—"the Real Stuff," in Elizabeth's phrase—because with a quick glance down the columns, you can identify the most frequent letter groups in the cryptogram and the letters that come before and after them. Letters in a given language are like children in a kindergarten class; they have affinities, cluster in cliques. In the lunch line, one kid likes to walk behind a second kid and in front of a third kid while a fourth sits off in the corner, eating from a paper bag. What you're really looking at when you look at this frequency table is a picture of "certain internal relations in the English alphabet," as Elizabeth and William would put it. You're looking at the structure of the underlying language itself.

Now you have some grip on the puzzle. You can begin to peel back the skin of the message, to see familiar shapes in the strangeness. Like with a crossword puzzle, there's no direct, guaranteed route to solving a cryptogram. The solver has to make educated guesses, plug in letters and see if they lead to recognizable words, backtrack and erase if a guess is wrong, try a new letter.

Elizabeth quickly got the hang of it, plowing through messages and counting letters, although it felt completely new and weird to her, a totally different way of looking at language than what she was used to. All her life she had celebrated the improbable bigness of language, the long-lunged galaxy that exploded out from the small dense point of the alphabet, the twenty-six humble letters. In college she trained herself to hear the rhythms of playwrights and poets, the syllables that slip from the tongue in patterns. Tennyson:

There lives more faith in honest doubt,
Believe me, than in half the creeds.
There LIVES more FAITH in HON-est DOUBT,
Be-LIEVE me, than in HALF the CREEDS.

But before, she had gone no further than chopping lines into meters. She left the words in their boxes, intact. Codebreaking required more drastic measures. Now Elizabeth had to shake the words until they spilled their letters. To rip, rupture, puncture, chisel, scissor, smash, and scoop up the rubble in her arms. To chip off flakes from the smooth rock of the message and place them in piles and ask questions about them. It involved a kind of hard-hearted analytic violence that she had never contemplated before. It was reaching into the red body of the text until the hands dripped with blood.

Ahhhh!

The first few messages she broke, real military messages, had been intercepted from the Mexican army. Like most military cipher messages, they were written in blocks of five letters, like TZYTV RGFQF MQFHC, in order to fuzz out the original lengths of the words and therefore make the messages harder for adversaries to break. Elizebeth counted the letters, drew her frequency tables, consulted materials on the frequencies of various letters and letter combinations in the Spanish language, scratched her guesses into the graph paper, and there, *right there*—she saw things that started to look like words. A lovely shape pried out of the murk, glistening.

The process gave her a sensation of power that was electric and new and made her want to keep going. It was nothing like working on the biliteral ciphers with Mrs. Gallup. Here there was no mystery, no squinting through a looking glass at the curls of italic letters and trying to sort them into categories based on vague criteria never fully explained. Here the method was sharp and clear, a series of small and logical steps that built toward a goal. "The thrill of your life," Elizebeth said later, describing how it felt to solve a message. "The skeletons of words leap out, and make you jump."

And she was never alone. That was the other thrill. She and William operated as a team. During the day they were never more than a few feet apart, handing papers back and forth, checking each other's work, asking questions when stuck, keeping up a friendly patter, "calling out" letters on their sheets in the "word-equivalent" alphabet commonly used by the U.S. Army: Able for the letter A, Boy for B, Cast for C, Dock, Easy, Fox, George, Have, Item, Jig, King, Love. If Elizebeth needed to read the ciphertext FVGEQ, she would call out, "Fox! Vice! George! Easy! Quack!"

It was demanding work. Each solution had to be checked. Errors corrected. A single miscopied letter could wreck hours of effort. You got tired and needed a friend to look at your page while you rubbed your eyes. Each learned to recognize signs of fatigue in the other and knew when to suggest that it was time for a break. The less you had to think about, the better and more accurate your work. Elizebeth and William used the same kind of pencil, the same kind of paper, and never deviated from these choices. They liked pencils with soft lead and big erasers, the eraser end seeing as much action as the lead end.

Cast! Easy! Jig! King! Opal! They called out letters all day long like teachers taking attendance at a strange school. *Pup! X-ray! Vice! Love! Sail!* The pencils at Riverbank were plentiful and free, black with white erasers, and doubled as advertising tools; a cipher alphabet was printed on each pencil in white letters, along with RIVERBANK LABORATORIES—GENEVA, ILLINOIS.

Mike, she called out, a smile playing at the corners of her mouth. *Zed. Rush. Fox. Zed.*

Watch, he called out, grinning. *Dock. Yoke. Pup. Easy.*

The paper they used was graph paper with a grid of quarter-inch squares. One letter per square. They never threw anything out. "Work sheets SHOULD NOT BE DESTROYED," the pair would soon write in one of several scientific papers about their discoveries. Worksheets "form a necessary part of the record pertaining to the solution of the problem. No work is too insignificant to discard, therefore it should be done well from the start."

Tare. Yoke. George. George. Able.

Unit. King. Nan. Zed. Boy.

No one told them how to set up this workflow, and no one told them they had to collaborate. They simply found, by trial and error, that collaboration made things go faster, that "a group of two operators, working harmoniously as a unit, can accomplish more than four operators working singly. Different minds, centered on the same problem, will supplement and check each other; errors will be found quickly; interchange of ideas will bring results rapidly. In short, two minds, 'with but a single thought,' bring to bear upon a given subject that concentration of effort and facility of treatment which is not possible for one mind alone."

Although William and Elizebeth solved their first batches of military messages using techniques they learned from Hitt's manual, they soon exhausted its teachings. Elizebeth filled the margins with her own notes in blue pen, comments about sentences she found imprecise, things Hitt had gotten slightly wrong, things he could have explained better or had left unexplored. (She underlined a sentence on page 85 and wrote next to it, "This is poorly expressed.") She and William had reached the cordon of what was known, the edge of the map. From here on, they would need to invent new techniques—to become scientists, explorers, pushing into a wild land.

One way of thinking about science is that it's a check against the natural human tendency to see patterns that might not be there. It's a way of knowing when a pattern is real and when it's a trick of your mind. Elizebeth and William had begun at Riverbank by looking for the false patterns of Mrs. Gallup. But now, over the next several years, they found ways of seeing true patterns. It was as if they had been tossed into a raging river of delusion without knowing how to swim and figured out how to save themselves from drowning, clinging to each other the whole time. This struggle made them stronger than they could have ever imagined. They climbed out of the river transformed, with new powers, shaking the water from their backs, and then took off at speed, racing across the mountains and through the swamps of an undiscovered continent.

Between 1917 and 1920, George Fabyan used Riverbank's vanity press to publish eight pamphlets that described new kinds of codebreaking strategies. These were little books with unassuming titles on plain white covers. Today they are considered to be the foundation stones of the modern science of cryptology. Known as the Riverbank Publications, they "rise up like a landmark in the history of cryptology," writes the historian David Kahn. "Nearly all of them broke new ground, and mastery of the information they first set forth is still regarded as the prerequisite for a higher cryptologic education."

The eight Riverbank Publications are commonly attributed to William alone, with two exceptions. Inside his personal copy of one paper, Riverbank No. 21, *Methods for the Reconstruction of Primary Alphabets,* William wrote in black ink beneath the title, "By Elizebeth S. Friedman and William F. Friedman." A second paper, *Methods for the Solution of Running Key Ciphers,* never included her name, but she and William always told colleagues it was a joint effort.

However, there's evidence that Elizebeth was involved with more than just the two papers. The original typewritten and hand-edited drafts of the Riverbank Publications are now held by the manuscript division of the New York Public Library, and her handwriting is all over them. William seems to have written a lot of the technical sections, with the drafts marked up by both of them, Elizebeth's comments interspersed with his, while Elizebeth wrote and researched the historical sections, which he edited in a similar fashion.

They worked as a team in most matters and the soon-to-be-legendary papers were no different. In a 1918 letter to Elizebeth, William referred to the early Riverbank Publications as "our pamphlets"—*our,* not *my.* And other Riverbank workers contributed as well: men and women, codebreakers and translators. The publications were "a piece of work that was done by the staff," Elizebeth said later. "No one person was mentioned as the sole conqueror or anything like that. Everybody worked together." This is as far as she ever went in claiming a piece of the credit. Today it's hard to know exactly what she did, because she wanted it that way. "Mrs. Friedman had a tendency to see that the record made little or no mention of her contribution to a number of their joint efforts," the custodian of the Friedmans' personal papers wrote in 1981 to a researcher interested in Elizebeth. "And therefore it will be difficult to get a clear picture of her exact role."

Why hide her role? Partly it was expected at the time, that the man was the scientist and the woman the helpmate, but Riverbank was a bubble world where the usual rules didn't apply. Fabyan had no trouble championing the work of women, as he proved with his zealous promotion of Mrs. Gallup. A more likely explanation is that Elizebeth was trying to help William win a battle with Fabyan over the copyrights of the Riverbank Publications. At first, Fabyan didn't even let William place his name on the covers of the pamphlets, only on the inside pages, and Fabyan registered the copyright under his own name. He said he saw no ethical problem because he had paid for the research. "It may be egotism on my part," Fabyan told William, "but so long as I pay the fiddler, I am going to have the privilege of selecting a few of the tunes."

It was hard enough for William—a credentialed scientist, a genetics Ph.D.—to get credit for the work. He and Elizebeth may have decided it would be doubly hard to convince Fabyan to share credit with her, too.

Whatever the case, the Riverbank Publications, and the breakthroughs they describe, still seem incredible today. Seven of the eight pamphlets were written in the space of two years, in a little cottage in the middle of Illinois, the cryptologic equivalent of Albert Einstein's annus mirabilis, when Einstein rewrote the language of light, mass, and time in the space of a single year, at age twenty-six, while working as a patent clerk in Switzerland, staring out the window of his office and bouncing ideas off a fellow clerk. This is the achievement that the NSA interviewer in 1976, Virginia Valaki, kept begging Elizebeth to explain: *How?* Elizebeth gave unsatisfying answers, noted in the transcript:

"That World War I leapt on, and so many things happened so fast. . . ."

"Nothing was ever as carefully executed as that. It was sort of on a day-to-day basis. You did what you could with what you had to do it with."

"I don't think I remember offhand. I was too busy either getting on this swing or getting off that one. ((Laughs.))"

"I feel no confidence whatever to speak on that point; wouldn't have the faintest idea what to say."

The likely truth: it only looked improbable in retrospect. At the time they didn't know what was supposed to be hard, and there was no one around to tell them. They didn't see themselves inventing a new science. They were playing the game day-to-day as best they could, as Fabyan always said. They were just trying to solve messages as they poured in and not get stuck.

The mail from Washington contained a frothy mix of messages from all over, a zoo of alphabets that had to be studied and classified. There were two main animal kingdoms of cipher, "transposition" and "substitution." A transposition cipher was like Scrabble, a jumbling of the same letters into a new order. A substitution cipher was a swapping of letters. Each kingdom contained a diverse multitude of beasts that had to be tamed in different ways, and there was always a time crunch, someone demanding a quick answer. Invention under pressure.

One day in early 1917 a heavyset man showed up at Riverbank on a mission all the way from Scotland Yard, the police headquarters in London. He had been referred to Riverbank by the U.S. Department of Justice. Fabyan barked an introduction at Elizebeth and William, and the detective opened a briefcase. Stacks of messages spilled out. He said the messages had been intercepted by British postal censors, and the recipients included as many as two hundred individuals in India, then a British colony. Scotland Yard suspected an attempt by Germany to spark a revolution among Hindu separatists, but no one knew for sure. All the detective knew were the names of a few of the suspects, which he told to Elizebeth and William.

The young codebreakers looked at the messages and found them "quite baffling." They were written in numbers, grouped together in shorter or longer blocks:

```
38425  24736  47575  93826
97-2-14
35-1-17
73-5-3
82-4-3
```

Elizebeth and William assumed from these groupings that the separatists were using three different codes. The blocks of five numbers looked like a simple type of codes based on a rectangular grid of letters:

	1	2	3	4	5	6	7
1	A	B	C	D	E	F	G
2	H	I	J	K	L	M	N
3	O	P	Q	R	S	T	U
4	V	W	X	Y	Z		

The grid turns a letter into a number (C is 13), which then has a number added to it, based on a prearranged key word. If the word is LAMP, the value of L (25) might be added to C (13), making 38. Elizebeth and William deduced the key word and solved the messages by analyzing frequent numbers and the intervals between them.

Thinking about the second set of numbers (97-2-14), which were confined to a single long message in the detective's trunk, Elizabeth and William noticed that the middle number was always either a 1 or a 2. This was a clue that the conspirators were using a specific, yet-unknown book to encrypt their messages and that the book had two columns of type, like a dictionary. The numbers likely pointed to words at certain locations in this mysterious book. For instance, in a sequence like 97-2-14, 97 meant the page number, 2 meant the right-hand column, and 14 meant the fourteenth word in the column. Applying similar logic to the third set of numbers in the detective's messages (73-5-3, 82-4-3), the young codebreakers deduced that the numbers pointed to *individual letters* within a different book possessed by the conspirators. The recipient of the letter, seeing the number 73-5-3, would turn to the seventy-third page of the book, go to the fifth line, and write down the third letter in the line.

Of course, this wasn't enough information to solve the messages, and the young codebreakers at Riverbank weren't sure they could go further: If the letters and words had been selected from specific books, and Elizabeth and William didn't know the names of the books or have copies of them, what was the use?

"For a time," William wrote, "it looked like an insurmountable task."

But they wrote down all the numbers in order, searched for repetitions, thought about it some more, and found a foothold. A Harvard professor had recently counted the words in a long English text, and the prairie codebreakers read his study. Of 100,000 total words, only 10,161 were unique, and just 10 words accounted for 26,677 of the 100,000: "the," "of," "and," "to," "a," "in," "that," "it," "is," "I." "You can't convey much intelligence using only these words," William wrote, "and yet you can't construct a long, intelligible, unambiguous message without using them over and over again."

Turning to the numbers that seemed to come from a dictionary, the codebreakers reasoned that a word at the beginning of
the alphabet, like "and," would correspond to a code group beginning with a lower number (1, 2, 3, 4) than a word that came
higher in the alphabet, like "the." This insight helped them solve
the most frequent words in the messages, and from there it was
possible to work out others: If 97-2-14 was YOU, then 99-2-17
must be a word close to "you" in the dictionary, perhaps "your,"
"young," or "youth." Elizebeth and William ended up solving 95
percent of the message in this manner, without ever seeing a copy
of the conspirators' dictionary. As for the last set of numbers, the
ones that seemed to refer to letters in an unknown book, they
used a similar process, matching frequent code groups with frequent letters and pairs of letters, and reverse-engineering the text
of the book as they went. Whenever they discovered a new letter
of plaintext, it told them more about the content of the book, and
whenever they pieced together a new line in the book, it told them
more about the plaintexts of the messages. One of the conspirators' notes read, in part:

> I challenge anybody who dares ignore the solid work done
> through our agencies. . . . Our men worked, suffered. Still
> suffering. . . . We have succeeded in laying foundation for
> future work . . .

The two young codebreakers ended up solving the whole trunkful of messages for Scotland Yard, revealing an intricate separatist
plot by Hindu activists living in New York to ship weapons and
bombs to India with the help of German funds and assistance:
dates, times, places, names. Several conspirators were charged in
San Francisco, and prosecutors summoned William to testify in
open court about how he broke the codes. Elizebeth wasn't asked.
She hated staying in Illinois while William went on an exciting
trip to the West Coast—she thought she deserved to be called as
a co-witness—"but someone had to stay behind and sort of oil the
machinery at Riverbank." She didn't speculate about why William
was chosen instead of her, perhaps because the answer was obvious: prosecutors thought the jury would more easily believe a male

expert. As it turned out, the trial erupted in spectacle: Before William had a chance to say his piece on the witness stand, an Indian man in the gallery stood up, pulled a handgun from his vest, and shot one of the defendants in the chest. He yelled a single word—"Traitor!" Then a U.S. marshal fired at the gunman over the heads of the shocked spectators, killing him. The shooter apparently thought the defendant had snitched to the government, betraying his friends by revealing the code. He didn't know about William, Elizebeth, and the science of codebreaking.

For the first eight months of the war, as incredible as it sounds, William and Elizebeth, and their team at Riverbank, did *all of the codebreaking* for every part of the U.S. government: for the State Department, the War Department (army), the navy, and the Department of Justice. And the broader scientific insights of the Riverbank Publications emerged directly from these day-to-day puzzles solved under wartime pressure. The pair would solve a cryptogram and realize they may have stumbled onto some more general method. Then they would test the method on additional examples, trying to see where it broke, what its limits were, aiming to strengthen a one-time solution into a universal principle and to share that knowledge with others.

It had long been known that the frequencies of letters in a cryptogram provide clues to its solution. Knowing this, cryptographers had invented many ways of obscuring the letter frequencies, making messages harder for adversaries to break. It was possible to encrypt a message with multiple cipher alphabets instead of just one (a poly-alphabetic cipher). It was possible to rely on a secretly chosen novel or dictionary to generate a code message, in the style of the Hindu revolutionaries. If the sender and recipient happened to select a key in their book that was exactly the same length as the message—known as a "running key"—the message became even harder to break. The War Department considered running-key messages to be indecipherable. And these methods could be mixed and matched to further frustrate the codebreaker. For instance, a plaintext word like "strawberry" could be turned into a block of code like WUBCW, then those letters transformed with a cipher into LWJIJ—a process of "enciphered code."

Each of these techniques placed a wall between the message and the codebreaker—sometimes a pane of frosted glass, sometimes a sheet of metal or stone. Elizabeth and William invented new tools for destroying these walls—hammers, corrosive acids, explosives. They learned to identify and solve several different kinds of substitution ciphers: straight alphabets, direct alphabets, reversed alphabets, poly alphabets, mixed alphabets. They developed general techniques of solving book ciphers without needing a copy of the book. They taught themselves to solve messages enciphered with running keys. Together in Engledew Cottage they strolled through cities of text with their wrecking kit, swinging hammers with glee, blowing up brick, melting steel, the sound of breaking glass echoing out into the prairie. Then they wrote the scientific papers, the Riverbank Publications, documenting exactly how they did it, and how other people could do it, too, if they followed the same steps.

This part was crucial. The test of a scientific discovery is if others can replicate it and get the same results. Mrs. Gallup had never passed this test. Elizabeth and William wanted to pass. They later wrote, "What Colonel Fabyan failed to realize, throughout his campaign to 'sell' Mrs. Gallup's decipherments, was that no demonstration, however good, can take the place of experiments which can be repeated and will produce identical results."

To drive home this point, William even invented a new word: "cryptanalysis," synonymous with "codebreaking." The new Riverbank methods were not magic but a species of *analysis,* similar to the analysis performed by a chemist or an astronomer or an engineer designing a bridge.

Serendipity still played a role in codebreaking. "Many times," the pair wrote, "the greatest ally the mind has is that indefinable, intangible something, which we would forever pursue if we could— luck." Epiphanies happened. Insights that seemed to come from nowhere, bolts from the blue, guesses that made more progress on a problem than days of dreary labor. Mrs. Gallup had always called this "inspiration." Elizabeth and William preferred to speak of "flexibility of mind" or "intuitive powers" because these phrases sounded less magical. Intuition, to them, was like a hard-earned internal compass, a grooved-in sense of how to move forward that came from patience, skill, and experience. It could be cultivated.

Starting here at Riverbank and continuing throughout their lives, people tended to describe the brains of Elizebeth and William in gendered ways, as if her style of solving puzzles had a starkly different texture than his. Elizebeth's was usually said to be the more intuitive mind, William's the more mathematical. He was supposed to be better with machines and she with languages—Elizebeth was rapidly picking up German and Spanish, and learning pieces of other tongues. There may have been some truth in it. But the reality is that they were both mathematical neophytes, even William. A future colleague of William's, Lambros Callimahos, a classical flutist and trained mathematician, idolized William to the point of copying his personal habits; upon learning that William liked to use tobacco snuff, Callimahos took up snuffing. But Callimahos recognized that whatever made William good had little to do with math. He described William as a man "cursed by luck," writing, "Even if he computed odds incorrectly, it didn't make any difference because he would forge ahead in his blissful ignorance and solve the problem anyway. On several occasions he told me that if he had had more of a mathematical background, he might not have been able to solve some of the things he did." If William had been older or better trained, "he could have been ruined. His definition of a cryptogram was simply a secret message that was meant to be solved, just that."

To those who had a chance to watch them both work, the minds of William and Elizebeth appeared equally amazing and equally incomprehensible. Their brains were Easter Island statues, stony and imposing. Colleagues resorted to mystical analogies. William was like a latter-day King Midas: "Everything he touched turned to plaintext." Elizebeth's gift for puzzles was "God-given," "an effect without a discernible cause." Who was the better codebreaker, William or Elizebeth? People gave up trying to figure it out. A raffish young army officer from Virginia, J. Rives Childs, met William and Elizebeth at Riverbank in November 1917; they taught him the science of codebreaking, and he went on to serve with distinction in the war. Childs found it impossible to tell if William was smarter than Elizebeth or if it was the other way around: "I was never able to decide which was the superior."

Elizabeth and William sometimes played into the stereotype that he was the mechanical male thinker and she the sensitive female thinker. It was a helpful shorthand for explaining the inexplicable.

There's a now-famous story that encapsulates how they thought about their own differing brains. One day during the war, a series of five short messages arrived at Riverbank from Washington. It was a test of sorts. The messages had been encrypted with a small hand-operated device recently invented by the British army to make their field communications more secure. The device was a kind of cipher disc, with two alphabets printed on rings that rotated with respect to each other, but with a twist: while the outer ring had the usual 26 letters, the inner ring had 27. The extra letter introduced a degree of irregularity, making it harder for a codebreaker to visualize the alphabets sliding against each other. The device also allowed the cryptographer to change the alphabets quickly and easily.

The British had already concluded that the device was unbreakable. So had experts in France and a few in America. But to be certain, an official in Washington had used the device to encrypt five test messages, using two alphabets of his choosing. He then sent the messages to the Colonel, to see if Riverbank could solve them.

William looked at the messages. He had been given a description of the device that produced them, but not alphabets. His only chance to solve the messages was to reverse-engineer the alphabets the Washington official had used.

He began with the assumption that the official wasn't an expert cryptographer: a safe assumption, because almost no one is. Therefore the official might have made any number of common blunders that people often make when trying to communicate securely. The strength of a cryptographic system usually has less to do with its design than with the way people tend to use it. Humans are the weak link. Instead of changing keys or passwords at regular intervals, we use the same ones over and over, for weeks or months or years. We repeat the same words (such as "secret") at the start of multiple messages, or repeat entire messages multiple times, giving codebreakers a foothold. We choose key phrases that are easy to guess: words related to where we live or work, our occupation, or to whatever project we're working on at the moment. A couple of human mistakes can bring the safest cryptographic system in the world to its knees.

It struck William that the Washington official, in preparing this important test of a cryptographic device, might have used key words related to the practice of cryptography. So William tried words like "cipher," "alphabet," "indecipherable," "solution," "system," and "method." After two hours of intense focus, he was able to piece together what he thought was the alphabet on the outer disc, which seemed to use the key phrase *cipher*. William now assumed that if the official had been careless enough to use a guessable key phrase in one alphabet, he had probably used a similar key phrase in the second alphabet too, on the inner disc. But this one proved tougher to crack. William tested all sorts of key phrases; nothing worked. He turned to Elizebeth for help.

"I was sitting across the room from him," she recalled, "busily engaged on another message":

> He asked me to lean back in my chair, close my eyes and make my mind blank, at least as blank as possible. Then he would propound to me a question to which I was not to consider the reply to any degree, not even for one second, but instantly to come forth with the word which his question aroused in my mind. I proceeded as he directed. He spoke the word cipher, and I instantaneously responded, "machine." And, in a few moments Bill said I had made a lucky guess.

Later, in writing and interviews, Elizebeth would try to explain the "springlike elasticity" of her mind in this moment. What led her to blurt out the word "machine"? Where did it come from? All she would say is that because the British device was small and hand operated, "it did not occur to [William's] meticulous mind to use the word machine. But to me it was a machine." Thanks to Elizebeth's guess, she and William were able to solve the five test messages in less than three hours.

William attributed Elizebeth's insight in this case to the fact that she was a woman. He later said in a lecture, "The female mind is, as you know, a thing apart." He appears to have made a joke about sex as well. He and Elizebeth had just gotten married when this story took place. William recalled in his lecture to a roomful of men, "I came to the end of my rope and said to the new Mrs. Friedman: 'Elizebeth, I want you to stop what you are doing and do something for me. Now make yourself comfortable'—whereupon she took out her lipstick and made a few passes with it." Imagine laughter.

To hear Elizebeth tell it, William was the brighter one. Before they were married and before they were a courting couple, she was already starting to praise his abilities in a way that minimized or overlooked her own, setting the pattern for the rest of her life, the moments when she would describe William to friends and to reporters as a man of history and destiny, "a wonderfully warm man, with the broadest of minds and intelligence," and even "the smartest man who ever lived." All the same, she was competitive by nature, and at times the two of them indulged a cheerful rivalry.

Once, in a dusty 1896 issue of the literary magazine *Pall Mall,* William and Elizebeth discovered an article about ciphers used by anarchist opponents of the old Russian czars. The article included a brief cryptogram at the end. In general, the shorter the cryptogram, the harder it is to solve, the same way a song is harder to identify by three notes of its melody instead of twenty. This "Nihilist" cryptogram consisted of only a few numerals and two question marks:

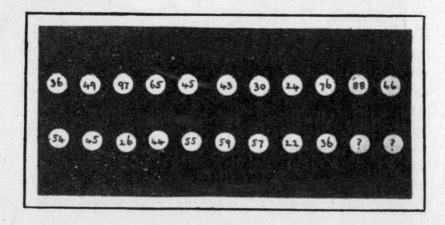

No solution was given. "The meaning of the cipher which now follows will never be solved by any one," the author wrote, concluding that the lock "has now closed and firmly shut its fastenings."

Naturally, William grabbed a pencil and began trying to pick the lock. "Well," Elizebeth writes, when William "met up with that message, he took the challenge and set his teeth into the tough nut with a snap. And would you believe it, he deciphered the message, short as it was, *and the key*, in 15 minutes!" The key relied on a single repeated word: "courage." The plaintext read: "He who fears is half dead."

It's a convincing piece of testimony to his greatness: William Friedman, the smartest man who ever lived. But instead of ending the story here, Elizebeth goes on: "Of course, when I learned that, I too had to try my hand" at the cryptogram. "I unlocked the forever-to-be-hidden secret in 17 minutes."

By the spring of 1917, William was in agony. He had known Elizebeth for eight or nine months now and he wanted her all the time. He was afraid to say it out loud because he didn't know if she would reciprocate. She had never called him anything but a friend. But he was sure he loved her. It was getting hard to sit with her all day and pretend to be thinking about work. In the moments when he appeared to be scratching away at a puzzle, he was really wondering what her hair would look like if he reached behind her neck and removed the pin, her beautiful loosened hair if he pulled her close.

He imagined a life with her, a house, children, and at the same time he could not imagine it. He knew that his Jewish family and friends in Pittsburgh would not approve of him marrying a non-Jew. The community there had always seen marriage between Jews and Gentiles as a kind of betrayal, a weakening of Jewish resistance to a hostile and bigoted American culture. When William was growing up, Pittsburgh's *Jewish Criterion* weekly newspaper made the case against intermarriage in repeated articles and editorials:

The glacial undercurrents of racial antipathy between Jew and non-Jew cannot be tepified by even the hottest fiercest rays of the sun of love! Statistics and the divorce courts prove this.

A part cannot become merged into a whole without ceasing to be a part. The Jews don't want to merge.

WILL THE JEWS COMMIT SUICIDE THROUGH MIXED MARRIAGES?

He feared what his people would say. But desire trampled the fear.

Soon enough, but not yet, he would tell Elizebeth what he thought of her: that "your soul and spirit and heart are as fine, sweet, and pure as your body is beautiful"; that the perfection of her body was matched only by the clockwork of her mind, "brilliant and quick and clever"; that she had him "skinned to a frazzle" in the brains department. He marveled at her ability to escape the bounds of a problem, to strike the flint of her thoughts against different rocks, against history and math and logic and against William, too, shooting out sparks, ribbons of flame. "You're lots smarter than I am in ciphers. You can soar away into the clouds and still remain planted firmly upon solid ground and reason. You can dream and be practical." He would tell her, over and over, that he couldn't express what was in his heart: "Oh Divine Fire Mine, I adore you, how futile are words!"

Divine Fire. A nod to the Bible of his father. The devouring God of the Old Testament, *whose fire is in Zion, and his furnace in Jerusalem.*

One of the mysteries of falling in love is that it makes you inarticulate and eloquent at the same time. You lose the ability to speak and write in normal ways (*How futile are words!*) even as you develop, with this person you love, assuming this person loves you back, a shorthand of glances and gestures. At first it seems like your beloved is "speaking in code"; later, maybe, it's like the two of you are *sharing* a secret code.

This feeling may have a deep scientific explanation. In the 1930s and '40s, before the digital computer was invented, a young scientist from rural Michigan named Claude Shannon wrote two papers that were like magic beans for the computing revolution, growing the great beanstalks of IBM, Apple, Silicon Valley, the Internet. As a graduate student at the Massachusetts Institute of Technology, Shannon realized that electronic circuits could be arranged to solve logic problems and make decisions, and that 0s and 1s could encode all the world's information, from a song to a Van Gogh. He didn't create the first computer, but he was one of the first to grasp the immensity of what digital computers could do.

Shannon, who would later work with William Friedman and other cryptologists on secret NSA projects, also enjoyed thinking about codes and ciphers. While employed at Bell Labs, he came up with the insight that the problem of communicating through a noisy system, like a phone wire, is almost identical to the process of enciphering and deciphering a message. In other words, according to Shannon, making yourself understood to another person is essentially a problem of cryptology. You reduce the noise of the channel between you (instead of noise, Shannon called it "information entropy") in a way that can be quantified. And the method for reducing the noise—for recovering messages that would otherwise be lost or garbled—is decryption.

Viewed through Shannon's theory, intimate communication is a cryptologic process. When you fall in love, you develop a compact encoding to share mental states more efficiently, cut noise, and bring your beloved closer. All lovers, in this light, are codebreakers. And with America going to war, the two young codebreakers at Riverbank were about to become lovers.

CHAPTER 5

The Escape Plot

Elizebeth with William in his army
lieutenant's uniform, around 1918.

To be your North Star—Billy Boy—I'd like to be!"
   She wrote "North Star" on the page of the letter without
knowing what it meant or how she wanted him to respond. It was
just a pair of words that captured a tug of attraction toward William
Friedman, a chemistry that made her curious: a small, persis-
tent tilt in his direction, like a plant bending toward a patch of sun.

Elizebeth first noticed it when she was home in Indiana watching her mother die, and writing to him at Riverbank, telling him—what? I don't love you but "I miss you infinitely." I am not sure I *love-you* love you but "I shall work for you" if you ask. I have dreamed about you but I don't remember what the dream was. "Anyway, Billy Boy, like me just a little bit always. I want you for the dear good friend you are, if nothing more. I want, oh, so much, for us both to 'achieve.' "

Before any feelings of passion, this is what Elizebeth expressed to William: a vague desire for the both of them to *win*. "Work hard on the letter tests—for my sake! You must win—because I want you to!" She saw his talent and was starting to understand the size of her own abilities. She sensed that the two of them were more powerful together than they were alone. *That* excited her, if not the thought of romance, a relationship, making love. When they kissed for the first time it didn't do much for her, as she would recall later in a poem:

> *There was a time when for my love I did not care*
> *The hot wooing, the passionate kisses*
> *left me cold.*
> *I yielded to him*
> *because he was good to me.*
> *And compassion led me to return his kisses*
> *when his longing eyes and eager heart spoke,*
> *"I wish you cared as I."*

She wasn't sure what to do when he started talking about marriage in early 1917. The ice cracked on the Fox River, spring picked the lock of winter, the water moved, and William Friedman spoke to her about the pros and cons of a potential union, obstacles and advantages, in a careful, unemotional tone, as if discussing a job opportunity. (He confessed later that he was simply trying to hold back the flood of his feelings, lest the dam burst and he embarrass himself.) He did not get down on his knees and propose, and his hesitation gave her the room to respond in kind, to examine the idea with detachment, to let the possibility of marrying him drift past at a distance like a harmless puff of cloud.

She could see a case for marrying him and a case against. If she did choose to marry, her family and neighbors back home would likely be confused—nice Quaker girls didn't marry Jewish boys in Indiana—but she had made more of a break with her family than William had. She had fought hard for the freedom to choose her own path.

Most of the time, talking it over that spring, Elizebeth and William agreed that getting married would be silly. There were too many barriers: family differences, religion, money. Incredibly, Fabyan was still paying them both the same salary as when they started, thirty dollars a month, despite their massive new responsibilities, and some months he didn't pay at all. If they did decide to get married, where would they live? In William's windmill? The prospect of being a married couple at Riverbank seemed absurd to both of them, as much of a fantasy as the perpetual motion machine and the messages from Bacon's ghost.

At the same time, they weren't sure they wanted to stay at Riverbank anyway.

The longer they lived here, the more concerned Elizebeth and William became about Fabyan's dark side, his need to control the people around him. He seemed to take special pleasure in humiliating William. Once, while both men were traveling in Washington, Fabyan demanded that William fetch a newspaper from a street vendor. William pointed out that papers were available at the hotel desk; Fabyan bellowed that he wanted one from the street. William obliged. Later, when William showed up to dinner in a freshly pressed evening suit, Fabyan, who was dressed shabbily, forced William to change into more casual clothes to match. "It just didn't go down to be treated like chattels," Elizebeth said. "We were sick and tired of Fabyan's scheming and dishonesty. Fabyan always came out ahead, and we always came out at the other end."

Fundamentally Elizebeth and William were two ambitious people. *Why should something with a risk in it give me an exuberant feeling?* She wanted to live a daring life, and he wanted to "make a mark in something," he told her. "Perhaps it will be Genetics." He expressed to Elizebeth "my ambition to know one little thing better than any other person, to be a pioneer in that field and to blaze new trails for the rest to follow. Why I feel that way, I don't know—it's just in me and will have to come out in some form or other." Riverbank had launched them on an incredible adventure, but now it was holding them back, and they sensed that if they were ever going to escape, they needed to do it together.

On a cool, rainy Monday in May 1917, they went missing from Riverbank.

It wasn't like them to skip work. The hours ticked away without them, the cows eating grass in the field, the fruit flies multiplying in jars. When the pair returned that evening, William was dressed in a dark blazer, light-colored striped pants, and a striped tie, and Elizebeth wore a simple gown of white lace. Colleagues gathered around and the couple shared the happy news: They had gone to Chicago and gotten married. A rabbi named Hersh performed the ceremony.

The wedding announcement ran on the front page of the May 23 *Geneva Republican*, next to a story about a Selective Service bill just passed by Congress, requiring men ages twenty-one to thirty-one to register for the military draft. "Mr. Friedman came to Riverbank soon after his graduation from Cornell University and was employed for some time on experimental work in the Riverbank greenhouses," the paper wrote of William. "He later took up the work in connection with the Bacon research studies." As for Elizebeth: "Miss Smith's home is in Indiana. She is a college graduate and a splendid reader. She and Mr. Friedman have lectured on the Bacon cipher before colleges, schools and clubs." The article didn't mention that Elizebeth was instrumental in convincing him to "take up the work."

She married him without being in love. She admitted this later in her diary, after picking up a novel called *The Prairie Wife* and reading the opening line of the book's woman narrator: "Splash! . . . That's me falling plump into the pool of matrimony before I've had time to fall in love!" Elizebeth recognized that sentence as one "I might myself have spoken." She married William because he was a good person, and he wanted it with such overwhelming intensity, and she trusted that the rest would come soon, certainly soon, because so much had already happened in the shortest time. It had been almost exactly a year since the Colonel met Elizebeth in Chicago and whisked her off to this patch of prairie. "I am learning to take things as they come," she wrote in the diary soon after the marriage ceremony, taking stock of how her life had been transformed in a flash:

> To be mangled and torn and castigated and macerated in soul—to wish passionately day after day only to die . . . and then to be brought, by a miracle, to a new place—to work that is absorbing, fascinating—to a place where I forget and find peace, glorious peace—and oh, miracle of miracles, to Love! Ah, Heaven is good! Truly, truth is stranger than fiction! I could not have believed it possible, but here am I. Is it possible I am to have them after all— Youth and Love, and Life?

The reaction to their marriage in Pittsburgh was just as they had feared. When William traveled back home briefly to tell his parents, his mother collapsed at the news that her son had married a shiksa. He told Elizebeth about it in a wire to Riverbank. She read it and felt sick. "I am cast into a whirl of remorse, pain, and sorrow for you," Elizebeth wrote to him. "Oh, Billy, Billy, what have we done?" She told her family in later years that when she visited her in-laws in Pittsburgh, William's mother would sit and weep. "You would have thought that Bill had committed murder," Max Friedman, one of William's brothers, later recalled. "If he had still been living in Pittsburgh he would have been ostracized."

But William didn't live in Pittsburgh anymore. He lived in Illinois, in a rich man's windmill. And now Elizebeth did, too. She moved from Engledew Cottage into the windmill. He made room for her journals and papers and books and brought her up the steps to the second floor. It was humid and cramped, and it smelled of soil, but it was theirs.

That summer the military was sending teams around the country to recruit volunteers. Fabyan invited the army to Riverbank in July. He ordered his employees to build a wooden stage at the highest point of the lawn next to the Villa, and a recruiting tent next to the stage. Three thousand people came from Geneva and surrounding towns, clogging the roads with horse-drawn buggies and automobiles, a prairie traffic jam.

A U.S. Army captain stood on the stage, tall and clean-shaven, with a hank of brown hair gelled to a stiff peak, wearing a uniform of olive drab wool with gold piping and black dress shoes. "Better to go and die than not to go," he cried. "Women, plead with your sons and brothers and sweethearts." The captain said that anyone who spoke of peace should be shot as a traitor. At the end of the speech, a boy from Elgin stood and walked to the recruiting tent. The crowd cheered, and more boys stood and followed him. A bit later, Fabyan invited the guests to tour his model trenches for a fee of twenty-five cents per person, to be donated to the Red Cross. He raised more than three hundred and fifty dollars. Men in bowler hats and women holding parasols stood at the edge of the trenches, peering down. Children ventured inside and played in the mud.

William did not volunteer for the army that day, but he was starting to think about it, partly out of guilt—he was a healthy male in wartime—but also out of concern for himself and Elizebeth, their future together. He wondered if he could use his codebreaking skills to get commissioned as an officer. The army paid more than Fabyan. It laid out clear paths to promotion. And there were army bases all over the country and the world. When the time came for the couple to leave Riverbank, they could leave with good prospects.

He began to pester Fabyan about it, asking the boss to reach out to his contacts in Washington and recommend him to the Army Signal Corps, Joseph Mauborgne's section. William said he wanted to go to France and apply his code and cipher knowledge closer to the fighting. Fabyan always waved him off. William was needed in his present position. He should forget about the army and concentrate on his work.

Frustrated, William took matters into his own hands, writing to Joseph Mauborgne and asking if the army had any use for his abilities. At the same time Elizebeth wrote to the navy to inquire about codebreaking positions. They waited for replies. Months passed. Nothing.

It wasn't until later that the Friedmans learned the truth. They heard it from Mauborgne and others who had been desperately trying to reach them the whole time. Fabyan was intercepting the Friedmans' mail. He had taken the job offers that arrived for them from Washington, put them in a drawer, and responded himself, informing Washington that the Friedmans were unavailable.

Also, one army officer who visited Riverbank for cryptologic training told William he discovered secret listening devices in the classrooms. Bugs. It seemed obvious that Fabyan didn't place the bugs to spy on the students. The students didn't know anything. It would have been pointless. The only logical explanation was that Fabyan had been spying on the Friedmans, in order to anticipate their movements and prevent them from ever leaving his Garden of Eden. *It's made honest bees out of them, this constant supervision:* Fabyan was surveilling his young employees as if they were two honeybees in his colony, under glass.

A tiny slip of paper fluttered down to Elizebeth. She was outdoors at Riverbank with William and Mr. Powell, the gentle University of Chicago publicity agent, the three of them working in the grass, the fresh air. She picked up the paper and saw a line of cursive written in light pencil. It was from William. "My dearest, I sit here studying your features. You are perfectly beautiful!! B.B." Billy Boy. She hid the note so Mr. Powell wouldn't see it, later pressing it between two pages of her diary. "My heart sang," she wrote there, "carolling bursts of ecstasy."

She wasn't pretending anymore or yielding to William out of kindness. She was the one throwing her arms around him in the cottage when Mrs. Gallup wasn't looking and pulling him into a kiss. "My Lover-Husband," she called him now:

TONIGHT MY LOVER-HUSBAND *and I made a tryst with the future.*
THE GOAL IS *set; will we win? We planned it all—cheek to cheek—facing the swelling power of the new moon—*
"WONDER-GIRL," HE SAID, *"It shall be all for You—only for You!"*
AS I HELD *him close and caught my breath in the intensity of hope, he said—"Dear Heart! You are not crying?"*
AND I REPLIED—*"No dear, only praying." And this was my prayer:*
*"*OH SPIRIT WITHOUT *and Within, keep me sweet! Keep me working on & on & keep me well—keep the Fire Burning!"*

Their work started to dry up in the summer and fall of 1917. Each new parcel from Washington was lighter, containing fewer intercepted messages to solve. Something had changed. Fabyan paced and fretted. He raged in the hell chair.

It turned out that the War Department had recently launched a codebreaking unit of its own, under the command of a twenty-eight-year-old lieutenant named Herbert O. Yardley, a scrawny Indiana native who had become entranced with cryptology after reading library books that told about the old black chambers of Europe. "Why did America have no bureau for the reading of secret diplomatic code and cipher telegrams of foreign governments?" Yardley asked himself. "Perhaps I too, like the foreign cryptographer, could open the secrets of the capitals of the world."

Fearless and charming, and a shark at poker, Yardley considered his new bureau to be an American black chamber, and he had no trouble convincing the War Department to let him solve messages that would have otherwise been shipped to Riverbank, seven hundred miles away. Known officially as MI-8, and based at the Army War College in Washington, Yardley's bureau had shattered Fabyan's near monopoly on American codebreaking.

Fabyan, aware that he was losing influence and power by the day, now came up with a plan to win it back. He knew the military needed many more codebreakers than it could locate and train quickly, both to work for Yardley's Washington bureau and for the American Expeditionary Forces (AEF) in France. America needed a codebreaking school, and here was River-bank, already set up as a university of sorts. He invited the army to send men to Riverbank for training. The army took him up on it.

The first students arrived in November 1917, four young lieu-tenants destined for the war front. They knew nothing about codes and ciphers. They were, as an NSA historian would later say, "as dumb as anyone just off the street." Fabyan asked Elize-beth and William to teach them.

The Friedmans had never taught a class before. They had no lesson plans and a grand total of one year of codebreaking experi-ence. There was nothing to do but to do it.

"What was taught was taught," Elizebeth later recalled, "and we taught it with what we had."

The first four students soon became eighty young officers in training sent from the Army War College, many accompanied by their wives. There wasn't enough space to house the visitors at Riverbank, so the Colonel booked the largest hotel in the nearby town of Aurora, and William and Elizebeth taught class there every day, lecturing in the morning and correcting problem sets in the afternoon. They started with the biliteral cipher of Francis Bacon and moved on to more contemporary methods of encryp-tion and decryption, using actual messages from the Spanish-American War and German intercepts from the first two years of the Great War. Mrs. Gallup sat off to the side during the classes, observing but not teaching.

The students knew that after graduating, many of them would be going straight to war, deployed to France as code and cipher officers with the AEF, and others would be assigned to similar work in Washington. This awareness of ordeals ahead made Riverbank seem that much sweeter. It was like being sta-tioned in paradise. Fabyan provided the students with daily box lunches with fresh food from the farm, organized outings into the countryside, and threw parties where the single men could mingle with local girls, including a lavish military ball that ush-ered the golden-haired daughters of Geneva into the arms of the uniformed officers. At least four of the officers' wives took the classes and completed the course. The Colonel commended the women for their "excellent work" in a letter to the War Department, though he did not list their own names but instead the names of their husbands.

On the last day of the course, in late February 1918, the students and instructors gathered outside the hotel for a photo, lining up in two rows that stretched from one end of the building to the other. In the photo, William, Elizabeth, and Fabyan sit front and center, William looking off to the right; the army men take stiff poses, some of them angling their heads 45 degrees to either side, some looking straight on. The significance of this curious feature of the photograph escaped almost all who viewed it at the time: *Each person stood for a letter of ciphertext in the biliteral cipher.* The ones looking to the side were the *b*-form of the cipher. The ones looking straight on were the *a*-form. United they spelled out the motto of Francis Bacon, the phrase chiseled into stone by the Colonel above the Acoustics Lab door: KNOWLEDGE IS POWER.

For the rest of his years, William would keep an enlargement of this photo beneath the glass surface of his work desk, glancing at it most every day of his life. It was a reminder of a more innocent moment, a time before two dark and interrelated forces began to draw boxes around his days, shaping his path and Elizebeth's, too. War was one. Justifiable paranoia the other.

William finally raised enough of a stink with Fabyan that the boss said he could leave Riverbank on the condition he return to Riverbank when the war was over. He entered the army as a first lieutenant in the signal corps, an officer but a low-ranking one. He was headed to France to ply his code and cipher abilities with the AEF. An army photographer snapped his official portrait in a darkened room with a lamp to the left, illuminating the left half of his face and body. In the picture, he looks serious and delicately handsome. His ears seem to stick out more than usual. He liked it and gave Elizebeth a framed copy as a gift. In May 1918 they said goodbye, Elizebeth smiling through her tears, and Lieutenant Friedman boarded a train to Chicago on the way to his destination, American General Headquarters (GHQ), in the farming town of Chaumont, France.

Elizebeth wanted to go with him. She saw no reason why she should not be allowed to serve in France as an AEF cryptologist. But the army told Elizebeth that "I, a mere woman, could not follow to pursue my 'trade,' " so she stayed behind at Riverbank, continuing to break codes in the Lodge with the other brain workers of the estate. In her diary she wrote original poems about war, exploring "the heartache of separation from the Dear One overseas" and recognizing that she needed to take care of herself, to preserve an inner mental space that was solid and clear ("a calm Whole, a unified peace"), and all the while reading the time-lagged stream of letters that arrived from William weeks or even months after he sent them, the envelopes stamped with the red mark of the AEF censor before crossing the ocean to the prairie.

She could tell from his letters that he missed being able to talk to her about a puzzle when he got stuck. For security reasons, he had to speak vaguely and omit all technical details. "The work is so hard," he wrote, "and the results so very, very meagre. Sometimes I fear that I haven't got it in me at all. I cannot explain to you—but just imagine yourself at work absolutely in the dark, up against the most baffling problem, with no data to base speculation upon, no guiding generalizations, except the most vague and unreliable—Oh, I tell you Honey, it's going to be an awfully hard task to make good."

At the same time it was clear from the letters that William's reputation was growing in the army, that some days solutions appeared to him "out of the clear blue," startling his colleagues. "On Saturday Col. M brought around a visitor, some Col—I don't remember his name. When he came to my desk he introduced me and said, 'He is our wizard on Code.' Dearest, I was quite embarrassed, and didn't know what to say." Col. M was his commander, Frank Moorman, and William made it a point to tell Elizebeth that Moorman admired the Riverbank Publications, hoping she would feel proud of her work on them. "Love-girl," he wrote, "yesterday at conference Major M passed around our R.K. pamphlet"—Riverbank No. 16, *Methods for the Solution of Running-Key Ciphers*—"and said that he went through it with much interest and that it was the best thing he had seen on the subject."

William told her not to worry about his physical safety. Chaumont was far enough from the trenches and the firefights that he felt there was no danger of needing the .45 pistol that he carried on his hip.

Like the other officers, William lived in a billet, the private home of a French woman he called Madame. It was so dark in the French countryside at night that during his first weeks he had trouble finding his billet and had to use cigarettes as torches. In the day he worked in a building behind GHQ known as the Glass House, surrounded by the other AEF codebreakers and radio operators. The Germans used a field cipher based on six letters: A, D, F, G, V, and X. A message might look like FAXDF ADDDS DGFFF, or DDFAX SDGVV AFAFX. William spent a lot of time fiddling with these six-letter nonsense messages, groping for the light cord in the darkened room of the cryptograms. The men in his section preferred to work alone, which he found baffling. "You know how much 'group work' counts in our business," he wrote to Elizebeth. "What can one person alone accomplish?"

He tried drinking French wine and didn't care for the taste. Lemonade was more to his liking. At the officers' club he kept to himself, nursing highballs in a plush chair in front of a roaring fire, except when the men dragged him into a poker game, which he always regretted, losing money each time he played. The Midas of codes had no talent for reading human faces. In this he was the opposite of Herbert Yardley, leader of the MI-8 codebreaking unit in Washington, who also spent time in France during the war. William met Yardley there for the first time and thought he seemed fake. "I must confess to considerable distaste for Y. Frankly, I didn't like him at all," he wrote to Elizebeth. "He acted like a wooden Indian."

Feeling lost and out of place, and wishing the army had permitted his wife to serve in Chaumont, he walked back to his billet in the dark at the end of the night and spent hours writing to her. He had placed a picture of her next to an oil lamp and each time he struck a match to light the lamp he looked at the picture and said, out loud, "Hello, you darling! Hello, Rita Bita Girl," then lit the flame and started to write, imagining he was back in bed with her at Riverbank, stroking her hair, talking in baby talk. He fantasized about spanking her. "Do you miss your Biwy Boy, my darling? Have you been naughty? Do you need to be spanked? You little 'imp.'" He said that was as far as he dared to go, with the censor reading every word, and promised that someday he would cable her "some real stuff which may burn the insulation of the wires."

During these months in Chaumont, four thousand miles from home, William became tormented by feelings of inferiority and romantic inadequacy that would never completely go away, gnawing at him for the rest of his life. He worried he was too unprosperous for Elizebeth, too interested in science instead of money, too effeminate. He apologized for "the many imperfections in my makeup." He asked for reassurance that he was a "good lover": "You have told me that, haven't you?" One day he happened to meet Colonel Parker Hitt, the Texas codebreaker now posted to France, and needed to crane his neck up to look at him: "He actually towered above me." He went to sleep at night with the windows open and often dreamed of his wife, a recurring dream where she was leaving him because she didn't love him anymore, then he woke up in a sweat and lunged for his pen and a piece of paper to copy it down: "You didn't yike me at all," he wrote, baby talking, "and I was all broken up." He begged her forgiveness for leaving her alone at Riverbank with "no money and a lot of debts" and promised to pay more attention to her happiness. "When we were together," he wrote, "I was particularly mean not to take you out more often, even though we couldn't 'afford' it, or even though there was lots of work to do. The Spring Time of love was ours—and I failed to make it all that it should have been for you. I owe it to you—and you shall have it all a thousandfold over when we are together once again. 'Afford it' or not."

Against these anxieties and regrets William possessed only one weapon: language, wordplay. Every day he felt he was losing a little more of his wife and every day he felt he must fight to win her back, so he labored over his letters, making corrections, emendations, fixing rare grammatical mistakes, turning the pages 90 degrees and adding sentences at right angles in the margins, trying to find the magic incantation of symbols to crush the globe flat and cheat the distance between them. He filled the pages with encoded messages of devotion he had every reason to believe she would understand. "This cable will read apparently harmless— but each letter and punctuation mark on it is but a group standing for a whole phrase which I wish I had the power to vary—but have not. The phrase is 'I love you!' It has two alternates—'I adore you!' and 'I worship you!' So if those tiny flashes of electricity can talk to you they will whisper to you over and over again, with an infinite permutation of expression my message of love for you." It was a lover's code:

A = I love you! / I adore you! / I worship you!
B = I love you! / I adore you! / I worship you!
C = I love you! / I adore you! / I worship you!
! = I love you! / I adore you! / I worship you!
. = I love you! / I adore you! / I worship you!

Elizebeth wrote long letters back to him. In one envelope she enclosed a lock of her hair.

Her letters don't exist today. It's likely she destroyed them after the war. Still, they left traces in William's: sentences of hers that he quoted, questions she asked that he answered.

A frequent topic was their future at Riverbank. Should they stay there or leave? What should they do about George Fabyan? The man was relentless; all through William's deployment, Fabyan had been writing him in Chaumont, asking that he return to Riverbank at the soonest opportunity. The Friedmans discussed this issue with caution, abbreviating Riverbank as R., George Fabyan as G.F., and the Bacon Cipher project as B.C. It occurred to them that Fabyan might have friends in the censor's office and they didn't want the rich man prying into the conversation any more than he already had.

Elizebeth tried to tell William in her letters that she no longer felt safe at Riverbank. She made a vague reference to Fabyan's "excesses," causing William to say he didn't understand: "How and where did you learn of these?" When William told Elizebeth he wanted to have children with her, she replied that it wasn't safe to have children at Riverbank. "You are perfectly right," he agreed. "When we are 'safe,' the children."

On September 21, 1918, she revealed something to him in a letter. All that survives is William's reaction: "Honey, I could have committed several crimes after reading what it had to say about that old nameless rascal. I was upset all day as a result. To think that he would do such a thing after all we have done." Elizebeth later confided to friends that Fabyan made sexual advances while William was in France.

William encouraged her to leave Riverbank if she was unhappy: "Honey, don't be afraid to take a step. You have ability and more brains than any other woman I've known. You can fill any job a woman can and many jobs that men fill."

The German lines collapsed in October 1918, British and American troops advancing and seizing territory. The roads around Chaumont began to fill with convoys of emaciated German prisoners of war. On November 10 at GHQ, a group of American soldiers huddled around a newspaper, laughing and shouting: The Kaiser had abdicated his throne. The war was over, the Allies had won. Three miles away the men of the AEF Gas Defense School blew up bombs and fired rockets in celebration, thunderclaps disturbing the sky. The dazed citizens of Chaumont wandered into the street and hung lanterns in their windows.

As William's colleagues drank and sang, he stayed indoors at his billet and wrote to Elizebeth, vomiting a great pent-up mass of insecurities and dreams onto the page. "Dearest Woman in the Universe," he began, "This is surely a fateful day." Then he made a series of promises, talking about what their lives would be like when he got home. He said he didn't want her to be consumed by housework. As a child he had seen his mother exhausted by her cleaning duties. "Home does not entail a spotless kitchen and a faultless parlor," William wrote. "Home does entail the presence of hearts that beat in unison—whether the shelter be a hovel or a palace." He was offering her the same freedom to pursue her intellectual ambitions that she had always extended to him—but did she really mean that? In her private heart did Elizebeth wish that her husband had more of a bank account and less of a brain? "Elsbeth, my Dearest, when you say that you want me to go on with my research work—blaze the trail and all that—do you realize that those chaps, poor fortunate-unfortunates, are usually not bank presidents? I should be happy, I think, with a fair share of the comforts and goods of this world, if I could continue with my studies, and unless I am seriously mistaken—and I don't think I am—you are not the woman to be hankering after life's luxuries and fineries. If you were, we would never have been attracted to one another."

That night, after 11 P.M., the oil in his lamp burned out and he went to bed. In the morning he learned of the Armistice and added a line to the bottom of the letter: "Honey, it's all over now."

At Riverbank she heard it from the news well before she received his letter, and she began a letter of her own, which William then quoted back to her. "The signing of the Armistice had one result—my indulging in thoughts, last night . . . dear, intimate things that burn one up with a fire of longing and ache of wanting you," Elizebeth wrote. "I must not again."

William replied, "What shall I say of the thrills that took possession of me on reading those words? I, too, have indulged in thoughts. . . . Ah, Dear One, when shall we too live them over again?"

He then broke some bad news: The army wasn't releasing him yet. He had to stay in Chaumont to write a secret history of the code and cipher work as a technical reference for future army use. He might be there for months.

This is when Elizebeth finally decided to leave Riverbank. She packed a bag in stealth without telling Fabyan and slipped onto a train for Indiana, reasoning that with the war over and no urgent messages to decipher, Fabyan could do without her for a bit, whether he liked it or not.

To pass the time in Huntington, she got a temporary job in the local library, a two-story building of limestone with a special room of materials about railroad engineering. She helped farmers find books and opened letters from the men in her life.

Some of these letters were job offers, eager replies to inquiries she had already sent. The Office of Naval Intelligence in Washington wanted Elizebeth to join its Code and Signal Section. Also, an officer in the War Department thought her Riverbank training would be "of the greatest value" in the MID; he was none other than John Manly, the Chaucer expert who used to argue with Elizebeth about the Bacon Ciphers and once shoved her on the shoulder. Manly now worked with Herbert Yardley. "Most of the work handled here necessitates a thorough knowledge of Spanish or German," Manly wrote. "Women who can *think* in either of these languages are needed as cryptographers at $1400 per annum."

William wrote to her, of course. He seemed as effusive and insecure as ever. He asked her if she knew how small an electron is, using that as the basis of an extended riff about the incomprehensible size of his love for her. He said he had gotten her a piece of lingerie in Paris, a silk teddy, custom sewn, with the help of an army captain who told him what measurements to use ("Can't two perfectly 'spectable married men get together on designing a perfectly proper—even if private—piece of woman's apparel?"), and he ruminated on their postwar future. Fabyan had demanded that William return to Riverbank at once: "You have had a long enough vacation," Fabyan wrote, "your salary has been going on and I do want you to get back at the earliest possible moment." But William worried that if he and Elizebeth did resume working at the estate, and Fabyan forced them to continue probing the Bacon Ciphers, it would destroy their credibility as cryptologists and make it hard to find other jobs. "I refuse to have anything to do with the B.C.," William wrote. "I think that whole business would be an excellent experiment for a psychologist. . . . Furthermore, I shall keep you away from it too. Nothing but unhappiness, and accusations, and unfruitfulness have ever come out of the whole business. Aside from our deep, and perfect love, the greatest treasure which life holds, we have found little else at R. but heartache, and argument, and unhappiness."

By the end of the letter he came around to the idea that they should leave Riverbank forever. "I don't want to flatter ourselves, but [Fabyan] is going to have one fine time trying to replace the Friedman Combination!" He signed off one letter with a love note in cipher, written using a type of transposition cipher called a "rail fence":

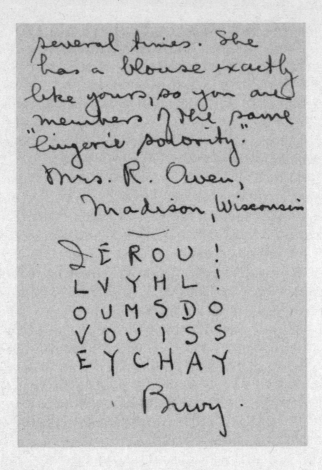

To find the hidden message, start on the upper left, with I. Read down that column, ILOVE, then start at the bottom of the next column to the right and read up, YOUVE, and then down again, up again, down, up.

George Fabyan also sent Elizebeth letters while she was in Indiana. She dug her nail into the wax and his black words uncoiled.

"I am wondering how you are and what you are doing," he wrote with a strained politeness that barely masked his fury, "and if your vacation has not been long enough to suit you." He liked to sign his name in a flourish of blue colored pencil, making the widest line possible, the tip round and blunt. Elizebeth didn't understand how anyone could bear writing with an unsharp pencil. It was barbarism.

He asked her in several different ways to come back, alternating charm with threats. He attempted a strategy of divide-and-conquer, suggesting that if Elizebeth committed to returning to Riverbank, it wouldn't necessarily bind William to do the same. (When Elizebeth relayed this to her husband in France, William was enraged: "Does he suppose you'd live at R. and I at Chicago!! . . . any man who attempts to sow dissension and create unharmony between man and wife, in that manner, is a scoundrel.") He tried to impress Elizebeth with his power, recounting a recent conversation with the head of MID in Washington, who had offered to hire all the women codebreakers at Riverbank, including her. "I told him that I would see them in hell before any girl whom I was interested in went to Washington under existing conditions," Fabyan wrote.

Finally, he tried being reasonable, addressing the Friedmans as a couple, through Elizebeth. "I am an old man going down hill; you are young people climbing up and it is for you to decide whether your opportunity lies at Riverbank or elsewhere." He said he wanted to talk it over with Elizebeth in person, in Chicago, given the "rather unsatisfactory" and slow nature of mail.

Elizebeth shot back in a letter, "I am inclined to agree with you that in most cases, correspondence is rather unsatisfactory. But with you I confess it has some advantages—for, you see, in conversation you insist on doing all the talking! Now I suppose you are going to retort, 'This, from a woman?'"

She finally got the letter she'd been waiting for in early February 1919: William was coming home. The army was done with him in France. "Won't our reunion be better than any honeymoon you can think of? I love you! I love you! I love you, love you love you!!"

He arrived in New York City two months later, in April, on a ship with other returning troops. Elizebeth went to New York to meet him, and they saw each other for the first time in a year.

They stayed in the East for a few weeks in rented rooms, thinking about what to do next.

Elizebeth knew she couldn't go back to Riverbank. Her husband, for reasons she didn't quite understand, had always found Fabyan at least slightly entertaining, but to her, the man was a scoundrel. As for William, he didn't want to be in the military anymore. He hated knowing that the army could send him anywhere in the world on a whim, separating him once again from his wife. The little he had seen of war convinced him there was no glory in it, once telling Elizebeth in a letter, "The War will not make better men or women out of us." If he could choose his own path, he confessed, he would unwind the last few years of his professional life and return to his first love, genetics. Maybe he could continue his plant and fruit-fly experiments at a university, or failing that, join a corporation and make some money.

Elizebeth agreed that William's "extraordinary gift of scientific analysis" should be properly appreciated and rewarded, and she encouraged him to get his discharge from the army, in April. After that they traveled so William could meet with potential employers. Elizebeth figured she would find work of her own wherever he landed. She noticed that the corporate executives who interviewed him were invariably amazed at his knowledge of codes and ciphers: "Everybody said, 'But where has this been all these years? Here we have this wonderful science all opened up for us now and where has it been hiding?'"

Strangely, however, no company offered William a job. Wherever the Friedmans went, a telegram would arrive from Fabyan, commanding them to give up the search: "Come back to Riverbank, your salary is still going on." The only way he could have known their whereabouts was if he had dispatched a spy. It was logical to conclude that Fabyan was threatening William's potential employers. "He had us followed," Elizebeth said. "He opened our mail."

Feeling defeated and not seeing any other options, the Friedmans told the Colonel they would return to Riverbank if he let them live in their own house in Geneva, gave them both a raise, and allowed them to question Mrs. Gallup's theory based on hard evidence. He agreed and welcomed them back home.

Fabyan didn't keep his promises. The raises never materialized. He continued to ignore and even suppress criticism of the Bacon Ciphers. When a famous type designer wrote a report showing how Mrs. Gallup misunderstood the printing practices of Shakespeare's era, Fabyan shoved the report into a drawer, even though he himself had paid for it to be written. (The Friedmans stumbled across the mothballed report years later in the Library of Congress.) Worst of all, he maneuvered to deny William credit for a crowning scientific achievement, his paper "The Index of Coincidence and Its Applications in Cryptography," written in 1920.

William had noticed that in any piece of English text, a letter sometimes appears directly atop the same letter in the line below—*d* on top of *d*, *w* on *w*, *q* on *q*. William discovered that the frequency of this "coincidence" could be measured, and it was distinct for each language, a kind of signature. In English, a coincidence happens exactly 6.67 percent of the time. Seven columns out of 100 contain an alignment. This insight married modern statistics with cryptology for the first time, and by doing so, kicked open a door that couldn't be closed. "The Index of Coincidence" and its offspring would lead directly to important feats of codebreaking in the Second World War. Instead of putting William's name on the cover, Fabyan had the paper published first in France; people got the idea that a French cryptologist was the author.

William and Elizebeth were enraged by "Fabyan's skullduggery"—Elizebeth scrawled this and other angry phrases in the margins of Fabyan's letters to William, annotating his duplicity, keeping a file of his lies—and within a year of returning to Riverbank they felt desperate to escape, reaching out to associates in Washington. This time they sent and received letters at their own address in Geneva, avoiding Fabyan's surveillance net. Joseph Mauborgne of the army leapt at the chance to hire the Friedmans and promised to create positions for both of them at once. "We feel that it would be a great misfortune if the Friedman family were to retire to some other kind of a job," Mauborgne wrote.

The Friedmans accepted the offer in December 1920. William was afraid to tell Fabyan, and he worried what the Colonel might do to his friend Mauborgne, too. William warned Mauborgne, "He is as powerful as he is ruthless." They all dreaded the rich man's reaction. "I expect a lively row when the news breaks upon

Colonel Fabyan's portly frame and expect that no little of his fury will be vented upon me," Mauborgne wrote to William. "Perhaps you had better fix up that side of it—if you can."

Elizabeth wanted to leave in the middle of the night without telling Fabyan. William found this overly cruel. She begged him to reconsider: If they wanted to escape, she said, they had to be just as tricky as Fabyan.

So together the Friedmans planned a clandestine operation: "our secret plot to be able to get away without getting our throats cut," Elizabeth called it. One morning they loaded all of their possessions into a car they had managed to borrow, cleaned out the house they'd been renting in Geneva, locked all the doors, drove to Riverbank, and located the Colonel. They showed him the car with all their luggage. They said they were leaving on the three o'clock train and their decision was final.

They thought he'd explode, turn red and scream, maybe try to restrain them. Instead, with an eerie calmness, Fabyan smiled and wished them well. It was so out of character that William assumed he must have already decided to seek his revenge at a future moment of his choosing.

There would be time to worry about that later. For now, they were giddy as they traveled east to Washington. They thought they were free. William believed that "after a very limited number of years," Riverbank "will disappear from the Earth and be but a black memory." He was grateful to escape and eager to work with his wife in a new city, keeping the Friedman Combination intact, as he had talked about during the war: "Oh, you are some partner, I'll say! . . . and to think you love poor me!"

Elizabeth's professional intentions were a little different. She thought leaving Riverbank might unshackle her from William to some degree. At times there, working at his side, she had felt pressure to compete, but the war had inflated William's renown to the point where she couldn't possibly keep up, and she figured that no one in Washington would expect her to match him. There was relief in that. As she put it later, "By the end of the war I was more or less known as a military cipher expert, but I was better known as the wife of my husband," who had "made a reputation so startling that I regarded the task of catching up to him as being altogether hopeless." But if she believed she would never again rival her husband, she was wrong. In the nation's capital, she was about to carve her own path, her own name, with a set of blades that would one day turn out to be just the right shape for dismembering the plans of Nazis.

PART II

TARGET PRACTICE

1921–1938

To work in this field, you have to become devious yourself. You have to think like a malicious attacker to find weaknesses in your own work. . . . Cryptographers are professional paranoids. It is important to separate your professional paranoia from your real-world life so as not to go completely crazy.

—CRYPTOGRAPHY ENGINEERING,
    FERGUSON, SCHNEIER & KOHNO, 2010

American cryptology was in a shambles after the First World War, a perilous mess. The codes used by the troops were already out of date, "completely inadequate" going forward, in William's opinion.

The problem was twofold: speed and security. Old pencil-and-paper methods of generating cipher messages were too slow compared to the speed with which dots and dashes of Morse code could travel by radio. "Military, naval, air, and diplomatic cryptographic communications had to be sped up." And because more vital messages were increasingly transmitted by radio and could therefore be intercepted and studied by the enemy, new techniques were required to protect that information from prying eyes. Security and intelligence are two sides of the same coin. Without one, the other is no good. You can intercept and decrypt the juiciest enemy secrets in the world, but if your own codes and ciphers aren't secure, you are defeating yourself, filling a leaky bucket at the top while secrets spill out the bottom.

Every powerful government on earth realized this now: the urgent need to invent new kinds of machines to make cipher messages, machines that were faster, easier to use, and dramatically more secure, all at once. "All the countries of the world were trying to develop something that nobody else could read and make sense out of," Elizebeth later said. "They were all playing with machines."

This epiphany marked the dawn of a peacetime arms race that would draw the battle lines of the next world war, and the Friedmans were plunged into the heart of it as soon as they arrived in Washington in the last days of 1920, fresh from their traumas at Riverbank, with no time to relax and take a breath.

Elizebeth and William both went to work for the government on January 3, 1921, three days after the first dry New Year's Eve in U.S. history, when 1,400 newly hired Prohibition agents ("dry agents") across the country surveilled the midnight parties, and the restaurateurs of Washington, D.C., spared no effort "to make the celebration, well, as festive as it could ever be under modern circumstances," one hotelier told the *Washington Herald*. That following Monday the Friedmans reported to the Munitions Building in Washington, a low hulk of concrete on the National Mall. Hastily constructed during the Great War, it teemed with fourteen thousand army and navy workers, including the staff of the Army Signal Corps, the part responsible for communications. The chief of the signal corps was Joseph Mauborgne, the kindly cryptologist, and the Friedmans now became his junior colleagues.

It was the first proper job they'd had in years, with regular paychecks and a nonderanged boss, and they felt happy, hopeful, and profoundly relieved. They had escaped from the clutches of a paranoid millionaire without getting their throats cut. Now they could live without fear.

William was placed in the army reserves at his prior rank of lieutenant and worked for the signal corps in that capacity, while Elizebeth worked as a civilian. He was twenty-nine, she twenty-eight. His starting salary at the army was $4,500 and Elizebeth's was $2,200, equal to $58,000 and $28,000 in today's dollars. This seemed like a lot of money to both of them after the poverty wages of Riverbank. And they were finally in a real city, which the young couple thought was just as nice as getting paid.

They could see friends, go to the theater, movies, the orchestra. Their first Washington apartment was a piano studio above a bakery. They woke each morning to the smell of fresh bread, and when they left for work, the owner of the apartment taught piano students there. Some evenings, when it was warm, Joe Mauborgne visited with his cello, and the three of them opened the windows and played together, Elizebeth on piano and William on violin, pedestrians stopping on the sidewalk outside to listen, not knowing that the musicians inside were three of the most experienced cryptologists in the country—almost the *only* experienced cryptologists.

The world of American cryptology was still tiny. There were only three codebreaking units in government, with fewer than fifty employees among them. The largest and best-funded unit, with two dozen people, was run by the former army lieutenant Herbert Yardley. After the war he had won funding from the State Department to launch a codebreaking bureau in New York City, in a four-story town house off Lexington Avenue. He considered it a modern version of the cabinets noirs of old Europe, the secret rooms in post offices where clandestine agents melted the wax seals of letters—an American black chamber. Yardley and his wife, Hazel, lived in an apartment on the top floor, and Herbert and his employees worked on floors below, reading the mail of foreign diplomats. They dealt in paper ciphers, and had some successes breaking the messages of Japanese diplomats, though Herbert wasn't skilled enough to go further, into the era of machines. The Friedmans knew the Yardleys socially and had dinner with them sometimes in New York, the two men chatting amiably across a deep chasm of professional rivalry and personal incompatibility. William was shy and monogamous, Yardley a boozy raconteur who during the war had kept a Paris apartment for his mistress. Although Herbert respected Elizebeth's intelligence, he once told William that she had "an edge on her."

The other two American codebreaking units were in Washington, one at the navy and one at the army, each smaller and more poorly funded than Yardley's black chamber.

The army is where the Friedmans started out, in a small windowless office in the Munitions Building. William smoked a pipe, Elizabeth smoked cigarettes. By the end of the afternoon the room resembled an industrial city seen from a distance, the couple's bodies like churches poking out of the haze. Together they produced the first scientifically constructed set of pencil-and-paper codes and ciphers in army history. There was still a place for "hand" or "paper" ciphers as opposed to machine ciphers. The best paper system was more secure than a weak machine, or a strong machine improperly handled, which is why inventors of cipher machines in the 1920s were struggling to make their proto-types easy to use, almost idiot-proof.

At the request of the army, William started tinkering with these machines, analyzing them, flipping them all around, search-ing for their soft and vulnerable places. He was the first Friedman to confront the machine era of cryptology, although Elizabeth would one day take on the machines as well, in the most spec-tacular way.

"There's all the difference in the world between machine ci-pher and paper cipher," Elizabeth explained later. When trying to break a paper cipher, with pencil and paper and brain alone, you had to depend on finding repetitions, patterns in the mes-sages. But the new breed of machines generated what seemed to be patternless winds of letters. "You can start from here and go to the end of the world and never have a repetition," Elizabeth said. In theory, the only way to read a message was to know the starting configuration of the machine's internal parts—the "key"—which only the sender and recipient would possess. What's more, the machines were designed to survive capture and study. Even if you got your hands on a copy and took it apart like a broken clock, examining each gear for as long as you pleased, you would still not be able to read messages produced by another such machine.

Many inventors of cipher machines were private citizens, and a quirky bunch at that: hucksters, dreamers, engineers, lo-tharios, thieves. William met Edward Hebern, a burly Califor-nian who had built a machine based on alphabet wheels that turned with the application of electrical current. Called rotors, these electrified wheels represented an important advance that would find more sophisticated expression in the German Enigma machine; rotors could be easily removed, swapped, and linked in a chain. William asked Hebern how he happened to think of this elegant concept of a wired rotor. Hebern replied, "Well, you see, I was in jail." William asked what for. Hebern said, "Horse thievery." William asked if he was guilty. Hebern said, "The jury thought so."

Hebern believed his machine was unbreakable and was trying to sell it to the navy. William placed the machine on his desk and thought about it. He stared at the small box for six straight weeks in 1923. He told Elizebeth he was "discouraged to the point of blackout." Then the solution occurred to him one night when he and Elizebeth were getting dressed for a party: "As I was tying my black tie, it suddenly came to me." It was the first-ever solution of a wired rotor machine. William mentioned the accomplishment to George Fabyan in a letter; Fabyan had let him borrow one of his cryptologic curios—a nineteenth-century cipher device—for a lecture, and William was returning the device. "P.S.," he scrawled at the bottom of the letter, "I busted a beautiful 'indecipherable cipher' machine recently and it is extremely important in many respects. . . . Gave the Navy an awful jolt! Best piece of analysis I ever did. B."

In contrast to Elizebeth, who never wanted to talk to Fabyan again and didn't write to him after leaving Riverbank, William kept a connection alive. He wrote regular letters to the tycoon from his signal corps office, sharing bits of personal news and asking for Fabyan to send him pieces of cryptologic literature from his unrivaled library at Riverbank. He did it partly out of professional fear; it was best to maintain cordial relations with such a powerful figure. But William also coveted the documents that Fabyan still controlled, the records of the Riverbank Cipher Division. He realized that the records could be used to write a true history of cryptology, a history that did not exist. William wrote to Fabyan, "It's a striking paradox that a subject which forms the basis of material which is exchanged every hour of the day over the whole face of the globe, under its waters, and through its ether—material that touches directly or indirectly every human being—lacks an authentic, detailed history. Somebody, somewhere, is going to write that history." He hoped it would be him, and he needed access to Fabyan's papers to do it.

After busting the Hebern machine, William moved on to the next supposedly unbreakable device, invented in 1924 by a German named Alexander von Kryha, who committed suicide in 1955. The Kryha cipher machine was shaped like a half-moon and contained two discs of alphabets, one a fixed semicircle, the second a circle that rotated against it. According to the inventor, it could encipher a message in 2.29×10^{82} ways, a number larger than the number of atoms in the observable universe. William was not impressed: "The number of permutations and combinations which a given machine affords, like the birdies that sing in the spring, often have nothing or little to do with the case." Other features were more important, like the method of selecting alphabets and the motions of the wheels. William conquered the Kryha and later demonstrated his mastery by solving a two-hundred-word message prepared by a lawyer in New York who believed the Kryha was unbreakable. William found the solution in a mere three hours and thirty-one minutes, including a fifty-minute lunch break.

In this new era of machines, William was showing that the human was still king. Until the invention of digital ciphers in the 1960s, the field of cryptology would be defined by heroic human attacks on physical cipher machines. These attacks would often be *aided* by machines specially built to speed the attacks—like the famous electromechanical "bombes" designed by the British codebreaker Alan Turing, and some of the world's first computers, monstrosities of wires and vacuum tubes that occupied entire rooms—but not necessarily. It was still possible at this point to defeat a machine with mere pencil and paper. The human brain could beat the machines, if it was the right brain, and if the owner of the brain was willing to accept the cost of victory.

The common saying about cryptologists, as William phrased it, was that "it is not necessary" to be insane, "but it helps." There was uncomfortable truth in the joke. To operate at the highest level of the field seemed to require the kind of pitiless attention and focus that turned some otherwise pleasant and well-adjusted people into zombies who stumbled down the stairs. Mental breakdown was a hazard of the job. During the coming war, several American cryptologists would crack under the strain. Captain Joe Rochefort, one of the navy's top codebreakers from 1925 until the end of the Second World War, suffered from ulcers. A slender, high-strung man, Rochefort recalled later that he would come from work three days out of four and lie in bed for two hours, unable to eat, because he felt so much pressure. The pressure didn't come from superior officers, or even from the urgencies of war; it came from within. "Here is a bunch of messages and I can't read them," Rochefort said. "Now what's wrong? It was this sort of a thing, you see. And this was sort of standard. You'll find people like this who maybe go into a trance, what looks like a trance when you're talking to them, and the first thing you notice is that they are not paying any attention to you at all, and their mind is on this other problem which they brought home from the office." Rochefort had to stop breaking codes for two years in the late 1920s for health reasons, but the navy pressed him back into duty. He knew William Friedman well. "There is no one that could compare with Friedman," Rochefort said, "no one at all."

There was one other cipher machine that William studied briefly: Enigma, invented by a young German engineer in 1918 and available at that time on the open market. It looked a bit like a typewriter. Under the cover of the Enigma were three rotors, wired wheels similar to the ones in the Hebern machine but invented independently and capable of more intricate movements, the electrical current crossing and doubling back in mazes.

William was impressed with the Enigma and its "clever inventor" but didn't make a serious effort to solve the machine. It was just a curiosity at this point, not a haunted piece of technology cloaking the schemes of fanatical killers. The Nazi Party had just been founded in 1920. Hitler was speaking to modest crowds in beer halls. William Friedman had no way of predicting the fatal battles around Enigma yet to come, and no way to know that Enigma loomed larger in the destiny of his wife than his own: that one day Elizebeth would conquer multiple Enigmas with pencil and paper.

By this point, she had quit her job. She wasn't working as a cryptologist anymore. After about a year at the army, Elizebeth resigned in the spring of 1922, saying she planned "to stay home and write some books." William had encouraged her to quit. For one thing, he was excited about her books—he admired good writers and thought highly of her as a prose stylist—and in another sense it was just the path of least resistance, the expected arrangement, for a young Washington wife to stay home. The decision seemed to clear the way for Elizebeth's ambitions and for the couple to start raising children. But as soon as she quit, Wiliam missed her at the office, complaining to a friend that Elizebeth was home "and I am all alone." Then she left Washington for a five-week vacation through the Midwest, visiting family and friends, and his loneliness shaded into panic. She sent William letters about how much fun she was having. "I drove 36 miles to Lake Erie and then swam and broiled ham on the shore for supper. After it got dark we told stories and sang around the fire—it was a glorious night with the moon making fairyland of the lake."

She mentioned that a certain midwestern stranger had been following her to social events, expressing romantic intentions. William tried to make her jealous in his replies. He described going for an evening walk with the attractive wife of a friend, an outing cut short by clouds of mosquitoes. Elizebeth wrote back, "My dear, I'm proud of you! If fate had only been gentle with you and spared the chiggers, what a nice Memory you would have." She added, "Hold as many hands as you can! Life grows short!" In another letter she asked, "Am I wicked to be glad you are missing me?" and signed off with a love note in a rail-fence cipher:

Plaintext: JE T'ADORE MON MAR, "I love you my husband" in French.

Not long after Elizebeth returned from her trip, the Friedmans decided to move out of the city, to the pine forest of what is now Bethesda, Maryland. They rented a house with a wraparound porch that looked out to pines, apple trees, a tulip grove, and a garden spattered with irises of purple, blue, red, yellow, and white. The interior was comfortable and a bit creaky, the forest dampness making the paint peel from walls. They got a dog and called him Krypto, after *kryptos*, the Greek root of the word "cryptology," meaning secret or hidden. Krypto was an Airedale terrier, tan on both ends and black in the middle, with the alert eyes of a hunter and a wiry coat. They added a cat to the mix, Pinklepurr, named after the A. A. Milne poem about "a little black nothing of feet and fur."

William rode the trolley to the Munitions Building in the morning, an hour spent in his head, thinking about alphabets. After Elizebeth resigned, the army had given him an assistant, a former boxer with cauliflower ears. The man's only skill was typing. If you would like to imagine the birth of the mighty National Security Agency, please visualize two men in a small room, one with a pug nose, pecking at a typewriter, the other a dandy in a suit and bow tie, smoking a pipe, wondering what his wife was up to at home, and if she was missing him.

"Now," Elizebeth wrote on her typewriter, "I wouldn't be a bit surprised to learn that there are some persons reading this book who if suddenly awakened from a sound sleep would not be able to recite the alphabet, so here it is:

A B C D E F G H I J K L M N O P Q R S T U V W X Y Z

"It is composed of 26 letters and their order is fixed. I couldn't for the life of me tell you why A comes first, B second, and so on, nor can anybody else tell you—and don't let them kid you into thinking they can, either."

Sitting with her typewriter by a crackling fire, Elizebeth worked on a book about codebreaking aimed at teens and curious adults, a "little book" to "afford you some amusement for leisure moments." The idea had been in the back of her head since Riverbank. She wanted to write about codes and ciphers with a light and whimsical touch nowhere to be found in other literature. She thought up playful analogies to explain cryptologic concepts. "Miss Transposition Merely Turns Her Clothes Around"; Miss Substitution changes into a new outfit. Adopting the tone of a mischievous schoolteacher, stern but kind, Elizebeth walked readers through sample problems and cheered them along: "You're just eating 'em up." "Bravo!" Meanwhile, she wrote a draft of a *second* book, a children's history of the alphabet, illustrated with her own drawings of hieroglyphs and cuneiform tablets. She had begun working on it at Riverbank. The alphabet, part of the backdrop of our lives, like the sky or electricity or advertising, but the one tool that makes all the others possible—she wanted kids to know that the alphabet is a miracle.

The goal with both books was the same: to share, explain, demystify, get people excited about the possibilities of words. This was the opposite of what government seemed to be about, in her experience—the dreary, smoky vault at the Munitions Building. She was having fun, developing a voice of her own, and she might have been perfectly happy here, writing books by a fire for the rest of her life, if men from the government had not begun to knock on her door, asking her to solve puzzles for America again.

Her skill was the force that pulled them. There were just so few cryptologists of her ability, or William's. They were like a binary star system in a void, twin suns rotating around each other, drawing lesser bodies by their light.

Having to choose between the two Friedmans, people almost . always approached William first. He was the man, the one with his name on publications, the one who had served in France during the war. A retired astronomy professor asked William to analyze radio signals collected from a device that printed the waves as black lines on a thirty-foot-long piece of film. He theorized that if aliens existed on Mars, they might be transmitting; William found no pattern in the signals (they were probably interference). He consulted on criminal investigations. A man sent a bomb to Huey Long, the Louisiana politician, along with a note full of hieroglyphic markings. William solved it. The warden of an Ohio penitentiary sent him a cryptogram smuggled into prison by the mother of a bank robber. William sent back the plaintext, which revealed a plot to help the inmate escape by placing bombs along a prison wall and exploding them on a Sunday during church.

In 1923 William gave expert advice to a congressional committee that was investigating political bribes written in coded letters. Top officials in the Warren Harding administration had taken cash from oil tycoons—the "Teapot Dome" affair, the biggest corruption scandal in U.S. history. His testimony caught the eye of a twenty-eight-year-old J. Edgar Hoover, then an FBI agent working the Teapot Dome case, and in the years that followed, after Hoover was named director, he asked William to consult on a few FBI cases. The bureau had no cryptanalytic section, no skilled codebreakers of its own. It had to rely on outside experts. When Hoover's agents arrested several of John Dillinger's gang members, notes scrawled with code were found in the pockets of the machine gunners. Hoover sent the notes to William. He solved them.

A wealthy businessman, *Washington Post* owner Edward McLean, hired William in 1924 to design a code for his personal correspondence, then refused to pay him. This was a man who had bought the Hope Diamond as a gift for his wife. She wore it around her neck at parties she threw for wounded veterans. And McLean stiffed William on his fee.

Elizebeth urged her husband to protest, but he let the matter drop, fearing the stereotype of the "money-grubbing Jew." America was growing more anti-Semitic in the 1920s, the successes of Jewish immigrants provoking ugly responses. The president of Harvard changed admissions rules to keep Jews out. Henry Ford launched an anti-Semitic weekly newspaper with a declaration that "the Jew is the world's enigma." Officers in the Military Intelligence Division of the War Department tracked intelligence reports about Jewish activities on index cards labeled "Jews: Race" and kept a central dossier on the "Jewish Question" that included documents like "The Power and Aims of International Jewry." The MID men were William's colleagues.

So he treaded carefully in Washington, lest he provoke an anti-Semitic reaction that seemed to always be near. He tried to get along, said yes a lot, loaned out pieces of his brain. Inevitably there was only so much of him to go around, and requests for his time spilled over to Elizebeth. "When they couldn't get him, I'd be offered a job," she said later. "That's the story of my life. Somebody asks for my husband and they can't get him, so they take me."

On the one hand, she found this insulting. It was like people were trying to use William's brain "second-hand" by hiring the woman who presumably enjoyed intimate access to it. "Sad for me," she said. But it also gave Elizebeth a way to demonstrate that she was a master codebreaker in her own right.

First it was the navy that wanted to hire her, in late 1922. They had lost a civilian cryptologist, Agnes Meyer Driscoll, a woman with a mathematics Ph.D. who had left for the private sector, partnering with the horse thief Hebern to launch a cipher machine factory in California. "I didn't want to work for the Navy," Elizebeth said, "but they were just sitting on my doorstep all the time and the only way to get rid of them was to go there for a little while until they found someone else." She took Driscoll's place and filled in for a time, designing codes for sailors. Then she became pregnant. Elizebeth left the navy after five months, again thinking she would never go back to the government, and gave birth to the Friedmans' first child, a girl they named Barbara, in 1923.

Elizebeth had always been hesitant to bring a baby into the world, her own childhood having been less than warm. She had watched her mother, Sopha, raise nine kids, exhausting herself in the process and dying of cancer, suppressing her own desires and dreams to the extent that she never said what they might have been. Elizebeth didn't want to sacrifice herself like that. Ever since she married in 1917 it had been her husband who pushed the issue of children and she who pushed back, saying she wanted to wait. William confessed in letters from France that he wished they had begun having children at Riverbank, before he deployed to France with the AEF. "Often I feel that in many ways we made a mistake," William wrote. "Our love will only be complete when there is a third—flesh and blood of yours and mine." She lacked this conviction; when Elizebeth did think about having children, it was an impulse that gripped her and then quickly passed. "Sometimes I wish I were going to have a child," she wrote in one letter to France. William replied that "a queer sensation" came over him when he read those words. A child, "the wonder of all mysteries!" Whatever Elizebeth wrote in response, she destroyed.

When children finally arrived, Elizebeth didn't put aside her ambitions, although the delivery of Barbara was a difficult one, leaving the mother laid up in bed for months with spinal pain. William made the mistake of mentioning Elizebeth's pain in a letter to Fabyan, who offered to prescribe some back exercises, spinal alignment being his long obsession. "If she were here I am quite sure we could help her," Fabyan wrote to William. "The general idea is to get the anterior curve in the back near the hips out, and one way of doing it is to lie on your back on a table and stick up your knees, bringing the heels as near the buttocks as you can." William replied that he thought she would be all right.

She hired a nanny, a black woman named Cassie (her last name is not recorded), and continued to work on the two book projects from home, sharing the cooking duties with Cassie and taking breaks to play with Barbara, the firstborn, followed by a second child, John Ramsay, three years later, in 1926.

The daughter and the son had opposite personalities and Elizebeth felt constantly perplexed by the size of the gap. Barbara was an affectionate, verbally prolific infant with blond ringlets who charmed all of the Friedmans' friends, loved books, and shared her dolls with Krypto the dog. John Ramsay didn't talk much, didn't like to be sung to, didn't seek out books, and swung wildly between moods, "his mirth excited by so little, his hilarity so whole-souled," Elizebeth wrote. "And then in a flash, fury and rage takes its turn." The only music he seemed to enjoy was whatever Elizebeth played at top volume on the Victrola.

She took a hands-off approach to parenting, hewing to a doctrine of no doctrines, in agreement with William. The Friedmans were determined not to consciously teach their kids anything or tell them what to believe, only to create a comfortable environment, pour in vitamins, "and let the rest take care of itself," as Elizebeth put it. She respected her offspring as autonomous, independent creatures and was captivated by their efforts to learn how to communicate. Some of Barbara's noises sounded like verbalized cryptograms. Elizebeth, always looking for insight into language, transcribed her daughter's babble inside a book Fabyan had given her, *What I Know About the Future of Cotton and Domestic Goods,* the one that contained only blank pages. "She strings together consonants impossible to an English tongue," Elizebeth observed. "Within the last week she says I don't know in answer to questions. When I first told her father of it, he scoffed saying the concept of not knowing a thing was impossible at her age. But as I pointed out, she does not use the separate words—it is rather one elided sound—IDONTKNOW, a perfect imitation." She copied Barbara's ejections verbatim and analyzed the text with a codebreaker's curiosity. *Pfnr-*

pfnh-hnwhwp. It seemed clandestine. There was structure in it, a pattern at the edge of legibility: *pfnr-pfnh-hnwhwp.* What did it mean? *IDONTKNOW.*

After two years living in the forest of Bethesda, the Friedmans decided to move back into the city, closer to William's job, and bought their first house together, at 3932 Military Road in northwest Washington, a newly constructed home with an elephant-shaped door knocker. The neighborhood was full of families and young oak trees, two hundred yards from the Maryland state line. Fathers tended to work for the military or the government. Lieutenants sat on porches, reading the newspaper. This is the house where Barbara and John Ramsay grew up—the Friedmans would stay at 3932 Military Road for more than two decades—and it's where government agents once again came knocking for Elizebeth, hoping to use her husband's brain secondhand.

One day in 1925, when her daughter was two, she answered the door to find a man in his forties with a boyish face and large ears. He wore a double-breasted navy coat and a shiny navy cap. He said his name was Charles Root, he was from the U.S. Coast Guard, and he needed her help.

Root might have had the most thankless job in America at the time. The coast guard was responsible for patrolling American waters, trying to catch the nimble "rum-running" boats that flouted Prohibition law and smuggled bootleg liquor from sea to shore. It was a cat-and-mouse game that the mice were winning. From the start of Prohibition, the coast guard's task had been laughably difficult—they owned just 203 slow small patrol boats to police five thousand miles of coastline—but recently the rumrunners had deployed shortwave radios and sophisticated codes to conceal their movements, giving them the decisive advantage.

One year earlier Root had launched an intelligence division within the coast guard, the first of its kind. He told Elizebeth he was desperate to get a handle on the rumrunners' communications systems. The coast guard happened to be years ahead of other government agencies in radio prowess. It had built several large radio towers along the East Coast to aid in search-and-rescue operations at sea, to help save boats and sailors in storms, and these same posts were able to intercept radio messages sent by the bootleggers. Still, no one in his office knew how to break the codes in the intercepted messages, and over the last several years hundreds of messages had piled up, unsolved and unread. Root said he wanted Elizebeth to come in and tackle the backlog. He asked her to consider breaking codes for America again. Sensing her reluctance to return to government, he pitched this as a temporary assignment, a ninety-day contract.

After thinking about it for a while, Elizebeth said she'd do it on one condition: that she be allowed to work from home.

The coast guard agreed. Root gave her a metal badge that said SPECIAL AGENT, U.S. TREASURY in gold letters. Once every week or so, she traveled to Root's office in the main Treasury building, flashed the badge, picked up an envelope of unsolved puzzles, took the envelope home, solved the puzzles, returned the answers to Treasury, and picked up a fresh envelope.

A block of pale granite across from the White House, the Treasury headquarters was guarded on its south side by a bronze statue of Alexander Hamilton, the Founding Father who designed America's system of money and gave birth to the forerunner of the coast guard in 1790 by launching a fleet of ten small "revenue cutter" ships to catch smugglers and pirates. Fifteen thousand people in Washington now worked for the Treasury Department, and another forty-six thousand in field offices across the country, doing all kinds of tasks: minting coins and paper bills; collecting taxes and customs duties; tracking the output of factories, the price of gasoline, the size of the annual wheat harvest.

Elizebeth didn't have anything to do with these bureaucratic and economic functions. She was involved with the side of Treasury that investigated crimes.

The department contained no fewer than six separate law enforcement agencies: the Prohibition Bureau, the Narcotics Bureau, customs, the coast guard, the IRS, and the Secret Service. The six agencies had broad authorities to probe financial fraud and most any product or person that moved illegally across a border—guns, liquor, drugs, migrants, counterfeit money. The Treasury detectives were known as "T-men" in the press, as opposed to the "G-men" of the FBI, part of the Justice Department. And although the G-men of the FBI tended to get the glory when famous gangsters went down, thanks to the publicity genius of J. Edgar Hoover, it was the T-men of the Treasury, more often than not, who made the cases. Treasury was the center of the fight against organized crime. It was T-men who eventually nailed Al Capone for tax fraud. It was T-men who caught the kidnappers of the Lindbergh baby.

The spiritual leader of the T-men was Elmer Irey, a soft-spoken father of two in glasses and a wool suit. Irey ran the IRS's Special Intelligence Unit. A bloodhound with a nose for money trails, he referred to the six agencies as "the six fingers of the Treasury fist." Al Capone wasn't much bothered by J. Edgar Hoover, but he feared Elmer Irey, who liked to relax in the evening by reading the first few pages of a mystery story and writing the solution in the margin. Reportedly he was never wrong.

Before long, the T-men would learn to see Elizebeth Friedman, this petite mother of two in low heels, as one of their most potent weapons.

The rum kings of 1925 were not gentlemen. The gentlemen had already been driven out of the game. During the first few years of Prohibition, it had been possible for independent sea captains to make a living by racing crates of booze from the Bahamas to the Florida coast, for the cash and the thrill of it, but those days were over. The men in charge of the liquor rackets now were mobsters, killers, associates of killers, and shadowy corporations with intentionally understated names, such as the Consolidated Exporters Corporation of Vancouver, whose rum fleet would have been the envy of many small nations: sixty to seventy boats of various sizes, from enormous "mother ships" the size of baseball fields to small speedboats. The mother ships functioned as huge floating warehouses, anchored as far as sixty miles offshore, and capable of holding up to 100,000 crates of liquor. Consolidated's ships spanned the U.S. East Coast and the West Coast and stretched into ports in the Caribbean and South America. The rum business was hemisphere-wide: "The whole half of the world," Elizebeth said, "was interested in thwarting the prohibition law."

It was a daunting thing for her to face. She got to work.

During her first three months with Treasury in 1925, Elizebeth solved two years' worth of backlogged messages. Captain Root was so appreciative, he asked his bosses for money to hire Elizebeth for good: "Mrs. Friedman is the only person available with the required skill and experience." She accepted a full-time job breaking messages for all six Treasury law agencies and continued to work from home, delivering envelopes of solved messages to the office, carrying envelopes of unsolved messages back to 3932 Military Road.

"Mrs. Friedman," went a handwritten note from the Office of the Chief Prohibition Investigator. "Please see what you can do with this and return it to us." Several cryptograms were attached, including this one, sent from Halifax, Nova Scotia, to an unknown ship on the East Coast:

> **AWJTSSK JQS GBQKWSK LYMSE EJBCG SPEC QPFYEYQD**
> **MYHGC PRPYC JWKSWE CWI PQTGJW EPFS VBSM**
> **AWAJASTCE HJJS.**
> 　**BLACKCAM.**

She saw that it was a simple mono-alphabetic cipher. Elizebeth penciled in the plaintext letters without much effort, revealing a note telling the rum captain to anchor near a New Jersey lighthouse.

> **PROCEED ONE HUNDRED MILES SOUTH EAST NAVISINK**
> **LIGHT AWAIT ORDERS TRY ANCHOR SAVE FUEL**
> **PROSPECTS GOOD**

At first all she did was solve these messages, unraveling the codes of the rumrunners one by one. But most of the messages weren't this easy, and with each passing month, the task grew more difficult. The rum codes of the bigger operators were more secure than any military codes she had encountered during the Great War, demonstrating "a complexity never even attempted by any government for its most secret communications," particularly the codes of the Consolidated Exporters Corporation. The syndicate used different code systems for different sets of ships. Some code blocks were three letters long, some five letters, some four, and the smugglers changed the codes every few weeks, meaning that a broken code didn't stay broken. Consolidated relied heavily on a multistep processes of enciphered code, the messages like little onions, each layer created with a different technique that Elizebeth had to peel back: Starting with the ciphertext block MJFAX, she might have to decrypt it into another five-letter block, BARHY, which corresponded to 08033 in a widely available book of commercial codes used by legitimate businessmen to save on telegraph charges, from which she subtracted 1,000 to get 07033, which matched up with the English word ANCHORED in a second code book. "If I may capture a goodly number of your messages," she wrote, "even though I have never seen your code book, I may read your thoughts."

As good as she was at solving the individual messages, Elizebeth's ambition didn't stop there. She wanted to build a broader, more comprehensive system for extracting intelligence from wireless messages, a system that didn't exist at Treasury or anywhere else in the U.S. government because radio was still such a young technology.

Radio intelligence requires cooperation and sharing of information, and because Elizebeth was the only senior cryptanalyst in Treasury, the only one who knew how to break codes, she became a link between people in law enforcement who didn't usually talk to each other. She communicated with the T-men at the listening posts; depending on atmospheric conditions, the listening posts could hear all the way to Germany, a portent of things to come. She kept in constant touch with T-men in the field who gave her lists of ships as they came and went in U.S. ports. She sketched maps of the naval routes and maps of the radio traffic. She helped train T-men to use direction-finding machines that they loaded in the back of pickup trucks, driving up and down the coasts to find hidden pirate radio stations used by the rumrunners. The smugglers essentially acted like foreign spies in wartime, trading intelligence by radio from hostile territories, and to read their thoughts, Elizebeth had to act like a spy hunter, a counterespionage or counterintelligence agent—skills she would ply against the Nazis in the early 1940s.

"I sort of floated around," she said. "It seemed I went here, there, and everywhere I was needed." It was a manic and exhausting period in her life. She traveled to the West Coast and consulted with customs agents in San Francisco about gangsters running rum boats off the coast. She exposed a ship called the *Holmwood,* which was sailing up the Hudson River with 20,000 cases of liquor, disguised as an oil tanker called the *Texas Ranger;* the ship was halfway to Albany, New York, its destination, when the coast guard and customs seized it on the strength of Elizebeth's decryptions.

The names of ruthless men appeared in her solved puzzles: mafiosi and underworld bosses, racketeers and semi-legitimate CEOs, some of them hidden behind veils of complicated corporate structures and webs of interlocking investments. There was a fleet of rum ships on the West Coast controlled by two Canadian brothers named Hobbs. A California gangster named Tony "The Hat" Cornero owned some of the Hobbs boats. Their rival, for a time, was the Consolidated Exporters Corporation, the unstoppable Vancouver syndicate. Consolidated seemed to have an unlimited stream of money, pumped in by the Reifel family, patriarch Henry Reifel and sons George and Harry, hotel magnates in Vancouver. The Reifels also had an American investor, a colorful Bostonian named Joseph P. Kennedy—the father of John F. Kennedy, future president of the United States. Elizebeth noted in one report that in 1928 the Hobbs brothers appeared to sell their interests "to Joseph Kennedy, Ltd., of Vancouver, large holders of stock in the Consolidated Exporters Corporation."

Elizebeth wasn't afraid of these men, and she didn't hate them, either. She was doing a job, and the job was frequently exciting, each case a detective story. Elizebeth flicked invisibly at the world, setting great chains of events into motion. She went to Houston, Texas, at the request of a federal prosecutor and solved more than 650 messages in 24 different code systems used by rum syndicates in the Gulf Coast. Many of Elizebeth's solutions became evidence in a Texas rum trial involving a one-legged cabdriver named Louis "Frenchy" Armatou. Several of the messages seemed to involve a rum ship not at issue in the trial, the *I'm Alone.* Elizebeth's solutions triggered an international manhunt for the ship's captains. One fugitive, Marvin Clark, went on the run and was gunned down by an unknown assailant in New Orleans, and the other,

Dan Hogan, was arrested and developed a vendetta against Elizebeth—"he was in a very mean mood"—that required her to travel to a federal hearing surrounded by a security detail. And all the time she wrote to her children, to Barbara and John Ramsay, sharing news of her travels. "Only woman on plane," she wrote on a pad of yellow paper during a cross-country flight, at a time when flight in America was still new and wondrous. "Co-pilot as attentive as if I had been a young and pretty girl. The country, so stunning with its rivers and streams, green wooded mountains, and even flat farm country laid out in such patterns that it seemed as if a directing mind from above had planned it." When she got home, she took the kids out for afternoon tea and did puzzle games with them as they sipped their cups.

When the stock market crashed in 1929, demand for liquor only increased, along with Elizebeth's workload. She worried about the family's finances in letters to William. "Did the bank call the loan?" she asked in one letter, then added, as if it were no big deal, "Broke into another system this morning, messages unearthed from a safe."

The volume of intercepts pouring into Elizebeth's coast guard office grew and grew until she felt "almost buried under the press of duties." The grind of it was staggering: two thousand messages per month demanded her attention, and some twenty-five thousand per year. Not all of these contained information relevant to law enforcement, but she had to analyze all of them to know which ones to transmit and which ones to discard. She begged Treasury for help but all they allowed her was a single clerk-typist, a woman. Despite these meager resources, Elizebeth reported to her superiors in 1930 that she had solved twelve thousand rum messages in the previous three years, "which covered activities touching upon the Pacific Coast from Vancouver to Ensenada; from Belize along the Gulf Coast to Tampa; from Key West to Savannah, including Havana and the Bahamas; and from New Jersey to Maine."

Until 1930, almost all codebreaking for the U.S. government's planetary war against smuggling was handled by these two tired and perpetually overworked women, Elizebeth and her clerk, but that year Elizebeth decided she'd had enough and wrote a seven-page memo to coast guard commanders, proposing that they create a "central unit" for codebreaking. It wouldn't do to stumble along with two people anymore; there needed to be a proper team. At a minimum, she said, the unit should have seven employees, two cryptanalysts and five cryptographic clerks, earning combined salaries of $14,600 per year. She argued that her accomplishments of the past several years "can be increased a hundredfold by a sufficiently staffed and thoroughly organized unit" striking at the heart of the smuggling trade from "the most promising point of attack": the interception and solution of encrypted radio messages.

While she waited and hoped for help to arrive, Elizebeth did her best with the meager resources at her disposal. To build an archive of the smugglers' words and stay on top of any shifts in the codes, she made a carbon copy of every solved message, and when the stack of copies grew to be an inch thick, her clerk-typist bound the plaintexts into a volume, a book. By the start of 1931, Elizebeth had generated thirty of these books, a *Britannica* of criminal clockwork. She was compiling a secret history of her age, using the very words of its hidden kings.

Elizebeth and William still had a chance to work together in these years. The collaboration flowed in one direction only, from her to him. William's army work was too secret now. She got the sense that her husband found the smuggling puzzles to be a refuge from problems facing him at work; for him the rum war was a lark. They passed messages and worksheets back and forth, drawing cipher letters in the boxes of sheets of grid paper, snippets of potential plaintext: "Position," "Landing boat," "Is there any indication there will be trouble?"

One evening Elizebeth came home from a League of Women Voters meeting to find William examining a rum message. He looked up and smiled. "Andrew says send glass eye," he said. Elizebeth often uncovered personal notes from the underpaid and exploited sailors on the rum boats, funny or sweet messages to their wives or children, and this was one that she had decrypted and William had apparently seen. The actual text read, "Andrew says advise wife to send reserve glass eye." In a different message to shore, this same Andrew had requested "a pair of shoes, size 15." He must have been a very large man. Elizebeth and William shared a laugh as they tried to picture it: a luckless giant on a rum boat, holes in his shoes, missing a glass eye, pining for his wife.

Elizebeth finally got some relief from her workload in July 1931, when Treasury cleared her to recruit a codebreaking team of her own, at long last, according to the specifications she had laid out in her earlier proposal. Officially the unit would live within the coast guard, but it would break codes for all six Treasury agencies. The department gave Elizebeth funds for three junior codebreakers and two stenographers, along with her first dedicated office space, a small block of rooms inside the Treasury Annex on Pennsylvania Avenue, a building that resembled a classical Greek temple and housed the coast guard, customs, and the Bureau of Narcotics. The new space was a large open workroom with desks and small offices on the periphery that fit two to a room. As part of the deal she got a pay raise, too, from $2,400 to $3,800 per year, and a fine new title: Cryptanalyst-in-Charge, U.S. Coast Guard.

It was the first unit of its kind in Treasury history, and the only codebreaking unit in America ever to be run by a woman—another pioneering moment for Elizebeth.

The first thing she did was hire and train a staff. There was no available pool of cryptologists in the commercial sector; you had to train your own. Scouring civil service lists of applicants who had taken math, physics, or chemistry, Elizebeth was unable to find names of women, so she hired young men for the junior codebreaker jobs, setting them to work on practice problems she had designed, a beginner's course in codebreaking.

She had never been the boss of men before and she worried that her new employees might not accept her authority as a woman, but this concern proved unwarranted, with one exception, the all-time top scorer on the math exam, a young mathematics Ph.D. from Columbia University who was supposed to be a prodigy. He refused to complete Elizebeth's exercises and babbled nonsensically about an "indecipherable cipher"; she decided "he did not comprehend the English language" and replaced him with a candidate who showed a more practical cast of mind. Elizebeth's first two successful hires were Hyman Hurwitz, a twenty-one-year-old electrical engineer from Dorchester, Massachusetts, who liked to tinker with radios, and thirty-one-year-old Vernon Cooley, from Kalamazoo, Michigan, who had taught schoolchildren there and also worked for a time in the factory of the Kalamazoo Paper Company. Both were given the title of Assistant Cryptographic Clerk. A third young codebreaker-in-training soon joined them, Robert Gordon, who was twenty-three and hailed from Waco, Texas.

The three men, "able, agreeable, and cooperative," treated Elizabeth with respect, and soon they all settled into a productive routine, learning each other's quirks and strengths, dividing up tasks as a team. Asking if Elizabeth experienced sexism was like asking if Marie Curie did. Cryptology was a young field. It hadn't yet sorted itself into rigid roles by gender. And Elizabeth was exceptional. She deployed overwhelming mental firepower against the pull of gravity. She was Cryptanalyst-in-Charge.

By the end of 1932, when Americans elected Franklin Delano Roosevelt to his first term as president, Elizabeth's team at the coast guard was pound for pound the best radio intelligence organization in America. They knew how to extract information from clandestine radio networks, map the hidden structure of the transmitters, and hunt the people using them. This set of skills would later make Elizabeth an important figure in the quest to destroy the clandestine networks of Nazi spies. Her fight against smugglers was like target practice for the coming fight against fascism.

Americans weren't yet afraid of Nazis. They were too worried about scraping up money for food. By 1933 the Great Depression had put 15 million people out of work. Fabyan told William in a letter that the mayor of Aurora, Illinois, had shut down the town for five days to stop runs on the bank. "The world is a mess, and everything is going topsy-turvy," Fabyan said.

Elizabeth took her daughter to hear FDR's inaugural speech on March 4, 1933, a viciously cold morning. They walked to the Capitol from their house. "The only thing we have to fear is fear itself," FDR said, focusing on his plans to restart the economy. He did not mention Adolf Hitler or the Nazi Party, who had taken power that January, deploying mobs of men in brown uniforms and swastika armbands to crush dissent. The international press covered him like a normal leader. Many Germans did not think he would really do the things he had said he would do.

After FDR's speech people stayed outside in the freezing wind, cheeks pink as raw beef, waiting to see the parade. Elizebeth and Barbara handed out League of Women Voters fliers to the spectators. Elizebeth was pleased to see her daughter playing the role of "ardent worker."

Eighteen days later, in Germany, in a vacant gunpowder factory northwest of Munich, the Nazis opened the first concentration camp, Dachau. The occasion was announced by Heinrich Himmler, *Reichsführer* of the SS, at a press conference.

One month after that in Washington, at the end of April 1933, Elizebeth prepared to leave the city on her latest assignment for the T-men. "I pack my bag," she wrote, "and hug my children a good-by which is to last for a week or a month or longer, I know not, and board a train with a prayer that the new fields will be not impossible of conquest." She kissed William. He asked her to be safe and not take unnecessary risks. If she had known what awaited her in New Orleans, she might never have gone.

"Please state your name," America's top Prohibition official said. He had solemn eyes and spoke in a slow drawl.

"Elizebeth Smith Friedman."

"What is your occupation?"

"I am a cryptanalyst."

"And what are the duties of a cryptanalyst?"

"A cryptanalyst is a person who analyzes and reads secret communications without the knowledge of the system used."

It was May 2, 1933, and Elizebeth was speaking in a witness box in a federal courtroom, about to play her part in a giant conspiracy case against twenty-three suspected agents of the syndicate she had been tracking for years, the Consolidated Exporters Corporation. They had recently expanded into the Gulf of Mexico, directing a fleet of eight rum-running ships from a pirate radio station in New Orleans. T-men intercepted at least thirty-two coded radio messages and mailed them to Elizebeth, and her solutions exposed the ribs of the scheme: the names of the rum ships (*Concord, Corozal, Fisher Lassie, Rosita, Mavis Barbara*); their system of sneaking crates of liquor into lonely bayou towns on small boats called luggers and then unloading the crates onto freight trains, covered in sawdust. The government considered it "the greatest rum-running conspiracy since Prohibition," and now Elizebeth had been summoned to this federal courtroom in New Orleans to explain her methods to a judge and jury.

She wore a pink dress and a hat with a flower pinned to its brim. Directly in front of her was a stack of yellow papers with her message solutions. She looked out at the courtroom, its wooden pews crammed with spectators and journalists. The defendants sat together like a sports team on a sideline, in suits; the previous day in court they had been quietly switching seats to make it more difficult for witnesses to identify any one of them. The accused ringleader was Albert Morrison, a rawboned man in his sixties with white hair. He went by the aliases Charles Cosgrove, M. Ryder, A. A. Brown, Harry Hale, J. J. Jones, B. M. McGregor, and "Mr. Burk." The government believed that at least three of the other defendants—Nathan Goldberg, Al Hartman, and Harry Doe—were Chicago associates of Al Capone.

The government had spent $500,000 and more than two years on the investigation, marking this as a case they could not afford to lose, which is why the lead prosecutor today was the chief of the Prohibition Bureau himself, Amos Walter Wright Woodcock, a methodical former army colonel. He had argued for years that the only effective way to enforce Prohibition was to bring "a steady attack" against major crime syndicates and leave the small-time moonshiners alone. Big fish, not little fish. Consolidated was the whale.

Woodcock looked at Elizebeth in the witness box. "Have you a message which we will identify as 6:07 P.M., April 8, 1931?"

She shuffled her papers and located the sheet containing this message, which began, in code, QUIDS, ABGAH, FLASH, SLATE, FABLE, SHOOT, BOWSKY. She was about to read the solution aloud when a defense attorney objected, claiming that her testimony was "incompetent, irrelevant, and immaterial." Eight other defense attorneys leapt to their feet, joining the objection.

One was Edwin Grace of New Orleans, who handled Capone's appeals in federal courts. Grace was joined at the defense table by Walter J. Gex Sr., the patriarch of an influential family from the nearby Gulf city of Bay St. Louis, Mississippi.

"I believe I am asked my opinion of the reading of this message?" Elizebeth said, turning to the judge. "This is not a matter of opinion. There are very few people in the United States, not many it is true, who understand the principle of this science. Any other experts in the United States would find, after proper study, the exact readings I have given these."

"I ask all that be excluded," Gex said. "I think it is very improper."

Woodcock continued to walk her through the messages, and Elizebeth read her plaintext translations to the court, one at a time, interrupted by additional objections from the defense, accusations that she must have gotten her information from federal agents and not from codebreaking ("That certainly is some information somebody gave to that lady"). The syndicate had used a system of enciphered code—words that stood for letters in a cipher that stood for letters in a code. To solve it, Elizebeth had to rewind each step. A message that looked like

GD (HX) gm ga HX (GD) R gm OB BT HR CK 25 BT BERGS
SUB SMOKE CAN CLUB BETEL BGIRASS CULEX CORA STOP
MORAL SIBYL SEDGE SASH (?) CONCOR WITTY FLECK SLING
SMART SMOKE FLEET SMALL SMACK SLOPE SLOPE BT SA
back to the word SLDGE its SEDGE instead of SLDGE HW

became in plaintext

SUBSTITUTE FIFTY CANADIAN CLUB BALANCE BLUE GRASS
FOR COROZAL STOP REPEAT TUESDAY WIRE CONCORD GO
TO LATITUDE 29.50 LONGITUDE 87.44

When it was time for cross-examination, Gex rose from the defense table.

"How shall I address you," he said. "Madam or Miss?"

"I am Mrs. Friedman."

"Before you could properly translate these symbols somebody

had to tell you it was symbols in reference to the liquor transportation?"

"Oh no," Elizebeth replied, innocently. "I might receive symbols related to murder or narcotics."

"The same symbols these gentlemen used to mean what you say, whiskey, beer, position, could not have been made up by people in code for transportation of women from Europe?"

No, Elizebeth responded, "not with the meaning given them here."

"I move that all of the testimony of this lady be stricken out," said another lawyer, Maxwell Slade. The judge overruled him.

When there were no further questions from the defense, Elizebeth reached for her handbag, stepped out of the witness box, exited the courthouse into the damp spring air, and returned to Washington. Four days after her testimony, the jury convicted five of the syndicate's ringleaders, including Albert Morrison, who received two years in prison. Woodcock gave credit to Elizebeth, telling her superiors that she "made an unusual impression on the jury."

And not just the jury. Reporters covering the trial were taken with Elizebeth, describing her in stories as "a pretty government scrypt-analyst or 'code-reader,'" "a pretty middle aged woman," "a pretty young woman with a filly pink dress," and "a pretty little woman who protects the United States." This was her first sustained encounter with the press and it left a sour taste. She was still young enough to resent being called a middle-aged woman and thought the kinder phrases were badly written. The newspapers wrote about her again the following year when she returned to New Orleans to testify in the convicts' appeal. Facing off against Edwin Grace, Capone's attorney, Elizebeth grew impatient with his attacks on the validity of her science and told the judge she could settle the issue quickly if she had a blackboard. A bailiff found a blackboard in storage and wheeled it into the court. Elizebeth stood with a piece of chalk and diagrammed the rum ring's codes on the board until the jurors were nodding their heads and Grace was muttering that this was highly irregular. "CLASS IN CRYPTOLOGY," one newspaper blared the next day. The headline made her queasy. She had not signed up to be a public figure. She hoped the attention would die down. To her horror, it was only just beginning.

William went dark as his wife ignited and lit up the sky. The army kept his projects locked behind thick metal doors. He wasn't allowed to talk about his work, and she knew enough not to ask. "He never put into words and never asked me to put into words," Elizabeth said. The two cryptologists were so connected, so attuned to the slightest crease of an eyebrow or curl of a lip, that their faces had a way of becoming mirrors. A grim look on his face was instantly reflected in hers. Because neither wanted to cause pain in the other, the most painless course of action was often to make their faces like masks and not speak about forbidden things, though despite these efforts at self-control Elizabeth could tell when something was troubling her husband: "Many times there was a certain grim look that came around his mouth."

What she didn't know at the time, and didn't learn until after the war, is that the army had asked William to break a series of ciphers used by Japanese officials to encrypt their diplomatic communications, an enormous undertaking that would come to define his career and consume the next decade of his life.

He wasn't doing it alone. Like Elizabeth he was starting to build his own codebreaking unit, hiring junior cryptanalysts and molding them to his needs and his vision. The army had never agreed to such an expansion before, but the plates of government codebreaking had shifted due to a fateful decision made by the newly appointed secretary of state, Henry Stimson. A former artillery officer in the Great War, Stimson thought the idea of reading other nation's messages in peacetime was immoral, and upon learning that the State Department was paying codebreakers in New York to read the mail of foreign diplomats—Yardley and his American black chamber—the secretary was appalled. "Gentlemen do not read each other's mail," he was supposed to have said. Whatever the reason, Stimson decided to pull Yardley's funding in 1929, and the chamber was forced to shut down.

Because Yardley had developed some expertise with Japanese ciphers, the closing of his bureau left America with no ability to break new ciphers developed by Japan, a nation with a growing military and ambitions of empire. Worried about being left in the dark, the army turned to William, and in 1930 he launched a new army codebreaking unit that would later become the nucleus of the National Security Agency. William called his new organization the Signal Intelligence Service, or SIS.

The first three people he hired for the new unit were male mathematicians in their early twenties: Abraham Sinkov, Frank Rowlett, and Solomon Kullback. He gave them adjacent desks on the third floor of the Munitions Building, in a vault behind a thick steel door, and began to train them with dusty books and sample problems. William handed Rowlett a book by a German cryptographer named Kasiski. How is your German? he asked Rowlett. Rowlett said it was not good. William suggested he study the Kasiski text anyway. Now here was a book by an Austrian military cryptographer named Figl, and here were some of Friedman's own works, the "Riverbank Publications," written with Mrs. Friedman during the war. "Do you speak French?" Friedman asked Rowlett, who was sorry to say that he did not.

One morning, when William felt the three young men were ready, he popped his head in the vault and said they should come with him immediately. They followed the boss down a staircase and through a long corridor, where he took a left turn into an area that seemed deserted except for a steel door with a combination lock. Friedman reached into his coat pocket, removed a card, examined a series of numbers on it, and manipulated the lock until the bolt swung free. The door opened; behind it was a second steel door with a keyhole. Friedman fished a key from his jacket and jiggled it in the lock. The inner door opened into a completely dark room. The men waited outside as the boss disappeared into the dark and struck a match. A smell of smoke drifted out of the vault. Friedman found the dangling cord of a lightbulb and pulled, revealing a windowless room full of filing cabinets. He switched on three wall-mounted fans, turned to his young employees, and said, with gravity, "Welcome, gentlemen, to the secret archives of the American Black Chamber."

Yardley's files. The cabinets from his bolted bureau. Friedman had obtained them on the theory the files would prove useful to future army codebreaking projects, particularly regarding Japan.

Over the next fifteen years, William and his three young colleagues, joined by others, would drive themselves to the brink of collapse while working as codemakers and codebreakers alike, struggling to solve Japanese cipher machines of unknown design and applying what they learned about the weaknesses of the foreign machines to design new kinds of secure machines for America. At heart it was all the same business, the business of dominating secrets.

William first taught his deputies about cipher machines by challenging them to solve different machines that he himself had mastered years before (the Kryha, the Hebern), offering gentle hints when the young codebreakers got stuck. Then he set them to work on their first mysterious and yet-unconquered Japanese machine, which began transmitting in 1930: *Angooki Taipu A,* meaning "type A cipher machine" in *romanji,* the romanized form of Japanese that was used for transmitting messages. The name *Angooki Taipu A* was the codebreakers' sole piece of information about the machine. No one in the Western Hemisphere had ever laid eyes on it. There wasn't so much as a drawing of one, much less a physical copy. Internally, William and his colleagues referred to it by a nickname, "Red." Later, in 1938, the Japanese replaced Red with a more sophisticated machine, a significant upgrade: *Angooki Taipu B*. This one the SIS codebreakers called "Purple."

Red and Purple were used by Japanese diplomats in Nazi Germany to communicate with the Japanese government back home, which meant that solving the machines would give the SIS a most intimate view of strategic thinking in both Japan and Germany. To attack each system—first Red, then Purple—the American codebreakers needed to build their own bootleg versions of the Japanese machines, reverse-engineering them based on nothing but educated guesses from analyzing the garbled messages they produced. It was a task akin to building a watch if you have never seen a watch before, simply by listening to an audio recording of the ticking and clicking of its gears.

Even as he tried to infer the shape of foreign machines, William was building an American machine of his own, to protect American communications: the Converter M-134, a multirotor device of innovative design. He filed a patent in hopes of selling it one day on the commercial market. The patent was granted but held secret for many years, essentially nullfiying his rights. This would become an enduring frustration for William. As hard as he tried to earn money on the side, security concerns always got in the way.

Thanks to powerful brainstorms from Frank Rowlett, the M-134 eventually evolved into the SIGABA, an Egyptian Sphinx of a cipher machine with fifteen rotors and an ingenious mechanism that Rowlett called a "stepping maze." Up to four rotors might turn at the same time, with a single key press, and the rotors could be inserted in reverse direction. The army and navy distributed 10,060 SIGABA machines across every theater of the Second World War. President Roosevelt used SIGABAs to communicate from his Hyde Park home and when he traveled on the presidential train. The SIGABA was like an American Enigma machine or Purple machine, only inviolate. No enemy codebreaker, whether German, Italian, or Japanese, would ever manage to break it, despite strenuous efforts; the Nazis ultimately stopped intercepting SIGABA messages altogether, since they could not be read. The machine ensured "the absolute security of army and navy high command and high echelon communications," William later wrote with pride, and "contributed materially to the successful outcome of the war."

"Never said a word to me," Elizebeth swore.

All through the 1930s and the first half of the 1940s, the only thing he revealed to his wife about his work is that it involved Japan. That was all. He didn't talk about Red or Purple, or his own designs.

People wondered later how this could have been true, how a husband and wife, lying in bed together at night, both of them cryptologists, could possibly resist sharing the dramatic details of their work. No: "My husband never never opened his mouth about anything." She had to guess at the mind of her husband by watching changes in his behavior. He came home in the evening and said little. He opened a silver snuff box and inhaled the black tobacco dust. He was almost never cross, only withdrawn. The kids picked up on it. They noticed there were times when friends and neighbors were welcome to visit, when William cooked steaks on a backyard grill and laughed a lot and "was hugging me warmly," as Barbara later recalled, and other times he seemed trapped in a fugue state and was unable to come to the door. In the morning he paced through the house. At night he couldn't sleep. The bed would shake at 3 A.M. and Elizebeth would see him out of a corner of her eye, heading downstairs to make himself a sandwich.

He was not as sick as he would get in later years. She was not yet seeing him unable to get out of bed, lethargic, shaking, acutely depressed, talking casually of suicide, carrying a length of rope in the backseat of his car. She still thought of William's affliction as "nothing more or less than exhaustion" and refused to call it mental illness, a rational choice given the wide stigma against the mentally ill and the poor treatment options (no antidepressants). The chief psychiatrist at the city's preeminent hospital, Dr. Walter Freeman of George Washington University, was an early adopter of electroshock therapy and the inventor of the ice pick lobotomy, a cruel and unwarranted procedure that involved jamming a sharp metal stick through the back of the patient's eyeball while the patient was awake and wiggling it around until a sufficient amount of brain matter was turned to goo. There was a fair chance that if William sought mental treatment in Washington, he would end up under the care of Dr. Freeman or one of his disciples (and sure enough, eventually, William did).

So Elizebeth did her best in the 1930s to cover for her husband in public and shore him up in private, lending her strength to a person who seemed unwell without letting on to friends that he was unwell, out of loyalty to her lover and also a simpler kind of self-interest. The Friedmans had always refused to acknowledge clear distinctions between his career and her career, his adventures and her adventures, because it all sprang from those same fertile years at Riverbank when they worked so closely together and ultimately forged a sacred alliance to escape, and that bond had survived through the decades: "Any story of my experiences is quite inseparable from that of my husband, whose wizardries in cryptanalysis are of international note," Elizebeth once wrote. Their relationship was progressive in this sense, a joint bank account of the mind. It didn't seem like sacrifice for Elizebeth to help her husband when he was down and weak. It was like treating a wound on one of her own limbs.

The closest she ever came to explaining the politics of their marriage was in a letter years later to Barbara, recommending she read a novel called *Immortal Wife*, about a real American couple from the nineteenth century, Jessie and John Frémont. John was an army colonel who mapped the wilds of California, Jessie the feisty daughter of a senator, and they collaborated all their lives, Jessie believing that "to be a good wife a woman must stand shoulder to shoulder and brain to brain alongside her husband." As a husband and wife they attracted fame and controversy. Jessie took the fragmentary journals from John's expeditions and turned them into rich narratives that enthralled Americans. The Frémonts made money and lost it all; the U.S. Army court-martialed John on a bogus charge of insubordination; he suffered a nervous breakdown, deteriorating into a "gray-haired and sunken-cheeked person . . . his face at war with itself." Some of the novel's more vivid passages describe the efforts of Jessie to rebuild John's shattered confidence:

> Everything that happened to John must of necessity happen to her; when two people marry they cease to be purely themselves but step into a new and expanded character, the character of their marriage. . . . She used every art and guile known to the heart of woman to nurse him to health. As they rode . . . spirited horses through the forests she challenged him to race with her, complimented him on how beautifully he sat the horse. Sitting before the warmth and bright red flames of the fireplace she played up the hours and episodes he had enjoyed most in their years together, in which he had appeared to the best advantage, filled him with her pride in his accomplishments. . . . She played the temptress, wearing her loveliest gowns, using her most delicate perfumes, shamelessly arousing his sexual love for her, the love that always had been such a strong and potent force between them. . . .

William also told his son to read this book if he wanted to understand his own parents: "In many ways it parallels my life with your mother," he wrote, calling Elizebeth "a remarkable woman" whose "own indominable spirit helped me climb up out a psychological morass that was pretty deep and distressing." The Friedmans connected with the tale of the Frémonts, this pair of American explorers menaced by the whims of the army and the nuances of neurochemistry. It made them feel, as all good books do, less alone.

Their lives had become so isolated in Washington. When they first met and fell in love, at Riverbank, secrecy was about exploration and connection, a joyous hunt for a hidden order. Now it was a force of loneliness.

They rebelled.

Not openly. Not by breaking laws or leaking secrets. Instead they figured out how to use cryptology to reach out to people, to resist the isolation that their cryptologic careers enforced.

They began with their own children, teaching them simple ciphers when they were seven or eight: A=B, B=C, C=D. Barbara wrote letters in this cipher from sleepaway camp. "EFBS NPUIFS BOE EFBS EBEEZ . . . Dear Mother and Dear Daddy. We went on a canoe trip. We went about 12 or 14 miles. We did it in 1 day too. I paddled 1/2 the way. Love, Barbara." William and Elizebeth replied, "XF BSF QMFBTFE . . . We are pleased to know that you can handle a canoe. It is lots of fun. You will be surprised to learn that Pinklepurr had her kittens on August 2."

The Friedmans also shared cipher letters with their friends. Each December they sent a holiday card in the form of a puzzle. In 1928, it was a "turning grille," a square of red paper perforated with circular holes, its four sides numbered 1 through 4, with a single left-facing arrow that said "TURN." When this square was laid atop a separate sheet of paper containing a 9-by-11 grid of letters, certain letters showed through the holes: FOR CHRISTMAS GREETINGS IN 28. Turning the overlay 90 degrees clockwise, new letters appeared in the holes (WE USE A MEANS QUITE UP TO DATE), then again with a third turn (A CRYPTO TELEPHOTOGRAM HERE), and a fourth

(BRINGS YOU WORD OF XMAS CHEER). Another year William drew a picture of a tree with happy faces and sad faces hanging from its branches like fruit. "Friedman's Wishing Tree," the caption said. "Individual fruits biformed, differentiated by upcurving or downcurving crevice." It was a message written in Bacon's biliteral cipher: a nod to the Friedmans' own past, to their creation story at Riverbank, the Garden of Eden where they fell in love and were expelled for the sin of learning to distinguish reality from delusion. The happy face was the *a*-form in the cipher, the sad face the *b*-form. The plaintext read, "Season's Greetings from the Friedmans."

They took these games further by organizing live puzzle-solving events that were famous in their social group throughout the 1930s. Some of these "cipher parties" were scavenger hunts that sent guests winging through the city. Elizebeth handed you a small white envelope. You tore it open to find a cryptogram. The solution was the address of a restaurant. When you arrived, you ate the salad course, then solved a second cryptogram to discover the location of the entrée. Other parties were hosted at the Friedmans' home with food cooked by Elizebeth. A shy army wife arrived at 3932 Military Road one evening with her husband and panicked when the Friedmans handed her a menu in code. "The first item was a series of dots done with a blue pen," she later recalled. "The 'brains' at the party worked over the number of dots in a group when it occurred to me it had to be 'blue points'— oysters—and it was! I had done my bit, and from then on I was quiet."

On a different occasion, Elizebeth designed a menu that listed one of the courses as "An Indecipherable Cipher." A guest wondered if this meant "hash," a cryptographic term for a string of text that gets scrambled once and never unscrambled, like a door that locks forever behind you (today hashes are used to protect Internet passwords). The guest was delighted when Elizebeth arrived from the kitchen carrying a steaming plate of meat-and-potato hash.

The Friedmans received so much praise for these parties that William thought he saw a chance to make money: If the army wouldn't let him sell his cipher machines because they were state secrets, couldn't he bake some of the same ideas into a mass-marketed board game? In a heat of inspiration he tried to design a Monopoly of codes and ciphers. The Crypto-Set Headquarters Army Game was a folding piece of cardboard with a red spinning wheel; players had to solve puzzles to advance tokens from the start line to the finish. A second prototype, Kriptor, featured two ivory-colored rotating discs printed with hieroglyph-like symbols. Players had to play "codemaker" and "codebreaker," trading secret messages.

He thought Kriptor had commercial potential and sent it to Milton Bradley, makers of Battleship and the Game of Life.

In the hands of the Friedmans, cryptograms were like poems or songs, a way of telling friends and family they were part of something wonderful, a shared language, and that they were loved. Given the immense secrecy of their profession, reaching out to people in these elaborate and whimsical ways, widening the circle instead of narrowing it, was a kind of defiance, although, ultimately, their most defiant act during these interwar years may have been the simplest.

They built a library.

Soon after moving to Washington, Elizebeth and William began to collect books and papers about all things related to secret writing, including documents that touched on their own lives in cryptology and that were not classified or restricted to government vaults. They stored objects on bookshelves in the den of their house. The library was as broad and curious as its creators. It contained books the Friedmans loved and books they hated; books they felt committed sins against cryptologic accuracy, good prose, or both; books that were centuries old but still contained relevant cryptologic knowledge; books they enjoyed but did not understand (William was beguiled and frustrated by the fiction of James Joyce and Gertrude Stein, who wrote sentences so garbled they might as well have been in code). The most valuable book dated to the sixteenth century, *De Furtivis Literarum Notis,* by Giambattista della Porta, an Italian cryptologist and scientific rival of Galileo. The Porta book, with a burgundy binding and text in Latin—an extremely rare 1591 forgery of the 1563 first printing, one of only three such copies in the world—

had arrived in the mail one day from Riverbank, an unexpected gift from George Fabyan. Elizabeth was moved by this gesture, because she had never known Fabyan to give anything away without demanding something in return.

All the rest of their books were collected cheaply. Wherever they went in the world, the Friedmans rummaged through used bookshops for ten-cent treasures. They salvaged books that other people discarded. Once, at the Munitions Building, William fell into conversation with a white-haired Civil War veteran. "He came up to me one morning as I was coming into work," William recalled later, "and he whispered, 'They're burning things today.' And I said, 'Such as?' And he showed me these books." They were a complete set of Union army cipher books, precious items that William rescued from destruction.

Some of the most important items in the library were books of unsolved puzzles and historical mysteries that the Friedmans thought they stood a chance of solving, if they ever got the time. In this sense the library was an archive of their dreams, a set of escape hatches into other, lighter-hearted kinds of lives that seemed possible within the world of secret writing, lives of academic exploration and scholarship. Elizabeth had begun taking graduate classes at American University in 1929, working toward a master's degree in archaeology. She was drawn to several richly illustrated books of Mayan pictographs that had never been decoded; she sometimes fantasized about quitting the coast guard and flying to Mexico, climbing through Mayan ruins in a straw hat, making sketches. William pondered his copy of the Voynich Manuscript, an illuminated book of uncertain lineage, written in a delicate looping script that corresponded to no known language, illustrated with pictures of flowers. He studied a book about the Beale Treasure, supposedly a stash of buried gold and silver ingots in Beale County, Virginia, its location revealed by an unsolved cryptogram. William once joked that he "worked on the cryptogram, on and off, but only in my leisurely hours at home, nights, Saturdays, Sundays, and holidays."

They also collected their own writings in the library. The archives of the Riverbank code section had remained at Fabyan's estate, so the Friedmans didn't own their earliest worksheets, but in the years since they had kept or copied every other nonclassified scrap of paper that had passed through their hands.

The Friedmans weren't necessarily doing this to document the history of American intelligence and tell a renegade story about its birth, although this would be the ultimate, magnificent result. Rather, the Friedmans built an archive because that's what the best intelligence professionals do. They become librarians. It's no accident that J. Edgar Hoover got his start in government as an eighteen-year-old Library of Congress clerk, a job that gave him "an excellent foundation for my work in the FBI," he later said, "where it has been necessary to collate information and evidence." The FBI was a library of human fingerprints and human deeds. Elizabeth's thirty bound volumes of rum messages were a library of the outlaw seas. The files of Herbert Yardley's American black chamber, now controlled by William, were a library of diplomatic intrigues from many nations, and William's SIS unit at the army was fast becoming a library of as-yet-unsolved Japanese communications, an archive pitched against a rising fascist power.

For all the harmless innocence conjured by the word "library," the Friedmans knew the truth: a library, properly maintained, could save the world—or burn it down.

They took their home library seriously enough to follow the practices of professional librarians. William made his own children sign a checkout slip if they wanted to carry a book from one floor of the house to another, and whenever the Friedmans acquired a book, they pasted a custom bookplate inside the cover, a rectangle of card stock designed by a professor friend who studied Mayan writing. The illustration on the bookplate showed a crimson warrior swinging an axe down upon the skull of a human. Pictographs spelled out a warning to book thieves:

> *Lay ca-huunil kubenbil tech same*
> This our book we entrusted you a while-ago.
> *Ti manaan apaclam-tz'a lo toon*
> It not-being you-return-give it us
> *Epahal ca-baat tumen ah-men*
> Is-being-sharpened our-axe by the expert.

On the wall of the library, to reinforce the message, William and Elizabeth hung an axe with a wooden handle and a black blade.

They were not trying to be mean or intimidating. They were showing their reverence for knowledge. Knowledge is power, as Francis Bacon once said and as George Fabyan and Mrs. Gallup taught them years ago. The Friedmans had taken this precept to an extreme, structuring their whole lives around it—an attitude of sharp curiosity and ruthless self-honesty that defined them. They had carried that little kernel of Bacon's philosophy along with them when they escaped Riverbank, and afterward they stayed true to Bacon's idea in ways their mentors at Riverbank never did, because to really live a life in search of knowledge, you must admit when you are wrong.

Mrs. Gallup never acknowledged that her method was flawed. She never moved on to a new project, never left the estate at all, and Elizabeth and William did not write to her about the Bacon ciphers, feeling it would be unkind to keep restating their skepticism. As the Friedmans climbed onward to brilliant careers in Washington, Mrs. Gallup remained in a cottage at Riverbank, peering through a looking glass at old books, expenses paid by Fabyan, until her death in 1934. She died believing she was correct, that Bacon was Shakespeare, that she had discovered proof, that history would vindicate her.

As for Fabyan, he never quite said the Bacon cipher project was hopeless, that the messages were not really there, only that the argument was not winnable, that he and Mrs. Gallup had failed to make the case. "I have no facilities or knowledge by which to prove the authorship of any volume of Shakespeare," Fabyan told the *Chicago Evening American,* which mocked him: "E'en Colonel Fabyan, the seer of Riverbank, now stands agnostic on the rock of doubt, chewing the food of sweet and bitter fancy, and says he does not know. . . ."

During the final years of Fabyan's life, William's attitude toward his old tormentor softened considerably. They wrote to each other like old friends. They gossiped about Herbert Yardley. Fabyan called Yardley an "ass." William loved that. He noticed that Fabyan seemed glum and tired. His health was failing.

He complained of a hernia and a prostate gland that had to be "reamed out" by a surgeon. He saw another world war coming and hoped it wasn't true. "With love to Elizebeth and the family, and trusting we are not going to get into another war," Fabyan wrote at the end of one letter, signing off, "Always the same old, GF."

George Fabyan died two years after Mrs. Gallup, in 1936, at the age of sixty-nine. The Friedmans heard the news in a letter from an old Riverbank colleague who still lived there. "The Colonel's interest in life had slipped a good deal in the last year," the colleague wrote. A bout of laryngitis had worsened into pleurisy of the lungs and a cascade of medical complications. Fabyan suffered greatly in the last ten days of his life, feverish on the bed in the Villa that hung from chains, the bed swinging to and fro with the violence of his coughing. He said that when he died he wanted his employees to shut off the light in the lighthouse, the one that flashed two white lights and three red lights in a continual pattern every night, the coded message saying *keep out, keep out, keep out*. It went dark according to his wish.

Fabyan left behind far less money than anyone would have guessed. The Great Depression, along with his lavish expenditures on the laboratories and their science experiments, had nearly wiped him out. In his will he gave $175,000 to his widow, Nelle, and provided for a monthly stipend of $150 to his loyal Scottish secretary, Belle Cumming, for as long as she might want to live on the grounds as caretaker. Nelle died of cancer three years later, in 1939, and in 1946, Cumming was killed in a gruesome accident along with two other women when an oncoming train struck their car as it crossed the Geneva tracks. County officials purchased the estate for $70,500 and added it to the local forest preserve. Riverbank was public property, the once-mighty kingdom now an empty set of buildings and fields.

All along in his letters to Fabyan, William had been coaxing the old man to give his papers to a library: his own personal papers and also the voluminous files of the Riverbank Department of Ciphers. In Fabyan's last letter to William before he died, he had mentioned destroying "a lot of old correspondence on ciphers because I did not know whose hands it might fall into." Now

William wrote to his widow, Nelle. What was to be done with the files of the Department of Ciphers? These records were part of the Friedmans' own history and America's history too, and the Friedmans wanted to make sure they were preserved.

The reply came that Fabyan, as one of his last wishes, had ordered many records to be burned.

William and Elizebeth guessed that he did it to prevent embarrassing revelations about his many deceptions and schemes. Fabyan had chosen to go out in a flame of self-immolation, a Viking funeral of documents.

It wasn't as bad as they thought. While Fabyan had destroyed some documents, others survived and would eventually make their way to the New York Public Library. But the Friedmans assumed the worst. To them it was a tragedy: loss of history, loss of knowledge. No one else in America particularly cared. A dying man had burned some papers about cryptology. Cryptology is a profession of secrets. Secrets staying secret is the norm. Officials only get riled up when the opposite happens—when secrets are leaked, published, disclosed. And by now, the Friedmans and their colleagues were struggling to manage the consequences of the biggest leak of cryptologic secrets in the country's history.

"A lie!"

At home, William turned the pages of a book with a black cover, facial muscles tightening, writing annotations in the margins:

"This is a patchwork of misstatement, exaggeration, and falsehood."

"Lies, lies, lies."

Six crimson lines divided the book's cover into seven black rectangles. Inside one of the rectangles, crimson letters read,

THE AMERICAN BLACK CHAMBER
HERBERT O. YARDLEY

Yardley.

The poker player and codebreaker had not reacted well to the closing of his State Department bureau by Henry Stimson, the secretary who believed that gentlemen do not read each other's mail. When the government put Yardley out of work, he had been unable to find a new job in the Depression. Down to his last couple of dollars, he decided to spill his secrets, for money and revenge. He published a book, *The American Black Chamber*, in the summer of 1931, which became a bestseller and international sensation. It also hit the U.S. government like an exploding volcano, hurling rock and lava into the stunned surroundings and forcing the Friedmans to navigate the smoking landscape it left behind.

The book claimed to be the true story of the black chamber, "a glimpse behind the heavy curtains that enshroud the background of secret diplomacy." He framed it as an act of patriotism, and he had a case. Yardley believed it was dangerous and naive to stop breaking the codes of foreign governments. It made America weaker, more vulnerable. He believed there should be a debate, but it was impossible to have a debate about the kinds of things the black chamber did if the public didn't even know that it had existed. He argued that because the bureau was now closed, there was no harm in revealing its activities.

There was enough truth in the book to scandalize U.S. officials. Yardley disclosed, for instance, how his country had been reading the diplomatic traffic of Japanese ambassadors, information that Japanese readers were surprised to learn; the book sold 33,119 copies there. He also tantalized with anecdotes of sex and deception. "A lovely girl dances with the Secretary of an Embassy," Yardley wrote in the book's introduction, promising more details within. "She flatters him. They become confidential. He is indiscreet." He devoted a full chapter to a bewitching female spy named Madame de Victorica, "the beautiful blonde woman of Antwerp," who worked for the British and French during the Great War and employed ciphers and invisible ink. Yardley claimed that an unknown enemy once enlisted a blonde spy to seduce him at a New York speakeasy: "It did seem to me that she showed a bit too much of her legs as she nestled in the deep cushions." They got into a taxi and went to her apartment, where she promptly fell asleep on the couch; Yardley, suspecting foul play, searched her desk and found a note that said, "See mutual friend at first opportunity. Important you get us information at once." The next night, thieves broke into his office and rifled through the cabinets: "I took it for granted that they had photographed the important documents which they required."

This probably never happened. Yardley's colleague later said that the story of the blue-eyed blonde was a "damned lie," and that the only things taken from the office were a couple of bottles of booze. "It was my booze and I think [Yardley] took it himself." As for Madame de Victorica, she did exist, but Yardley embellished her biography. He admitted to friends that he fictionalized parts of the book. He compressed time, invented dialogue, added "bunk" and "hooey," and made no apologies: "To write saleable stuff one must dramatise."

William Friedman found this loose attitude intolerable. Truth was truth and anything else was fuckery. He also disagreed with Yardley's assertion that there was no harm in telling old stories. William thought that some of Yardley's revelations might startle adversaries into boosting the security of their communications, particularly Japan, thereby making the jobs of American code-breakers more difficult.

In the end, though, William resented that Yardley was telling a story about cryptology that William wanted to tell himself. William had always been a stifled writer, unable to publish what he would have liked. At Riverbank the obstacle was the ego of a rich man, George Fabyan, who insisted on taking credit for William's work, and now the reason was national security. He couldn't tell stories of his army feats because, unlike Yardley, he wasn't willing to violate his promise to keep America's secrets.

Yardley had actually sent William a copy of *The American Black Chamber* as a courtesy, signing his name on the inside cover. Beneath the signature, William wrote, "OMNIS HOMO MENDAX," which means, in Latin, "Every man is a liar." Then he began to write his own alternate history of the events and concepts Yardley had described, inside Yardley's book, in the margins: "A lie! Which can be so proved to be. See papers attached. Exhibit 1." *He attached exhibits to another man's book.* He underlined sentences, bracketed paragraphs, tagged words with asterisks, spangled pages with exclamation marks. Revenge by

annotation. Not content with his own annotations, he then circulated the copy of Yardley's book among four of his colleagues
in the army and MID, and they added their own annotations in
their own handwriting, a chorus of jeers and boos. William created a numbered key in the front of the book so that future readers could keep track of the different voices.

The fallout from the publication of Yardley's book was lasting
and broad. Intelligence bosses and lawmakers grew newly anxious
about the security of cryptologic information and moved to crack
down on future disclosures. Yardley became persona non grata in
U.S. intelligence circles for the rest of his life; exiled from his old
haunts, he made friends in Los Angeles and wrote screenplays for
Hollywood movies. In 1933, Congress passed a law specifically to
prevent Yardley from publishing a book of codebreaking yarns
focused on Japan; the new law, called "An Act For the Protection of Government Records" and derided by Yardley as the "Secrets Act," declared it a crime to reveal secrets about cryptologic
information. But stories about codes and ciphers only increased
after Yardley opened the gates. He had proven that there was a
market for them, particularly yarns about Yardleyesque women
who dealt in secrets, and when editors looked around for such a
woman, they did not have to look far.

Elizebeth Friedman was attractive. She was a mother. She was
American. She was testifying in open court against the criminal masterminds of her age. And unlike Yardley's women, she
was verifiably real in every detail. "Widespread interest in the
romantic stories of beautiful female spies, secret codes and ciphers which Yardley had told caused editors from this time on
to become keenly aware of the news value of such stories," went
a confidential memo later circulated by the U.S. Navy to warn
about the dangers of cryptologic publicity. "Consequently, when
in 1934 magazine and newspaper accounts broke concerning Mrs.
Elizebeth [*sic*] Smith Friedman, a Coast Guard Cryptanalyst, a
number of similar incidents followed."

"I'll confess, Mrs. Friedman, I was thunderstruck the other day
when I met you for the first time. I simply wasn't prepared to find
a petite, vivacious young matron bearing the formidable title of

Cryptanalyst for the United States Coast Guard. How did you ever get interested in the highly technical science of codes and crypts?"

"I never thought of my job as terribly unusual until the newspapers stumbled upon what I do for the government, Miss Santry."

Margaret Santry was a radio reporter for NBC. In 1934 she launched a series of interviews with "First Ladies of the Capitol," mostly socialites and the wives of congressmen, and in May she asked Elizebeth to speak about her career on a national NBC broadcast. Elizebeth brought her children to the NBC radio studio in Washington; commercial radio was less than a decade old, still slightly wondrous, and she thought the kids might enjoy seeing the technology. Barbara was ten, John Ramsay was seven. The NBC staff let the kids watch their mother from the glassed-in control room adjacent to the sound booth, analog indicator needles twitching back and forth as Santry peppered Elizebeth with questions.

"Does the habit of thinking in code ever creep into your family life?"

"I guess it's bound to—it's so much a part of our life," Elizebeth said.

"And are your children experts in code, too?"

Elizebeth replied that Barbara was "quite an expert for her age" and had sent her parents messages in cipher two years earlier, when Elizebeth and William sailed to Spain for an international conference about radio transmissions.

Santry asked, "And I suppose she wants to be a cryptanalyst like her mother when she grows up?"

"No, she wants to be a professional dancer," Elizebeth said. As for her son, John Ramsay "states he wants to be a code expert when he grows up," but "at present he is in the boats and guns stage."

"How do you solve the business of running a home and a family—and an important job—all at the same time, Mrs. Friedman?"

"Oh, it solves itself rather nicely; especially when I have such a grand housekeeper to look after things at present. I never really made any definite plans for a career, Miss Santry—it just happened."

This NBC interview was one of the rare media appearances that Elizebeth enjoyed; she preferred speaking with female reporters, and NBC treated her children with kindness. Most of the time she hated dealing with the press. Depending on the reporter, she found the experience irritating to unbearable, and altogether the articles and radio shows represented a real threat to her livelihood.

Codebreaking is a secret profession. Its practitioners aren't generally supposed to talk about how they do what they do, as demonstrated by the shock at Herbert Yardley's disclosures. Elizebeth wasn't anything like Yardley; he was motivated by money, and she was motivated by doing the job that the government was asking her to do. Elizebeth only discussed active cases in public when a prosecutor called her to testify in court, and at all other times she limited her comments to closed cases, and then only with the full authorization of the Treasury publicity office. She testified for a reason: to put bad guys in prison.

Yet Elizebeth was so talented at this task, and the trials so spectacular, that the resulting waves of publicity threatened to wash away her career, in an era when government was growing concerned about the security of secret information.

The publicity coincided with a rise in the drama and stakes of her cases. When Prohibition was repealed in 1934, destroying the market for bootleg liquor, several of the rum rings made a nimble transition to smuggling drugs, mainly opium and the drugs derived from it, heroin and morphine, which were refined by pharmacists and criminal gangs in the port cities of China. Elizebeth adapted. Bags of heroin were smaller than crates of booze, easier to conceal, which only made codebreaking more essential—the only reliable way to discover the drugs was to swipe the details of their location from the smuggler's own lips.

The drug networks were global in scope, spanning more lands and languages than the rum syndicates. Elizebeth coordinated the investigations with T-men in offices all across America, with members of the U.S. State Department, and with police inspectors in foreign countries. She worked with translators to read encrypted notes in Portuguese, Spanish, French, Italian, German, and Mandarin. She was able to take messages written in blocks of letters or numbers and trace these figures back to specific Mandarin words in commercial code books designed for Chinese merchants. Reporters and readers seemed amazed that she could break a code in a language she did not speak, but that was the power of having a system, a science. "The whole deciphering science is based on what we call the mechanics of language," she explained on NBC radio, making it seem easy. "There are certain fixed ways in which language operates, so to speak; and by studying the known elements and making certain assumptions, one can arrive at a result that usually does the trick."

Global heroin rings went up at the stroke of her pencil. Smuggling ships were skimmed off the ocean like fat from a simmering pot of soup stock. Each new feat startled loose a flock of news articles that sang about Elizebeth's previous feats and added one more verse to the ballad of her growing legend. She decrypted a stack of intercepted letters and telegrams exchanged between a member of Shanghai's fearsome Green Gang of criminal warlords and two brothers in San Francisco, Isaac and Judah Ezra, the dissolute twin sons of an upstanding Shanghai pharmacist. They had been smuggling opiates from Yokohama, Japan, to San Francisco on the passenger liner *Asama Maru,* the drugs hidden in barrels of tree-nut oil. The messages spoke of "wyset," "wysiv," and "wyssa" in various quantities; Elizebeth determined that "wyset" meant cocaine, "wysiv" was heroin, and "wyssa" was morphine. She alerted the T-men in San Francisco, who searched the *Asama Maru* when it arrived and found 520 tins of smoking opium, 70 ounces of cocaine, 70 ounces of morphine, and 40 ounces of heroin. Informed that their code had been broken, the Ezras quickly confessed and were each sentenced to twelve years in prison.

EZRA GANG FALLS IN TRAP OF WOMAN EXPERT AT PUZZLES, went a headline in the *San Francisco Chronicle*. SOLVED BY WOMAN. The press had a way of praising Elizebeth and condescending to her at the same time, professing amazement at the capabilities of the female brain. She had "supplied the Federals with enough dynamite to break up the ring," the paper said, and now "the Ezra boys have twelve years to figure out another code she can't break."

When the reporter asked her how she broke the Ezras' code, Elizebeth declined to say: "We have to keep our ideas secret so that we do not give other smugglers any new ideas." Worried that she was attracting too much attention, that T-men would feel slighted by her fame and that Treasury chiefs would fret about smugglers getting wise to her methods, she tried to get the press to stop writing about her. She begged reporters to credit the coast guard as a whole instead of one woman for solving the cases. She wrote apologies to colleagues and bosses, complaining that reporters would not leave her alone: "The mystery-lure of the words code and cipher, coupled with a woman's name, invariably inflames the news reporters and they start on the trail of a story."

Nothing worked. She had chosen a profession that continually immersed her in lurid realities. "She is entrusted with more secrets of the crime world and of federal detection activities than any woman in history," reported *Reader's Digest* in 1937, in a five-page feature that declared Elizebeth "Key Woman of the T-Men" and was mailed to more than a million subscribers. "When one of the Treasury Department's enforcement agencies gets the scent of a new international enterprise in smuggling, dope running . . . there is one unofficial order that sticks in every agent's mind: 'Get some of the gang's correspondence and send for Mrs. Friedman.'"

There were people in the U.S. military who didn't like that Elizebeth was talking about codebreaking in criminal trials and didn't much care that prosecutors were forcing her to do it. Her testimonies showed that she knew how to break many different kinds of codes, including codes in foreign languages, which "could lead to only one conclusion on the part of espionage agents—decryption of other nation's codes was in progress behind the scenes," according to the confidential 1943 memo. After the publication of "The Key Woman of the T-Men," the army's inspector general stormed into William's office and demanded to know why William had been mentioned in the article, as Elizebeth's husband and as the army's top codebreaker. Officials at both Treasury and the War Department had reviewed the article before publication and confirmed that it contained no classified information, but the inspector had not been consulted. William had no idea what to say. "I lost my tongue completely," he wrote, "and failed to ask for permission to sleep in the same room and/ or bed with my wife." Journalists mentioned him in articles about Elizebeth because the two of them were married and shared a profession. If it was a problem for two cryptologists to be married, what was the solution?

The troublesome publicity came to a peak in the first months of 1938, when the buzz from the *Reader's Digest* profile dovetailed with a new round of press from a flashy drug trial in Canada. At the request of the Royal Canadian Mounted Police, Elizebeth flew to Vancouver to analyze messages discovered in a raid of a Chinese merchant's home and business. Police seized thirteen Luger pistols, four hundred Mauser clips, almost one hundred machine gun drums, numerous cans of opium, and two dozen coded cablegrams in a safe. Working with RCMP translators, Elizebeth solved the messages, exposing a smuggling ring that traded Canadian weapons to Hong Kong in exchange for drugs. The cablegrams were written in English ciphertext, four letters per block, which Elizebeth turned into Arabic numbers, that corresponded to Mandarin characters in a Chinese commercial dictionary, that stood for Mandarin words, that were translated into English words:

UUOO AMAS ANAG USOG UKUU IUUI AEIY thus became "Cable three thousand select fully Wat list," Wat Sang Co. being the name of the drug dealer's Canadian company. She found that the smugglers referred to opium as "ginseng" and "groceries," and guns and ammunition as "hams," "presses," and "tails."

Her testimony at the Vancouver criminal trial helped convict all five defendants, and a fresh wave of articles and headlines appeared. CANADA SMASHES OPIUM RING WITH U.S. WOMAN'S AID. WOMAN TRANSLATES CODE JARGON. The February 15, 1938, issue of *Look* magazine included Elizebeth in a feature on "outstanding" women "in careers unusual for their sex," along with a female deep-sea diver, a female conductor of a symphony orchestra, and a silversmith. She had not provided *Look* with a photo or an interview, but the magazine somehow got its hands on a gauzy black-and-white photo of Elizebeth in a white robe. *Detective Fiction Weekly* printed fourteen pages about Elizebeth's career written by a newspaper editor who moonlighted in gangster fiction. "Lady Manhunter," he titled the piece, "A True Story." A telegram arrived in Elizebeth's office from a magazine writer in New York. "PLEASE COOPERATE BY ANSWERING QUESTIONS BELOW BY RETURN SPECIAL DELIVERY TO ME," the man wrote. "FOR WHAT DEPARTMENTS DO YOU DECIPHER MESSAGES? HOW MANY HAVE YOU DONE? WHAT TYPES? SENT BY WHOM? HOW DO THEY FALL INTO YOUR HANDS? . . . LIKES, DISLIKES, SUPERSTITIONS HAVE YOU, ANY FURTHER ANECDOTES OF HUMAN INTEREST, HUMOR, OR UNUSUAL EXPERIENCES WOULD BE APPRECIATED." Across the bottom of the telegram, in large letters, Elizebeth scrawled in disgust, "Ad Absurdum!"

She was now the most famous codebreaker in the world, more famous even than Herbert Yardley, the impresario of the American Black Chamber. And she was more famous than her husband, too—a reversal from the longstanding pattern.

All their lives, William had been the celebrated one, the master, the genius, and she the dutiful wife, supporting him from the shadows. But Elizebeth was now said to be the true genius in the family, the driving force of the duo. A rumor spread that she had taught her husband everything he knew about cryptology. People started to approach William at military functions and regale him with stories about Elizebeth's adventures, as if he were unaware of them. He thought the inversion of the narrative was hilarious, "a scream"; it had been silly before when everyone thought he was the only genius in the couple, and it was silly now. The Friedmans had always considered each other equals. He didn't mind the attention swinging to Elizebeth. "When people introduce me and then say that my wife is also etc & is really better at it, I invariably assent, with a real smile," he told her in a letter. He was proud of her.

Elizabeth, of course, didn't think she was a genius; actual geniuses never do. Codebreaking to her was about teams, systems, cooperation.

She told people she never wanted to see her name in print again. In the final months of 1938, she got her wish. The U.S. government gave her a new assignment that was every bit as clandestine as her previous missions had been public. Soon the mark of her pencil, once celebrated on front pages, would become one of the most closely held secrets in America.

William left Washington for Hawaii in October 1938 on a secret mission to the Pacific. The army wanted to deploy the cipher machine he had invented, Converter M-134, and start using it to protect "highly secret communications" between its headquarters in Washington and army installations in San Francisco, Hawaii, and the Phillipines. William's job was to carry two M-134s to each base, install and test them, and train army staff in their operation. He boarded a troopship, the *Republic,* in New York City, with a trunk containing six bulky cipher machines and some books of essays and poetry, and the ship steamed south to the Panama Canal and passed from there into the Pacific.

He was at sea during the atrocity known as Kristallnacht, the Night of the Broken Glass, when Nazi mobs in a thousand German towns murdered at least ninety-one Jews and set fire to synagogues. These were pogroms like the ones in Russia that had terrorized William's kin. The mobs attacked with stones, bricks, wood poles, and automobiles driven through plate-glass windows, destroying Jewish stores, Jewish hospitals, Jewish nursing homes, Jewish kindergartens, Jewish cemeteries, Jewish Scrolls of Law. On a busy Berlin street a cheering crowd hacked apart a grand piano with hatchets.

In Washington a few weeks later, Elizebeth fell ill. A creeping fatigue glued her to the bed for long periods. Not wanting William to worry, she concealed the details from him, only telling her sister, Edna, who assumed that her sister was simply overworked. Edna didn't know about Elizebeth's new mission, and William didn't, either. He was on the ship, cut off. His airmail letters arrived in Washington within a few days, but they were expensive.

Regular-mail letters took weeks; telegrams had to be short. The *Republic* had a radio transmitter for military use only. William was sometimes able to sneak on and send Elizebeth messages at her Treasury office.

From a few words in her letters and telegrams, he got the sense his wife was struggling, that something was wrong, but she didn't spell it out. He was sick of these distances and wished they could go back to the time when they were together all day long, solving puzzles side by side. On the ship he daydreamed about Riverbank. "I can almost forsee the time when you and I will be working together again in an office," he wrote to Elizebeth. "It would be a joy to have you alongside me again—to recapture those days when we worked side by side and stole kisses in the vault at Engledew—such passionate kisses on my part, and sometimes on yours."

Arriving in Hawaii in late November, William began tests of the cipher machine and bought gifts for his family: hula skirts for Elizebeth and the kids, and a glass vial of Shalimar perfume for Elizebeth. Then he sailed to San Francisco for more tests at a different army facility. He was back on the ship, heading homeward, by Christmas Eve. The crew of the *Republic* arranged a Christmas dance for passengers. He sat in the smoking room reading a book while others danced. After a colonel entered and shot him a dirty look for being a prude, William put the book down and danced a waltz with the daughter of the Cuban minister to Japan, a young beauty with a slight accent, "as brunette as can be."

He was still thousands of miles from Elizebeth when the new year dawned, and skimming along at fourteen and a half knots. During these first days of January 1939, Japanese pilots dropped bombs on civilians in the Chinese city of Chongqing. The cardinal of Munich praised Hitler's "simple personal habits." The Nazis announced the immediate deployment of hundreds of new U-boats, a smaller and swifter generation of submarine. A physics professor in a basement laboratory at Columbia University wrote in his diary, "Believe we have observed new phenomenon of far-reaching consequences"—nuclear fission, the splitting of uranium atoms into lighter elements, achieved with an atom smasher.

The motion of the ship disturbed William and he slept irregularly, conking out for hours in deck chairs at midday, sun cooking his pale skin, then twitching with insomnia at night, wandering the ship alone. He watched Hollywood movies in the evening, above deck, where the ship kept a film projector and passengers gathered in the open air. During Cecil B. DeMille's *Cleopatra,* the moon rose fat and white between the mountains near Monterey and the cryptologist's eyes wandered from the movie screen to the sky. He played shuffleboard with two congressmen. He fretted about the state of his patent applications and the low tide of his bank account. He wondered what was happening with the board game he sent to Milton Bradley. No one had gotten back to him. He learned later the game had stumped the firm's designers. It was one thing to design a fun code game for a party with friends and quite another to deliver a game in a cardboard box to a stranger.

On the ship he began a letter to Elizebeth that ultimately ran to thirty pages. He wrote on transparent tracing paper because it was thin and he could fit more pages in an envelope, saving on postage. Each day he added to the letter. His cursive handwriting smeared on the delicate clear sheets and the writing on the reverse showed through to the front. "I hope these sheets don't drive you wild to read," he wrote. "I wanted to get on as much as possible and use this paper to reduce the weight." He was trying to express a feeling that seemed too large for a single page or a moment's thought. He copied love poems by Tennyson and Thackeray from one of the books in his trunk. Thackeray wished to be a violet, plucked by his belle to live for an exalted hour "shelter'd here upon a breast / so gentle and so pure." He watched a romantic comedy starring Helen Hayes, *What Every Woman Knows,* about the shrewd wife of a British politician who helps his career behind the scenes, inspiring William to tell Elizebeth, "Well, that wasn't exactly news to me, my Darling. For I've known for a long time that you are the one in back of me and responsible for what little I've done. Had it not been for you I'd have been sunk long ago by unsolved infernal conflicts, by windy storms of emotion, by failure to keep up the fight when things seemed not worthwhile. . . . I know how much I owe to you—for love, for wisdom, for courage, and common sense."

Floating above the blue-black depths soon to be lethalized by German U-boats, he often stood at the top deck's rail looking out at the miles of ocean. Billions of droplets exchanging secret histories every fraction of a second. Trillions. Actual unbreakable code. One night when it was hot and he couldn't sleep he drank a glass of cold tomato juice and ate a few crackers and went up to the top deck. The place was deserted except for two or three men in chairs along the rail, smoking. He sat and pulled out the diary, flipping to the next empty page. He was a man on the verge of mental breakdown, writing a letter to his wife, who was also close to collapse, at a moment when the world was rearranging itself, and they would both have to give more of themselves than they had ever given before.

"The ocean," he said, "is as calm as a bowl of warm milk sitting on a table."

PART III

THE INVISIBLE WAR

1939–1945

CHAPTER 1

Grandmother Died

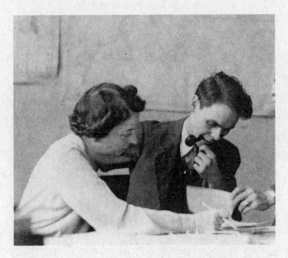

Elizebeth Friedman, U.S. Coast Guard
Cryptanalyst-in-Charge, and a junior cryptanalyst,
Robert Gordon, puzzling out a problem together, 1940.

Lights out 'cause I can see in the dark . . .
—FUGAZI

The Second World War did not begin with a gunshot or a bomb. It began with a feat of deception involving elements long familiar to Elizebeth Friedman—a code phrase, a radio station, and a murder. The men responsible were Nazis, and they belonged to the same part of the Nazi state that would soon attract Elizebeth's deep attention.

At 4 P.M. on August 31, 1939, in a hotel room in a small Polish town four miles from the German border, a Nazi officer named Alfred Naujocks dialed a number in Berlin. Someone in Berlin picked up. A high-pitched voice said, "Grossmutter gestorben." "Grandmother died." Naujocks hung up. He went to gather his team, the six operatives he had brought across the border. "Grandmother died" was the signal to execute a preplanned mission at 8 P.M.

The mission was to provide Germany with an excuse to start a war. Hitler had already decided to attack Poland, to seize his neighbor to the east, but he did not want to appear as the aggressor, so a pretext was needed, a simulated attack on German forces that would allow Hitler to claim he was acting in self-defense and create confusion about where the truth really lay.

This is where Naujocks and his colleagues entered the picture. They would invent the proof of Polish aggression.

They belonged to the SS, the chief instrument of Nazi terror. They were the men in black, the storm troopers, numbering 250,000 by 1939. They wore the death's-head symbol on their uniforms, the skull and crossbones. As individuals they were like any other large group of humans, containing multitudes: opportunists, idealists, fanatics, scholars, mediocrities, petty crooks. But as a collective they became "the guillotine used by a gang of psychopaths obsessed with racial purity," in the words of the historian Heinz Höhne. It was SS men who built the concentration camps and managed the ghettoes and trains that herded and transported Jews and other minorities into the camps to be enslaved, tortured, and killed. They were the guards at Dachau and Auschwitz, the murderers of millions. They were the mobile killing units, the Einsatzgruppen, that swept in behind the advancing German military, shooting resisters and Jews. They were the Gestapo, the ruthless Nazi police. They were the Nazi intelligence men, the ones who spied on their fellow Germans to keep them in line, and they were the spies who worked undercover in other nations, extending the reach of the regime until it encircled the world. And they were not meant to be confronted or understood. One SS leader bragged that the organization was "enveloped in the mysterious aura of the political detective story." Elizebeth would spend much of the war trying to penetrate this veil.

After the SS men in Poland received the code phrase over the phone on August 31, they waited until just before dusk, then drove two cars through the pine forest toward their objective, a Nazi radio station that transmitted propaganda broadcasts. The plan was to pose as Polish insurgents, take over the station, and broadcast a message denouncing the Führer. They stopped near the station and met a Gestapo captain to pick up what they had been calling "Canned Goods": the unconscious body of an SS prisoner, a forty-three-year-old Catholic farmer named Franz Honiok. He had been shot, sedated, and dressed in a Polish uniform. His face was smeared with blood. Naujocks carried him to the steps of the station and left him there slipping away from his fatal gunshot wounds, then stormed into the broadcast area with his team, aiming a revolver at the staff: "Hands up!" One of the SS men spoke Polish. He grabbed the emergency microphone used for storm warnings. "Attention, this is Gliwice," he shouted in Polish, pretending to be an insurgent. "The radio station is in Polish hands." He called for an uprising. The men fired bullets into the ceiling to simulate an armed struggle.

Thousands of radio listeners heard the gunfire and the burst of Polish. Two hours later stations in Berlin were spreading news of the "Polish attack." The BBC in London reported that "Poles forced their way into the studio." And while diplomats around the world tried to get a fix on the truth, the Nazis were massing at the border. At dawn on September 1, 1939, the morning after the incident in Gliwice, the Wehrmacht sliced into Poland, forty-two divisions all at once, with one and a half million men. It was the middle of the night on the East Coast. President Roosevelt woke to a ringing phone at 2:50 A.M. He picked up the receiver at his bedside and heard the voice of William Bullitt, the U.S. ambassador to France. Bullitt was calling with the news from Warsaw: Germany had invaded Poland. The president sat up. "Well, Bill, it has come at last. God help us all." He lit a cigarette and started making calls, waking up the cabinet secretaries.

Later that morning Roosevelt called a quick press conference. The first question was, can America stay out of the war? Roosevelt said, "I not only sincerely hope so, but I believe we can."

He was speaking for most of the country. In the days that followed, as Britain and France declared war on Germany, the U.S. public met the news with relief. Europe would defeat fascism on its own. The fight was across the ocean, far from U.S. shores. Nazis seemed a safe distance away.

In private, though, Roosevelt and his advisers were planning for the worst.

For years now, they had been thinking about the possibility of direct Nazi attacks on the United States. It was obvious that no effective invasion could be launched from Germany itself; ships were too slow, and airplanes couldn't carry enough fuel to cross the ocean, drop bombs, and return home.

But there was a catch in this argument, an unnerving loophole: *What if the Nazis got control of South America?*

South America. It was neutral ground—for now. No government there had declared a position in the war. Brazil, Argentina, Chile, Paraguay, Bolivia, and Ecuador were content, for the moment, to sell beef and raw metals to the combatants and to measure the political winds.

But this would certainly change as the war evolved and politicians cut deals. Hitler had already shown an ability to destabilize foreign governments. That's how Austria had fallen to the Nazis, and Czechoslovakia, too. And President Roosevelt was convinced that if Nazism took root in South America, even in just a few places, it would pose a clear and present danger to U.S. cities like New York.

South America was very big: the land mass of Brazil alone was slightly larger than the entire continental United States. South America was also very close: as Roosevelt put it in a 1940 speech to Congress, "Para, Brazil, near the mouth of the Amazon River, is but four flying-hours to Caracas, Venezuela; and Venezuela is but two and one-half hours to Cuba and the Canal Zone; and Cuba and the Canal Zone are two and one-quarter hours to Tampico, Mexico; and Tampico is two and one-quarter hours to St. Louis, Kansas City and Omaha."

If Britain fell to Hitler, the thinking went, Nazi ships could move west and set up bases in South America, seizing its rich resources, the metals to make war machines and the food to sustain armies, and then U.S. coastal cities would be within reach of Nazi bombing raids. Some officials dissented from this view, namely at the State Department, which considered a Nazi invasion from South America to be unlikely, but the Reich's rapid military victories and the erratic behaviors of the Führer had caused many to revise their sense of what was possible.

There was another sound reason to worry about German influence in South America. Millions of Germans were already living there as colonists. They had emigrated in waves since the late nineteenth century, seeking land and work, 140,000 arriving between 1919 and 1933 alone, many fleeing the same desperate economy that fueled the rise of the Nazis.

The Germans had left a cold country of worthless currency and sailed into the crystalline seas hugging a warm and open continent. A "bewildering abundance" met them ashore. "Everything is violent—the sun, the light, the colours," the Austrian exile Stefan Zweig wrote of Rio de Janeiro, Brazil's glinting capital. "The glare of the sun is stronger here; the greens are deep and full; the earth tight-packed and red . . . Rather than encouraged, growth has to be fought, so as to prevent its wild power from overwhelming the efforts of mankind." In Rio the green leaves of palm trees burned white in the midday sun as if radioactive. Men wore suits of white linen that became soaked by sudden downpours and thunderstorms. Women strolled topless on Copacabana Beach. A cream-colored luxury hotel, the Copacabana Palace, faced the beach and the crystalline blue waters of the Guanbara Bay; Fred Astaire and Ginger Rogers danced in the hotel's ballroom in the 1933 Hollywood musical *Flying to Rio*.

Some of the arriving Germans decided to stay in Rio; others filtered into the surrounding Brazilian provinces, sparse rural lands of forest and cattle, and still others scuttled south along the coastline to the fast-growing industrial city of São Paulo, which Zweig likened to Houston, Texas, because of its abrupt rise from nothing: "There are times when one has the sensation of not being in a city, but on some gigantic building site." Farther south, another great city of the continent beckoned to the Germans: Buenos Aires, Argentina, a polyglot metropolis of four million, a chaos of automobiles and bookshops and neon lights and cobblestone streets where tango music descended from the open windows of brothels. The nation had grown rich from cattle and wheat raised in the Pampas, the flatlands to the west and south of Buenos Aires, where tens of thousands of Germans lived alongside the *gauchos* and their horses.

Wherever Germans settled in South America, they built German schools (two hundred in Argentina alone), German businesses, German radio stations, German newspapers, and transportation links back to the homeland. Zeppelins floated people and cargo from Berlin to Rio, and two airlines, Condor and LATI, connected South America and Europe. Condor was owned by Germans, LATI by Italians. A visiting U.S. consul reported "a fair sale for German Bibles" across three Brazilian states and that 20 percent of all residents spoke only German; parts of southern Brazil became known as Greater Germany. "The German spirit is ineradicably grounded in the hearts of these colonists," wrote a German physician, "and it will undoubtedly bear fruit, perhaps a rich harvest, which will not only prove a blessing to the colonies, but to the Fatherland." A German visitor to Brazil reported with pride, "Surely to us belongs this part of the world," and the Nazi ambassador in Buenos Aires, Baron Edmund von Thermann, believed that German Argentines must show "complete subservience" to "the ambitions and desires of the home country. Germans naturally count on these prosperous nuclei to assist eventually in the rebuilding of a new Germany."

A small percentage of German immigrants brought fascist politics to South America, starting local Nazi clubs and chanting Nazi songs, but these groups were small and disconnected, stagnant ponds of fascist fervor. The bigger rivers of fascist sympathy in South America coursed through the local populations. It was a time of protests, marches, fantasies of revolution. Right-wing parties and radicals on the continent found inspiration in Nazism. Followers of a Brazilian movement called Integralism raised their hands in Nazi-style salutes, wore uniforms of green (the men were "Green Shirts," the women "Green Blouses"), and goosestepped through the streets of Rio. In 1938, a throng of Argentine youths marched into the Jewish quarter of Buenos Aires, chanting anti-Semitic slogans, "stripped to the waist like Mussolini,

mustachioed like Hitler," writes one historian. "When enraged
Jews attacked them, police arrested the Jews."

Similar movements were gaining followers in Chile, Bolivia,
and Paraguay, and powerful local officials poured fuel on the
fires. Across the continent, men who dreamed of leading their
own regimes had risen to the top of police and military hierar-
chies; many had gotten their training from German officers. A
group of Paraguayan officers formed a secret lodge, the *Frente de
Guerra,* to organize an ultra right-wing revolution; their motto
was "Discipline, Hierarchy, Order." The chief of the Paraguayan
national police, wishing to honor the dictators of Germany and
Japan, decided to name his son Adolfo Hirohito. In Argentina, a
young military instructor with a Cheshire-cat smile, Juan Perón,
was studying the leadership styles of Mussolini and Hitler and
found much to admire. One Argentine general, Juan Bautista
Molina, displayed so much zeal for National Socialism that even
Thermann, the Nazi ambassador, found it "embarrassing."

Hitler appreciated this wellspring of sympathy in South
America. His strongest affinity was for Argentina, which had
protected German interests in the First World War while osten-
sibly remaining neutral. In June 1939, three months before he
invaded Poland, Hitler met with Argentina's ambassador to Germany. Writes the historian Richard McGaha, "Knowing that war
was going to break out soon," the Führer "cryptically stated that
he hoped Argentina would stay neutral and that neutrality could
be the basis of a closer relationship." Then Hitler launched into
a tirade about America and England, saying that "the U.S. was
the worst-governed country in the world," that Roosevelt wanted
war "at the instigation of the Jews, who controlled industry and
the press," and that England was "a paper tiger with its little fleet
and meager air force."

In his mind the Führer had already added South America to the
Nazi column. If he decided not to invade at this time, he would
simply annex the continent after defeating Europe. As Baron
von Thermann put it, "Once the war were decided in Germany's
favor, her domination of Latin America would follow without
much effort." This was the Nazi attitude, and it meant that the
war in South America could not be a hot war, a war of soldiers and

sailors in recognizable uniforms, a war of battleships and mortars and planes and bombs. Instead it promised to be a war of languages and secrets, codes and conspiracies, masks and seductions, wireless transmitters and cipher machines—the type of war where everything depended on the invisible flashes of energy radiating from a radio coil hidden on a farm or beneath the floorboards of an unremarkable house.

The term of art for an intelligence operation that must remain entirely concealed is "clandestine." If a clandestine job is successful, no one ever knows it happened. It is invisible. The war in South America would be the Invisible War.

There was a school in Hamburg where SS intelligence officers trained combatants for this war. Male party members were selected to receive a basic course in espionage tradecraft. They were taught to write letters in secret inks. The SS had developed a disappearing ink that actually looked like ink, bluish in color and carried in a regular ink bottle; a message written with this ink would turn invisible after a few minutes and could only be unmasked with a certain reagent. They learned how to operate a German-invented "microdot" camera that shrunk documents to the size of the dot above an *i*, allowing espionage reports to be concealed in otherwise innocuous letters, and they were shown different methods of writing messages in cipher, including an ingenious system for exploiting a popular novel, any novel, to generate garbled text.

For the purposes of spying, this hand technique was often preferable to cipher machines like Enigmas, which were bulky, harder to transport, and more incriminating if discovered. A novel aroused no suspicion. One in common use was *All This and Heaven Too,* a period potboiler about a French governess falsely accused of murder. *Would the unlucky Henriette Desportes manage to clear her name? Or would the conniving Parisian judge dispatch her to the dungeon?* German men abroad pressed their noses to the book, eyes wide, turning the pages quickly, underlining words—no, these were not Nazi spies, these were simply readers under the spell of a story, needing to know what happened next.

The SS instructors taught students how to transmit text in Morse code and how to operate shortwave radio transmitters and receivers. Radio technology had made giant leaps since the heyday of rum-running. A shortwave transmitter of moderate power could now fit in a suitcase. The transmitter was a small metal box with vacuum tubes on the inside and dials affixed to the cover, and the antenna was a long wire looped into a tight coil.

Portable transmitters in hand, the novice agents were dispatched to begin their espionage careers for the Führer. U-boats delivered some of the spies onto alien shores, and others parachuted from planes or sailed on neutral ships under phony names, sometimes getting caught by customs inspectors or police along the way, their radios confiscated. The SS issued all foreign spies two kinds of suicide drugs to ingest in case of arrest. The first was a tablet that caused death by heart failure within ten minutes, and the second was a powder that resulted in "a slow process of general collapse over a two-week period" when rubbed on the body.

If a spy managed to arrive at his destination with the radio intact, he unfurled the wire antenna and established contact with the fatherland, tapping out an encrypted message in the dots and dashes of Morse, the signal aimed at a receiving station in Hamburg or Berlin. Sometimes it worked, and the spy could be heard in Germany—there were no atmospheric disturbances, and the signal squeezed through the crowded frequencies—but storms and interference often fuzzed out the radio pings, making it necessary to build more powerful stations, which required a higher level of expertise. A *Funkmeister* was needed: a technical leader, a radio wizard, able to piece together clandestine radio transmitters in foreign lands. And this is why, in 1941, the Nazi SS dispatched its most capable *Funkmeister*, Gustav Utzinger, a twenty-six-year-old man with short brown hair and a chemistry Ph.D., to South America.

Elizebeth Friedman's next mission for America became the biggest secret of her life. She would never speak in detail about what she did between 1940 and 1945, even as an old woman, and the records of her work, the documents that now make it possible to tell the story, were classified after the war and locked away for a generation, unsealed only after her death. In the 1950s and 1960s, when she gave speeches or interviews about her career, she freely shared anecdotes about various colorful adversaries of the past—the millionaire George Fabyan, the rumrunners, the drug smugglers— but she skipped the Second World War entirely. These were the years when she disappeared into "a vast dome of silence from which I can never return," she said.

The one time she seems to have alluded to her wartime mission, briefly and vaguely, was in 1975, during an interview with her husband's biographer. She uttered a few words and then the transcript cut off.

"The spy stuff," she said. "That's what I did."

It wasn't Elizebeth's conscious decision to spend the war chasing Nazi spies. It was yet another "pure accident" in her career. This is how it always happened: She put her ear to the ground here and there to learn how the pieces of the world fit together. She figured out how to hear a new sound. Then men in uniform showed up at her side, asking questions, wanting to listen over her shoulder. This had been true at Riverbank two decades earlier, when "the world began to pop and things began to happen," as she put it once; it was true in the 1920s and '30s when she shone a floodlight on the American criminal underworld; and it was proving true again now, in early 1940, when she and her team identified a new and sinister set of voices in the intercepts furnished by the listening stations.

The basic rhythm of her typical weekdays had not changed since the early '30s. She was still working in her coast guard office at the Treasury Annex building near the White House, serving as chief of the Cryptanalytic Unit that she had founded in 1931 and nurtured ever since. Her three junior codebreakers, Robert Gordon, Vernon Cooley, and Hyman Hurwitz, the ones she had originally recruited and trained, were still with her, and a handful of women clerk-typists had also joined the team as support staff. Elizebeth, Gordon, Cooley, and Hurwitz often worked together at a long table in the office, analyzing the ever-replenishing piles of cryptograms that arrived from the coast guard listening stations, chewing the ends of their pencils, maps of the world pinned to the wall behind them, the clack of the clerks' typewriters filling the room.

Outside the door, they could hear the muffled noise of T-men going this way and that, customs men, narcotics men, IRS men, coast guard men. They pressed their foreheads to the intercepts, Elizebeth perhaps wearing a simple white high-collared dress, Gordon smoking a pipe in a suit and vest, chomping on the pipe and frowning at a page. Sometimes Elizebeth would stand up and disturb Gordon's cloud of smoke as she walked to a shelf to look at a piece of cryptologic literature or to examine one of the cipher machines she kept there in case she should encounter a message that had been generated by one. She had an Enigma machine on the shelf, an old version that had been freely available in the 1920s. She also had a Kryha there, the semicircular German device that William had once mastered.

Elizebeth reported to the chief of the coast guard communications section, a salty vice admiral named John Farley, and Farley reported to the secretary of the Treasury, Henry Morgenthau Jr., an old friend of President Roosevelt from a prominent Jewish family. Morgenthau was the kind of person Elizebeth tended to get along with—polite, educated, pragmatic—although Elizebeth came to dread phone calls from his devoted personal secretary, Henrietta Klotz, who had a habit of calling Elizebeth's office at 4:28 or 4:29 P.M., one or two minutes before the 4:30 close of the day, and making what Elizebeth called "rapid-fire dictator-sort of requests," demanding that Elizebeth and her team solve some difficult problem in an impossibly small amount of time. Morgenthau would usually phone Elizebeth the next day and reverse Klotz's order with bashful apologies.

Morgenthau needed Elizebeth to be happy. He now depended on her to perform one of the department's wartime functions. Smuggling wasn't what it used to be—the war had disrupted the drug networks and made business perilous—so Elizebeth's Cryptanalytic Unit had shifted its attention to British and German ships. The Treasury was responsible for enforcing U.S. neutrality laws, and foreign ships along the East Coast needed to be monitored for any violations that might cause diplomatic controversies. At Morgenthau's request, in 1938, the unit began to analyze the wireless messages of British cruisers and German merchant vessels. Elizebeth broke the codes of Nazi captains as they tested the limits of U.S. neutrality and provoked tense confrontations. In December 1939, a German freighter flying the swastika flag pulled suspiciously close to Florida shores and was chased by U.S. Army planes and a nearby British cruiser, *Orion*. Elizebeth decrypted the German captain's panicked messages home. It was the first gunfight of the war in American waters:

AM TRYING TO RUN INTO AMERICAN HARBOR PORT EVERGLADES OR MIAMI CODE DESTROYED

THE CRUISER HAS TRAINED HIS GUNS AGAIN HE IS RUNNING SLOWLY FORWARD

CRUISER NAMED ORION

THREE AMERICAN ARMY PLANES HOVERING OVER US

"Exciting, round-the-clock adventures," she said later about these episodes. But an even more intense mission was yet to come.

While monitoring these radio signals for her Treasury bosses and solving the puzzles that were given to her, Elizebeth started to detect a new import to the messages. In January 1940, with Hitler preparing to invade Scandinavia, dozens of mysterious encrypted texts piled up in Elizebeth's office all at once, apparently transmitted by several different unregistered radio stations and intercepted by U.S. listening stations.

At first, the messages looked similar to the thousands of smuggling messages she had solved before. They used the same kinds of call signs and similar frequencies. But after a brief period of confusion, Elizebeth realized that the messages hadn't been sent by smugglers at all. The plaintexts were in German. They contained sensitive information about the routes of U.S. and British ships and the capacities of U.S. factories. And according to the bearing fixes, the signals originated from unknown radio stations in Mexico, South America, and the United States.

It soon became clear that the stations had been built by Nazi spies to share sensitive information with their bosses in Germany, transmitting and receiving dots and dashes of encrypted text at the speed of light. A pair of stations exchanging wireless signals formed a "circuit," and each circuit was protected by a different code or cipher that had to be broken before the messages could be read.

These were clandestine circuits, meant to stay invisible, and it became Elizabeth's goal to pry them out of the dark while remaining invisible herself—an essential part of the job. She knew that if the spies discovered that she was breaking their codes and reading their messages, they would switch to more secure codes, and she wouldn't know what the spies were saying until she could break the new codes, which might take weeks or months. A spy who speaks in a broken code is "the goose that lays the golden eggs," as William put it once. If you want to keep gathering the eggs, you must not frighten the goose.

For this reason, Elizabeth's Cryptanalytic Unit "was probably even more secret than other [codebreaking] organizations," the NSA concluded after the war, "because it dealt with counterespionage." Counterespionage, counterintelligence—these are the formal terms for what Elizabeth was beginning to do. She was counterspying on foreign spies, serving as America's eyes and ears in the invisible world of fascist espionage. Today there are large sections at CIA and FBI that perform foreign counterintelligence, teams of American professionals who spend their days trying to monitor the activities of Russian and Chinese spies, but in 1940 there was almost nothing, and Elizabeth had to act with extreme caution every day. It was essential that her Nazi targets never learn that she existed.

The first few batches of eggs fell smoothly into her basket. As soon as Elizabeth began to analyze the clandestine circuits in 1940, she realized that the spies were relying on different kinds of hand ciphers, variations of tried-and-true methods. Some were familiar systems from the rum days, adulterations of commercial codes like the ABC code and the ACME code. These were solved in a snap. The key for one circuit was found to be 3141592, the first seven digits of the mathematical constant pi. Elizabeth called this circuit "the pie circuit." Sometimes the Germans sent the key at the start of the message and in groups of three or four letters instead of five, indicating that there was something special about these letters and giving away that they were a key.

When an unfamiliar system was encountered, and nothing was

known about the speakers "to provide an entering wedge," Elizebeth and her teammates tried to start with something small and
simple. For instance, if they determined by a routine sort of check
that they were dealing with a transposition system, with the letters
mixed up instead of swapped out, they would look for common
German words in the messages, like *zwo*, "two," which is a useful
word to a codebreaker because it contains two low-frequency letters, *z* and *w*, which makes it stick out more. (The names of numbers
were often spelled out in messages to eliminate potential confusion
from dropped letters due to radio interference.) Another technique
that often helped was to take multiple messages and stack them on
top of one another, creating a "depth" of text that made it easier to
identify patterns as opposed to analyzing one message at a time:

1	E	A	W	I	Z	T	Z	N	X	O
2	I	E	U	R	Y	R	X	F	E	H
3	U	I	U	H	Z	F	E	N	N	X

Here, Elizebeth was able to look at row 1 and anagram the letters, Scrabble-like, to make the word *zwo:*

1	Z	W	O
2	X	U	H
3	E	U	X

Now the columns were in a different order, and this new order
gave a clue to the structure of the underlying cipher that allowed
her to break it.

Essentially, Elizebeth's goal was to look at these daunting
mountains of nonsense and chart a route up the slope in small dis-
crete steps, each of which was like a little game—not quite child's
play but not totally unlike child's play, either. And the games
grew more intricate as the months went on and the coast guard
codebreakers followed the intercepts.

Several sets of Nazi spies were using book ciphers similar to the
ones that Elizebeth and William had long studied but with new
twists. For instance, on January 1, 1940, she received her first intercept from a wireless circuit that linked Mexico with a radio tower

in Nauen, Germany. The messages contained only eleven letters of the alphabet: *N, R, H, A, D, K, U, C, W, E,* and *L.* One message began

UHHNR LNDAL NURND WCNCK NRHLN DNRAN CHNDR UNDEN

Relying on intuition and experience, Elizebeth made a few quick assumptions. *N* was the most frequent letter. She guessed it was being used as a "word separator"—a space bar. She also guessed that because there were only eleven letters in the messages, one of which was a space, the letters must stand for the numbers 0 through 9. But which letters stood for which numbers? If she was correct, the spies might have used a key word to determine that. Elizebeth and her colleagues tried to find the key word by anagramming the eleven letters:

WACKELND RUH

WAHL DRUCKEN

ACH RUND WELK

DA LUNCH WERK

DURCHWALKEN

There it was: *Durchwalken,* a colloquial German word meaning "to give a good beating." This was probably the key:

D U R C H W A L K E N

1 2 3 4 5 6 7 8 9 0 -

Now Elizebeth was able to turn the letters of each message into numbers, using *N* as the separator:

UHHNR LNDAL NURND WCNCK NRHLN DNRAN CHNDR UNDEN

255-3 8-178 -23-1 64-49 -358- 1-37- 45-13 2-10-

Cleaning up the numbers, the line became

255-38 178-23 164-49 358-1 37-45 132-10

This looked like a book cipher to Elizebeth; the numbers probably corresponded to locations in some unknown book owned by the spies. After translating the letters of several messages into numbers, she saw that some number combinations appeared more frequently than others: 1-1, 132-10, 343-2, and 65-12. The coast guard codebreakers underlined these frequent combinations, and "after a little experimenting the following was produced":

65-12	132-10	373-2	301-21	285-25	343-2
B	E	R	L	I	N

65-12	375-2	132-10	321-2	132-10	343-2
B	R	E	M	E	N

BERLIN and BREMEN, two German cities. (In some cases, a letter like *R* was linked to a few different number combinations.) These frequent letters gave her a start, and when able to solve the code in full, Elizebeth identified the names of two known Nazi agents in Mexico, MAX and GLENN, who would appear in other messages in the future, linked to agents in the United States and South America. The two Nazi spies were reporting to Berlin on the movements of U.S. and British ships, making those ships vulnerable to U-boat attacks.

Elizebeth solved their book cipher without needing to see the book and did the same with messages that used other books: *The Story of San Michele,* the memoirs of a Swedish physician; *Soñar la vida,* a spy story by a female Mexican fascist; *O servo de Deus,* a Portuguese novel. One Nazi spy proposed using the 1936 novel *Vom Winde Verweht*—in English, *Blown Away by Wind,* i.e., *Gone With the Wind*—and asked Berlin to locate a copy. Berlin replied that *Blown Away by Wind* was unavailable in Germany and another book would need to be chosen.

Several Nazi agents, Elizebeth discovered, were using a copy of the romantic novel *All This and Heaven Too* and a sophisticated process that generated messages full of garbled letters instead of numbers. Each spy had been assigned a unique identification number, such as 7. To encrypt a message, the spy would take that day's date, add the number of the day and the month to his identification number (for a January 10 message he would add 1 + 10 + 7 = 18) and turn to the resulting page in the novel (page 18). The first words of the first line became part of that day's key—the key for transforming plaintext words into blocks of nonsense according to a Scrabble-like method that jumbled the letters by stacking them into columns. The rest of the key was taken from the first letters of unindented lines going down the page.

To solve the messages, Elizebeth first had to deduce that *All This and Heaven Too* was the novel these particular spies had chosen. To do this she went through the same process of reverse engineering that she and William applied in 1917 to solve the Hindu messages. Then she bought her own copy of *All This and Heaven Too* and kept it on her coast guard desk, allowing her to easily ungarble any new message sent with that system, flipping through the novel and underlining or circling the pieces of the daily keys in red pencil. Here is how she marked up page 15, where the novel's fictional heroine is deciding whether to become the governess for a hot-tempered Parisian family and move into their home:

<u>Yet she did not dread</u> the thought of entering it. The difficul-

ties (i)t presented would at least be stimulating. One would not

(p)erish of boredom in a place where charges of gunpowder might

(l)urk in unexpected corners to explode without warning. She felt

(o)ddly exhilarated—almost, she thought, as if she were about to

(s)tep upon a lighted stage filled with unknown players, to act a

(r)ole she had had no chance to rehearse beforehand. She must find

(t)he cues for herself and rely on her own resourcefulness to speak

(t)he right lines. Henriette Desportes's heart under the plain gray

Elizebeth wrote the letters of the key horizontally on a piece of graph paper and used it to fill in the German plaintext.

Her basic puzzle-solving style hadn't changed from the smuggling days, and it remained effective: a process of trial and error with pencil and paper, deduction and experimentation, granules of eraser dust swiped away with a flick of the palm. Her scrap papers still looked like the scrap papers of a person doing the newspaper puzzle page over Sunday-morning tea; she wrote no equations, only numbers and letters grouped and stacked in rows, columns, squares, rectangles, and more exotic shapes. This approach worked for her because over the previous twenty-five years, encountering tens of thousands of messages, Elizebeth had solved so many different kinds of puzzles that she knew how to find shortcuts, to identify patterns in fields of text that were like signatures telling her what to do next. She was a kind of human computer in this sense. Today, if you want a computer to recognize certain patterns, you can train it through a process of "machine learning." How do you get a computer to recognize a picture of a cloud, for instance? You feed it a lot of pictures and say, essentially, *This here is a cloud*, and *This here is not a cloud*. After the computer gains enough "training data," it's able to look at a new image, do some math, and say, *This is almost certainly a cloud*. By 1940, Elizebeth's brain had probably accumulated more training data about codes and ciphers than any other brain on the planet. She had just seen so many damn clouds. It's why she was able to make inspired guesses about puzzles. She may not have been writing equations, but she was thinking mathematically.

This is also why, in 1940, when Elizebeth encountered her first Enigma messages from a German Enigma machine, she didn't feel overly intimidated.

Enigma was a straightforward idea expressed in a diabolical device. In the simplest sense, it was a box that cranked out poly-alphabetic ciphers. Remember the secret messages that eight-year-old Barbara Friedman sent her parents from summer camp? A=B, B=C, C=D. That's a MASC, a mono-alphabetic substitution cipher. One cipher alphabet encrypts the whole message. Enigma was *poly* instead of *mono*, using multiple cipher alphabets per message.

Poly-alphabetic ciphers date to the sixteenth century and can be written by hand with the aid of pre-printed grids of letters or sliding strips of paper. Instead, Enigma did the job with three or more rotating alphabet wheels connected to electrical wires. The wheels lived inside a box with a typewriter keyboard on the outside, the keys arranged in a familiar order, starting with

Q W E R T Z U I O. Above the keyboard was a "lampboard" of the same twenty-six letters in the same order. When a writer pressed a key, such as Q, a different letter, perhaps Z, would illuminate on the lampboard—the cipher letter, lit by a small battery-powered bulb. Later, the recipient of the message, operating his own identically configured Enigma, would type Z, and Q would light up, decrypting the message letter by letter.

With each key press, an electrical circuit was completed, and Enigma stepped the right-hand wheel, shifting it one letter forward. Once the wheel stepped through all letters, it stepped the middle wheel by one letter, then the left-hand wheel. The motion was similar to a car odometer—after you drive 9 miles, the right-hand number flips to 0, and the next number to the left flips to 1— and it generated a seemingly random, nonrepeating sequence of 16,900 cipher alphabets before the three wheels returned to their starting positions.

Crucially, no letter could be enciphered as itself. If you pressed *j* a million times, you would never see *j* light up on the lampboard.

Although this was a known limitation of the machine, it seemed to pale in comparison with Enigma's flexibility. The wheels could be arranged in different orders (1-3-2, or 2-3-1), the alphabet rings on the wheels could be set at different starting positions on the wheels, and the starting letter of each wheel, as seen through a small window on the box, was another variable. The choice of variables comprised the machine's key—the starting configuration used to encrypt all messages on a particular day, week, or month, depending on how often the key was changed.

How many possible keys existed? Depending on the model of Enigma, the number of keys might be as large as 753,506,019, 827,465,601,628,054,269,182,006,024,455,361,232,867,996, 259,038,139,284,671,620,842,209,198,855,035,390,656,499, 576,744,406,240,169,347,894,791,372,800,000,000,000,000.

Each one of these keys produced a unique set of 16,900 alphabets before repeating.

All of this seemed to make the job of a codebreaker impossible. There were too many possibilities to comprehend, and then there were possibilities about those possibilities, and possibilities about those possibilities about those possibilities. Clearly, shortcuts had to be discovered, and by the late 1930s, finding these shortcuts—and conquering Enigma—was the biggest problem facing Allied intelligence. After Polish mathematicians made some early breaks into the device, the Germans kept changing its design and how it was used, so the battle over Enigma was ongoing, a cryptologic arms race. The machine had been clunky at first, weighing as much as one hundred pounds, but subsequent versions grew lighter and more compact. The German navy, the Kriegsmarine, first adopted them in 1926 and installed Enigmas in ships and U-boats, followed by other branches of the military, embassies, and intelligence services. In 1936, the Nazis banned all commercial sales of Enigma and began to improve the machine in secret, adding additional components and subtleties intended to make Enigma codes absolutely unbreakable. Different Nazi organizations developed their own variants. Germany withheld knowledge of these alterations from the enemy, as if Enigma were a submarine or a bomb.

To extract useful intelligence from an Enigma system, Elizebeth Friedman (or anyone else) needed to accomplish two separate and immensely difficult things. First, the machine itself had to be "solved," its inner workings deduced and mapped—the motions of its wheels and the maze of wires controlling them. This required some leap of human ingenuity, some feat of mathematical deduction or inspired guessing. Then, once the wiring was solved—the part of the system that generally didn't change—the keys had to be recovered, which changed at different intervals (month, week, day) depending on the practices of different Nazi services. If you found an Enigma key in the morning, you might go to bed at night and get locked out again in your sleep, and the next day you had to find the key again if you wanted to read the new day's messages.

There were too many Germans using too many Enigmas with too many shifting keys to ever recover the keys by hand, so codebreakers needed to build machines of their own to assault the enemy's machines, giant electro-mechanical contraptions and some of the first digital computers, too. Automation. Polish codebreakers were the first to solve Enigmas and automate the process of recovering keys. They built "bombes" that mirrored the Enigma rotors, ticking through possible alphabets until they found ones that might fit. Later, the British mathematician Alan

Turing discovered how to make bombes dramatically more powerful, based on mathematical principles and previously solved bits of text known as "cribs"—a crib might be the name of a Nazi officer, the time of day, or "Heil Hitler." His solutions were essentially search algorithms, ancestors of the Internet search algorithms of today. Turing's biographer calls these "search engines for the keys to the Reich." It was anti-Nazi Google.

The British codebreakers worked at Bletchley Park, a mansion in the countryside outside of London. Bletchley grew from a handful of people in 1938 to thousands by 1945, the bulk of them women, recruits from the Women's Royal Navy Service who operated the bombes, among other jobs, and were billeted in large country houses.

The Enigma codebreaking program would come to be known as ULTRA; Enigma decrypts were stamped with the imposing phrase TOP SECRET ULTRA as a reminder to handle them with the utmost care. Later, America would join forces with the British, assembling its own ULTRA factories in Washington and sharing the burden. But early in the war, when Elizebeth and her coast guard unit analyzed their first Enigma machine, ULTRA was a strictly British franchise. There was no one to tell the Americans what to do. They had to invent their own method.

At first, Elizebeth didn't know that she was dealing with an Enigma at all. Enigma cryptograms look like lots of others, generic blocks of nonsense letters. In January 1940, coast guard radio monitors began intercepting one to five messages per day with the call signs MAN V NDR and RDA V MAN. Elizebeth wasn't able to make heads or tails of the first twenty or thirty messages that were intercepted on this circuit. However, after accumulating a greater "depth" of messages, sixty or seventy, she was able to write them one on top of another on a worksheet and see the letters in a new way by gazing *down the columns*.

Enigma is poly-alphabetic. It creates a new cipher alphabet with each key press. That's the beauty of the machine. But if an Enigma user types a number of messages using the same starting position of the rotors, the first letter of each message will use the same alphabet—and the second letter of each message will use the same alphabet, and the third letter. In other words, any individual message is full of alphabets, but if a codebreaker lines up the messages in a tower, *each column in the tower is mono-alphabetic*—one alphabet:

```
1   2   3   4   5   6   7
D   X   J   X   L   H   N . . .
L   W   S   X   I   Y   F . . .
M   H   O   S   S   L   C . . .
```

The letters in the first column here, D L M, all use the same alphabet. And the letters in the second column, and so on.

With only three messages, there isn't enough information to help the codebreaker. The "depth" is too low. There need to be more floors in the tower. At greater depths, closer to twenty messages and beyond, letter frequencies become visible:

```
1   2   3   4   5   6   7
D   X   J   X   L   H   N . . .
L   W   S   X   I   Y   F . . .
M   H   O   S   S   L   C . . .
M   A   P   A   C   T   Y . . .
F   P   W   S   G   S   C . . .
Y   Q   A   S   A   C   W . . .
N   S   H   W   U   F   C . . .
F   U   W   X   G   S   P . . .
M   B   D   W   X   U   O . . .
O   P   O   D   Y   X   L . . .
A   J   Y   S   X   F   D . . .
M   W   S   X   E   C   C . . .
```

M appears four times going down column 1. In this column, *M* might be equivalent to the letter *E*, the most frequent letter in German as well as English. In other columns, different letters might be equal to *E*. And now the codebreaker can use tried-and-true methods to fill in plaintext letters and piece together the adversary's words.

In this way, the technique of "solving in depth" can take a hard problem and turn it into a simpler problem. The trick is often to get the messages aligned in depth in the first place. If the Enigma user changes the starting position of the rotors from message to message, the floors of the tower have to be staggered to track with the shift in the starting position.

Figuring out how to align messages in depth is a subtle art. It can be done with clever guesswork and trial-and-error, and it can also be done by applying the principle of the Index of Coincidence, William Friedman's fundamental insight about the relationships between letters that sit in towers of text. Elizebeth tended to use both approaches in her work, but luckily, in this case, she didn't need to align the messages, because the senders had made a mistake by using the same starting position for all the messages. The messages were *already* in depth. Before long, then, the coast guard codebreakers were able to identify frequent letters in the columns and use those letters to piece together the plaintexts for most of the first batch of messages.

The words seemed to be in German.

Elizebeth and her colleagues still didn't know what type of cipher they were dealing with, so now they decided to write down the alphabets for many of the messages they had solved in depth, the ciphertext equivalents of the plain letters, to see if a pattern popped out. They quickly noticed that no letter was ever enciphered as itself: an *A* never meant *A*, a *B* never meant *B*. This suggested an Enigma.

They went to the shelf in their coast guard office and picked up their old commercial Enigma machine.

The codebreakers had already solved most of the messages, but now they wondered if they could solve the machine itself—the wiring. Knowing the wiring makes it easier to solve new messages. Without the wiring, they would have to repeat the laborious process of solving in depth every time the key changed. Their challenge now was to use the text they had recovered, the plain letters and the cipher letters, to work backward toward the unknown machine, almost like a police detective analyzes the spatter pattern of blood at a murder scene, starting with the red evidence and rewinding back to the moment of the crime, deducing from the crusts of blood the speed and angle of the knife.

Unbeknownst to the coast guard, groups of British and Polish codebreakers working on the Enigma problem had already discovered methods for working backward from the text to the machine. The Poles had done it with an algebraic approach, the mathematics of permutations, and one of the brilliant Bletchley codebreakers, a linguist and scholar of classical literature named Dilly Knox, had relied more on pattern recognition and a kind of alphabetic grid called a "rod square." But the coast guard didn't know about these approaches, and so, working in isolation, the codebreakers had to grope toward their own method. They poked and prodded and turned the wheels; they wrote alphabets on sliding strips of paper and moved the strips against one another, thinking.

It seemed to Elizabeth that there must be a fixed relationship between the alphabets she had already discovered by solving in depth—the plaintext letters and their cipher equivalents—and the motions of the Enigma's wheels. To test this hypothesis, she drew a number of diagrams that visualized the relationships between letters at each position of the machine. She wrote new kinds of towers of letters on the worksheets that were more like X-rays than photographs, probing more deeply into the identities at the heart of the Enigma, and immediately she saw clear patterns, hints of order and regularity.

Certain letters repeated vertically on the page, like LL and HH, and also pairs of letters, like SJ and EM. Elizabeth and her colleagues realized that these letter groups were telling them something about the spacing between pairs of wiring contacts on the Enigma's rotors. The maps were whispering secrets about the physical intricacies of the machine. Building upon these "remarkable results" over the following days, filling more worksheets to the brim with letters, drawing more towers and analyzing the patterns that appeared, the codebreakers managed to solve the wiring for all three wheels of the unknown Enigma. Then they were able to reveal the full plaintexts of all unsolved messages from the radio circuit.

The codebreakers now realized two slightly disappointing facts: The plaintexts seemed to contain no Nazi secrets; later the codebreakers learned that the messages had been sent by the neutral Swiss army, which sometimes used Enigmas to communicate in German. Then the coast guard shared the wiring diagram of the Enigma with William's codebreaking team at the army, in case it might be useful to them, and the army reported back that the diagram corresponded exactly to the wiring of a commercial version of Enigma.

Elizebeth had hoped that she was mastering a new kind of Enigma entirely. Still, it was a significant achievement. "This recovery of wiring assumed to be unknown was achieved without prior knowledge of any solution or technique and is believed to be the first instance of Enigma wiring recovery in the United States," her team wrote in a secret technical memo after the war. As far as Elizebeth and her codebreakers could tell, and they were hardly prone to bragging, they were the first Americans to solve an unknown Enigma.

Until this moment, cipher machines had always been William's territory, not Elizebeth's, but her solution of the commercial Enigma showed that she had a similar aptitude for solving machines, and this initial headfirst dive into the pool of Enigma codes would lead her to deeper waters later in the war. Of course, she didn't know this in early 1940. Demolishing that first Enigma was just work. She was confident enough in her abilities that solving an Enigma seemed like a reasonable and normal thing that she might accomplish with her team on a given week. She didn't brag or make a big deal. Anyway, there was no time. New puzzles were arriving at the coast guard all the time, new codes to break, along with increasing demands for assistance from outside agencies.

All along her plaintexts had been circulating through other parts of government. Each time her unit solved a message, a clerk typed the English solution on a fresh sheet of paper, a decrypt, and gave it to the coast guard chief of communications, Vice Admiral Farley, for dissemination. Depending on the content of the decrypt, the vice admiral might send a copy to navy intelligence (OP-20-G), army intelligence (G-2), the State Department, British intelligence, or the FBI. The decrypts were like blood cells in the veins of government, delivering the vital oxygen of raw intelligence, and as different intelligence agencies realized that Elizebeth had tapped into a trove of information about Nazi spies, they inevitably asked the coast guard for more decrypts. In the 1920s, she had complained about government men "appearing on my doorstep," wanting her to solve puzzles. They were still appearing on her doorstep, but now, instead of relatively anonymous T-men, they were some of the most powerful spymasters in the world.

J. Edgar Hoover liked to eat dinner at Harvey's restaurant on Con-necticut Avenue, next to the Mayflower Hotel, a five-minute walk from his suite of offices in the Department of Justice headquarters, a gargantuan gray edifice near the National Mall. Harvey's had separate dining rooms for men and women. The ladies' dining room was on the second floor, accessible by a separate entrance at street level. The first floor was the gentlemen's restaurant and bar, with waxed floors and rich leather banquettes. It was one of those places in Washington where men of influence slurped oysters and let their guard down for an hour or two.

The FBI director's face was beginning to acquire some of the first creases and pouches that would characterize the eventual marble busts of him. He was forty-five years old, one of the few immutable objects in an ever-changing city. He wore white shirts, double-breasted Brooks Brothers suits, and a hat with a brim that could be turned up or down. His agents wore the same uniform. A pink expanse of forehead separated his bushy black eyebrows from his thinning hair. The large neat desk back at his office, a corner office on the fifth floor of Justice, contained a radio, usually a vase of fresh flowers, and a framed copy of "Penalty of Leadership," the text of a Cadillac advertisement from 1915. It read in part, "When a man's work becomes a standard for the whole world, it also becomes a target for the shafts of the envious few."

At Harvey's he usually ordered steak or roast beef and a Caesar salad. He ate at the same table every time, the most secure in the room, almost invisible from the door under a stairway. A reporter once watched Hoover sign twenty autographs during a single dinner at Harvey's. He liked to sit there with his chief deputy, armed bodyguard, and longtime companion Clyde Tolson. It was a table for four with just two chairs. There was always a bottle of wine waiting for Hoover at his table when he arrived—part of a ritual that he performed here.

William Friedman dined at Harvey's on occasion. There were times at the restaurant when the cryptologist sensed motion in his peripheral vision, when a shadow darkened the white cloth. He turned his head and saw Hoover standing there with the bottle of wine. Without saying a word, the director nodded and poured wine into the cryptologist's glass. He had respected William for years and appreciated his periodic assistance with FBI cases, with the little encrypted notes written by criminal suspects that William would solve in his free time and send back to the bureau.

Hoover was almost certainly aware of Elizebeth Friedman. But he would not yet have had many chances to cross her path. She wasn't allowed to eat in the gentlemen's dining room at Harvey's. There were a lot of male enclaves like this in the city, inaccessible to her. And Hoover was a chauvinist of the old school. When he first took charge of the bureau in 1922, there had been three female agents. He got rid of them. The next two female agents wouldn't join the bureau until after his death in 1972. He argued that women weren't agent material because they couldn't be taught to shoot guns. Female clerks and secretaries at the bureau had to wear skirts and weren't allowed to smoke at their desks as the men could. One of his least favorite people was Eleanor Roosevelt, who wrote him a mildly indignant letter after the FBI conducted an intrusive background check on one of her friends. Hoover compiled a secret dossier alleging she was a communist. "When a woman turns professional criminal," he wrote once, "she is a hundred times more vicious and dangerous than a man." Women at Hoover's bureau were only deemed fit for "boring clerical functions," according to the memoir of one longtime agent. "It was perfectly all right to bullshit 'em and ball 'em: Just don't tell 'em any secrets."

But by 1940 Hoover had gotten himself into a jam serious enough to require the technical assistance of a woman.

It had long been the FBI's job to disrupt espionage rings within U.S. borders. Any Nazi spies operating in America were Hoover's quarry. However, he didn't seem to be very good at catching them. He had built the bureau's name on its flashy investigations of jazz-age gangsters, men who enjoyed attention and went out in public with entourages. Counterespionage was another discipline entirely, a matter that required a certain finesse, and the bureau's first sizable Nazi spy case, in 1938, had ended in a public-relations disaster.

That year in New York, the FBI arrested a Chicago man of Austrian parentage, Guenther Rumrich, along with two associates suspected of spying for Nazi Germany. Then an FBI agent named Leon Turrou made the mistake of tipping off Rumrich's collaborators that an indictment was coming. They panicked and fled the country.

Newspapers mocked the FBI for letting Nazis slip through its fingertips, and U.S. intelligence agencies that had long resented the FBI found new reason for their scorn. Over the years, Hoover's insatiable hunger for publicity had caused a lot of bad blood; in the press he repeatedly claimed sole credit for investigations to which other agencies had contributed but were not free to discuss. The head of army G-2, George Strong, one of William Friedman's superiors, despised Hoover, and the navy OP-20-G chiefs couldn't stand him, either. Henry Morgenthau at the Treasury hesitated to even speak the director's name in meetings, and Hoover thought of him as "that Jew in the Treasury." When British intelligence officers started to arrive in Washington in 1940, hoping to forge links with U.S. agencies, they were shocked by this toxic atmosphere of mistrust and quickly traced the cause to Hoover. "J. Edgar Hoover is a man of great singleness of purpose, and his purpose is the welfare of the Federal Bureau of Investigation," a group of British operatives later wrote. "It was once remarked of a well-known Oxford scholar that, while he had no enemies, he was hated by all his friends. Something of the same kind would express the feelings towards the FBI of its fellow U.S. agencies."

For a man as vain as Hoover, and as publicity-obsessed, and as intensely disliked by rivals in his own government, the bungled Rumrich case represented both a personal black eye and a threat to the FBI's future authority. Somehow he needed to salvage the bureau's reputation, to prove that it was capable of catching fascist spies, and in 1939, he proposed a bold plan to do just that.

Hoover knew that the concept of "hemisphere defense" had become a fixation with Roosevelt and military chiefs: Guarding the United States meant guarding *the entire Western Hemisphere* from Nazi encroachment. In other words, it wasn't enough to fortify U.S. defenses. South America must be protected as well. Roosevelt talked about hemisphere defense in speeches, arguing that "no attack is so unlikely or impossible that it may be ignored," and Secretary of the Navy Frank Knox raised the specter of Nazi planes taking off from South American airfields in the night and dropping bombs on "our own women and children in our teeming seaboard cities." Seeing an opening, J. Edgar Hoover pressed Roosevelt to dramatically expand the FBI's jurisdiction. For the sake of "the common defense of the Western Hemisphere," Hoover argued, the FBI must be allowed to operate beyond U.S. borders. He demanded the authority to send men into South America, "to seek out and identify agents of the Axis operating in all the Americas, to ensure the ultimate safety of the United States."

Hoover got his wish in June 1940, with a presidential directive that represented a historic expansion of the FBI's power. For the first time, the bureau was free to dispatch agents into other countries. He created a new division called the Special Intelligence Service (SIS) and began recruiting agents for duty in South America.

Their mission would be to find and monitor the secret mail drops and radio stations used by the spies; to map the structure of their organizations and communications networks; to determine the true identities of the enemy agents; and to cooperate with local State Department officials and police in arresting the spies, seizing the radio stations, and destroying the rings.

A tall order. The first five SIS agents were dispatched to the continent in September 1940, one each to Peru, Uruguay, Argentina, Brazil, and Venezuela. Pale and corn-fed, they stepped off their planes into the lacerating sun of another continent. They wore snap-brim hats and looked like detectives that South Americans had seen in newspapers and movies. The agents knew little about codes, ciphers, or radio, these crucial tools of their adversaries, and didn't speak the local languages. The SIS man sent to Brazil had been given a crash course in Spanish. When he arrived, he realized, to his frustration, that the language of Brazil was actually Portuguese.

Hoover's men in South America were so unprepared that they

had almost no chance of catching the spies through old-school gumshoe tactics: interviewing associates, recruiting confidential informants, developing leads. They needed to know what the spies were saying to one another in private. They needed codebreaking. And this was exactly the problem.

To break codes, you need intercepts and you need codebreakers to solve the intercepts. The FBI had neither. It had no intercepts because it had no listening stations; when the bureau wanted intercepts it was forced to obtain them from the coast guard and the Federal Communications Commission (FCC). And when the FBI got these intercepts, it couldn't read them, because the FBI had no codebreaking unit. What it had instead was a Technical Research Laboratory, essentially a crime lab, a place where bureau technicians analyzed bullets, fingerprints, threads of fabric, and blood samples.

All of this spelled trouble for J. Edgar Hoover. At the very moment he was launching a hemisphere-wide hunt for spies who communicated in code, his bureau had no ability to discover what they were saying.

Around this time, Elizebeth received an unusual order from her Treasury bosses: They asked her to visit FBI headquarters. She was to teach codebreaking to an agent named W. G. B. Blackburn, an employee in the Technical Lab. Elizebeth proceeded to train Blackburn in codes and ciphers, much as she had trained her own junior colleagues, and Blackburn established a small Cryptographic Branch at the FBI, which would grow to the size of a handful of employees over the next several years, all of them codebreaking novices.

This still wouldn't do for Hoover's purposes. The Invisible War demanded a level of technical firepower and prowess that his Technical Laboratory simply did not command. What he required was the full assistance of a mature codebreaking organization, whether they wished to help him or not. He needed Elizebeth and the coast guard.

She read pacifist poetry. It resonated. She thought of her kids. Barbara was in her last year of high school and planned to attend college at Radcliffe, and John Ramsay was a fourteen-year-old freshman at Mercersburg Academy, an elite boys' school in rural Pennsylvania. He wasn't young enough to be safe from a military draft. War would scatter her family. She also worried about the fate of her team at the coast guard. She had built this little organization and it was good and she wanted to protect it from disruption. Codebreaking is delicate work. You have to look at the page and get all the letters aligned just right, then you have to look at your team and get all the people aligned just right, so that the flow of intercepts and records and ideas and solutions becomes as efficient as possible.

Elizabeth escaped Washington for a week in June 1940, traveling to Mexico on a quick vacation with daughter Barbara and sister, Edna. It was the last time in the next five years she would get a break, a chance to pause and look around and spend time with the women closest to her. They drove a beat-up rental car through the farmlands of Oaxaca and the mountain ranges of Puebla Cordoba, descending into canyons on the backs of burros. Elizabeth wrote to William, "All Mexico is so full of resounding cockcrows, piggrunts, burro-brays, and church bells that all sleep is intermittent, at best." The two sisters had a great time and woke up early each day; Barbara wanted to sleep in and complained that the altitude made her knees wobbly. She was a gorgeous girl of seventeen now, six inches taller than her mother, confident and voluptuous. One day, when they were all on a plane above Oaxaca, Elizabeth happened to fall asleep in her seat, and when she woke, she saw Barbara up in the cockpit, next to the pilot. *Wait, what kind of airline was this?* Are girls just allowed to ride in the cockpit without their mother's permission? Isn't that unsafe? She felt like a mom.

The news of the war got rapidly worse while she was in Mexico. She had to stop reading the papers in the morning because it was too depressing. Nazi tanks were said to be plowing through the French countryside on the way to Paris. The Mexican papers seemed to think America was bound to join the war. The peso was rising, eating into Elizebeth's meager trip budget of fifty dollars. She airmailed William a letter about the rising cost of goods. In his reply he begged her not to spend more money than was absolutely necessary, "or we shall never never climb out of this morass of debt."

She wasn't sure if William was okay. He sounded sad and mopey in his letters. He said it had been rainy in Washington, and in the evenings he had been sitting alone with a pencil and a pad, listening to the rain on the roof, writing a technical paper on cryptology. He told her, "There won't be anybody [to] read this thing, I imagine, at least not for some centuries," and added a lament about the shackles of secrecy: "I wish I could write about forbidden subjects. What a story could be told."

By the time she got back to Washington—to home, husband, and job—the Nazis had entered Paris, hanging the swastika flag from the Arc de Triomphe.

CHAPTER 2

Magic

One day in September 1940, inside the windowless vault of William Friedman's army codebreaking unit, one of the two female codebreakers on the team, Genevieve Grotjan, stood at her desk and let the men know that she might have found something.

The men called her Gene. She was twenty-eight and quiet and wore rimless glasses. She had a background in statistics; she would later teach math as a professor at George Mason University in Virginia.

Gene Grotjan had been looking at raw Purple intercepts for hours, weeks, months when she called the men over. The team had not been able to penetrate the garbled Japanese text on the intercept sheets that had been streaming into the Munitions Building. But now Grotjan thought she had noticed two patterns that others had missed—subtle cycles of repetition, loops of letters in the text, much like the ones discovered by the coast guard in Enigma messages. Frank Rowlett, one of William's deputies, came over and looked at Grotjan's worksheets. Then he looked at her. He got the sense that her eyes were beaming through her glasses, that she was struggling to contain her emotions. Others started to crowd around the desk. Rowlett started jumping up and down.

"That's it!" he shouted. "That's it! Gene has found what we've been looking for!" Another man busted into a funny little dance. He threw his arms up in a victory pose. "Whoopee!"

Grotjan was a modest person. "Maybe I was just lucky," she said later in an NSA oral history. "I perhaps had a little more patience" than some of the other workers. She didn't become as animated as Rowlett because "I regarded it more as just one step in a series of steps."

William Friedman heard the commotion and shuffled into the room from his nearby office: "What's all this noise?"

Rowlett showed Gene's worksheets to the boss, and William agreed within seconds that they revealed a loose thread in the code. There was still more to do, but it was clear that the thread, if pulled, would allow the team, with great grinding effort, to re-cover the daily keys and consistently read the Japanese messages.

The men were bouncing and laughing with the excitement of the discovery. Friedman seemed almost sad. "Suddenly he looked tired," Rowlett later recalled, "and placed his hands on the edge of the desk and leaned forward, resting his weight on them." Rowlett knew that Friedman had been under a lot of stress, work-ing sixteen-hour days for weeks, months, years. The younger man pulled out a chair for his elder. Friedman sat quietly for a few mo-ments. Everyone was looking at him, waiting for his reaction. He turned to the codebreakers. "The recovery of this machine will go down as a milestone in cryptologic history," he said in a formal, distant voice. Then he left the room.

In codebreaking, the larger the success, the more it must be sup-pressed. Any leak might reach the enemy and cause them to switch to a new code system, destroying the value of the break. Heroes celebrate briefly and in secret. Someone went and got Cokes.

A few minutes later, the group dissipated, and everyone re-turned to their desks to explore the new textual terrain they had unlocked—everyone except Rowlett, who went looking for Fried-man. The young man was still shaking, still full of adrenaline, wanting some kind of catharsis. The boss's lack of enthusiasm confused him. Rowlett found Friedman in his office, "sitting at his desk, studying some notes he had made on a pad. When I entered the room, he sat quietly, merely looking questioningly at me."

Over the next hours and days, the team kept applying pressure to the hairline fracture in the code until it shattered and the first bits of plaintext revealed themselves. Five days after Grotjan's discovery, on September 25, 1940, the team produced their first full decrypted message. It was a big moment. William and the rest of the codebreakers had never been able to look at the Japanese machine, or touch it. They had never seen a drawing of it, or a patent illustration, or a photo. Yet they now understood how it worked and how to recover the daily key for a given set of messages. People had reverse-engineered cipher machines and devices before, but nothing at Purple's level of complexity. Today historians of cryptology believe that in terms of sheer, sweaty brilliance, the breaking of Purple is a feat on par with Alan Turing's epiphanies about how to organize successful attacks on German Enigma codes.

Once William and his colleagues fine-tuned their bootleg Purple machine, the one they were building to help them read Japanese messages, they demonstrated it for the unit's commanding officer, typing out a sample of ciphertext and then decrypting it as he watched. A sheet of fresh plaintext inched its way out of the machine, and after the man grabbed it and looked for a few seconds, a smile lit up his face, and he congratulated the codebreakers on their "magnificent achievement." Then he rushed off to get *his* commanding officer, Joe Mauborgne, William's old friend. When they returned together, the first officer pointed to the machine and said to Mauborgne, "Last night your magicians completed the reconstruction of the new Japanese cipher machine," and the codebreakers repeated the demonstration for Mauborgne. "By God, it really works beautifully!" Mauborgne said.

Your magicians. It really did seem like that. Like magic.

MAGIC became the top-secret moniker for these Japanese decryptions, for the astonishing fountain of secrets that would keep gushing up all through the war, secrets of Japanese strategy and Nazi tactics that flowed across Japan's encrypted circuits and were tapped by the U.S. Army (and later by the U.S. Navy too, which solved Japan's naval cipher), giving Allied planners the drop on their foes. The first handful of decrypted messages turned into 20, into 100, into 1,000, piped directly from William's unit through the corridors of power. MAGIC beguiled all who touched it. Men read the daily "MAGIC Summaries" with bulging eyes and could not quite believe they were reading the authentic words and orders of imperial Japan. It was almost too good to be true. The president read MAGIC, and Army Chief of Staff George C. Marshall, and Secretary of the Navy Frank Knox, and, eventually, Prime Minister Winston Churchill in Britain, who insisted on getting MAGIC raw, unsummarized by his generals.

MAGIC led directly to bombs falling on imperial ships at Midway and other decisive naval battles. It caused the deaths of hundreds of thousands of Japanese and saved the lives of unknown numbers of Allies. MAGIC changed the war. It was also one of the great secrets of the war, exactly like ULTRA, the Enigma codebreaking program. These were tremendous military advantages that could not be revealed to the enemy lest the enemy get wise and cut off the stream of intelligence. The advantage "would be wiped out almost in an instant if the least suspicion were aroused regarding it," George Marshall wrote later in a classified letter that captured the value of MAGIC in the Pacific War against Japan:

> The Battle of the Coral Sea was based on deciphered messages. And therefore our few ships were in the right place at the right time. Further, we were able to concentrate our limited forces to meet their advances on Midway, when they otherwise would certainly have been some 3,000 miles out of place. We had full information of the strength of their forces in that advance, and also of the forces directed against the Aleutians, which finally landed troops on Attu and Kiska. Operations in the Pacific are largely guided by the information we obtain of Japanese deployments. We know their strength in various garrisons, their rations, and other stores available to them. And what is of vast importance, we check their fleet movements and the movements of their convoys.

The triumph over Purple would turn out to be William's last hurrah as a hard-core codebreaker, his final death-defying climb. From now until the end of his life, he would serve America as an inventor of cipher machines and an architect of intelligence institutions (and ultimately a critic of them as well). He had reached his peak. Elizebeth, though, was still climbing, and she couldn't see him up there, across the gap between their two towers, starting his descent. She couldn't share this victory with him, because on the day he and his team broke Purple, a historic achievement that had required all of his battered brain, all he had learned in his unexpected life of exploration with the woman who meant everything, he said nothing about it to her when he came home. He did not seem different to her than he did on any other evening. He said hello and asked what was for dinner.

Heavy air attacks on London had begun that month. The Blitz. On September 7, 1940, shortly after 5 P.M., a thousand German planes appeared in the sky above London. It was a bright blue afternoon. The planes arranged themselves in vertical formations, fighters and bombers. The fighters had bright yellow noses and tails. The bombers were black and set their sights on industrial facilities that lined the Thames River. The bombs destroyed factories, shock waves and oily smoke rippling out. British Spitfires gave chase to the German planes. "The sky seemed full of them," one British pilot later said, "packed in layers thousands of feet deep. They came on steadily, wavering up and down along the horizon. 'Oh golly,' I thought, 'golly, golly. . . . ' "

The bombings of London continued for fifty-six straight days. Sirens and shelters, blackouts at night. The Axis was growing bolder in the final months of 1940. Japan invaded Vietnam, expanding its empire in East Asia. The Nazis confiscated the private radios and telephones of Jewish families and cordoned off the Warsaw Ghetto with barbed wire, trapping 400,000 adults and children, most of them Polish Jews.

America didn't want war. Both major political parties still supported neutrality. The aviation pioneer Charles Lindbergh argued in popular radio speeches that it would be foolish and hypocritical to fight Germany. He said America had no standing to accuse the Nazis of aggression and barbarism because America had sometimes been aggressive and barbaric itself. Later he argued that American Jews were a "danger to this country" on account of their "ownership and influence in our motion pictures, our press, our radio and our government." Lindbergh became the public face and champion of an antiwar group called the America First Committee. "America First," a campaign slogan of Woodrow Wilson, had been adopted by the Ku Klux Klan in the 1920s. Within a year the America First Committee was holding rallies at Madison Square Garden.

The worsening of the war in Europe, combined with U.S. reluctance to fight, was about to drag Elizebeth into the orbit of a highly motivated and capable group of British spies. The British were afraid. They knew they didn't have the money, the people, or the weaponry to sustain a long fight against the Nazis. They needed America to join the war. Their survival as a nation depended on it.

In the early summer of 1940, British officers began to arrive in America on a covert mission. Some went to Washington, making the rounds of embassy cocktails and dinner parties, looking for all the world like bright young chaps out for a good time, and others worked in the heart of New York, in a Fifth Avenue skyscraper, the thirty-fifth and thirty-sixth floors of Rockefeller Center. The group included Ian Fleming, a handsome lieutenant with blue eyes and a smart blue naval officer's jacket, and twenty-three-year-old Roald Dahl, a tall, elegant Royal Air Force fighter pilot who looked a bit like Gary Cooper. These guys would both become famous fiction writers after the war; Fleming invented the character of James Bond, and Dahl wrote children's books about chocolate factories, flying peaches the size of zeppelins, and foxes who outwit monstrous humans. For now, though, Fleming and Dahl were spies. Dahl was a particularly good spy. He seduced actresses and heiresses in Washington, gathering gossip in bed, and he charmed the president and the first lady, becoming a regular guest at their Hyde Park, New York, home, where they spoke so freely with the young pilot that he had difficulty maintaining his composure: "I would do my best to appear calm and chatty," he later wrote, "though actually I was trembling at the realization that the most powerful man in the world was telling me these mighty secrets."

They called their organization British Security Co-ordination, an intentionally boring name meant to deflect scrutiny. BSC was really one of the most fantastic associations of men and women ever created. It had one thousand members who worked toward a single goal: ending American isolationism and pushing America into the war, by any means necessary. One BSC recruit had been told, "All I can say is that if you join us, you mustn't be afraid of forgery, and you mustn't be afraid of murder." BSC planted anti-Nazi information in the American press, some of it false, through relationships with columnists like Walter Winchell. BSC staged protests at rallies for isolationist politicians and dug up dirt on their pasts. It used sex to steal information, sending gorgeous female spies to seduce enemy diplomats and swipe documents. And BSC also hoped to apply British radio expertise to catch enemy spies operating in the Western Hemisphere, which put BSC in direct conflict with the formidable American who had already claimed that ground and was not eager to give it up.

Early on, the British tried to strike a deal with J. Edgar Hoover. Ian Fleming and his superior officer went to FBI headquarters one day in June 1941 and met with the director in his corner office. The white dome of the Capitol was visible from the window, beyond a set of stone columns at the back of the National Archives, the central repository of government records. Fleming and his colleague explained that they wished to partner with the bureau and share intelligence on the Nazi threat. Hoover listened politely, "a chunky enigmatic man with slow eyes and a trap of a mouth," in Fleming's description. Then Hoover said he couldn't help; U.S. neutrality rules prohibited him from giving aid to any combatant nation.

This was true, but it was also an empty excuse. Hoover didn't want the British operating in America because he saw them as a rival to the FBI. The British didn't care one way or the other. They needed a friendly American spy agency as a partner, and if Hoover wasn't willing to be that agency, for whatever reason, they would find another one, even if they had to create it from scratch. And so they did. They planted the seed that eventually grew into the CIA. Behind the scenes, the British argued to U.S. officials that the FBI was ineffective. The FBI had "no conception of offensive intelligence as we know it," wrote Captain Eddie Hastings, a retired Royal Navy officer, now working for BSC in Washington; according to Hastings, America needed a new agency capable of "offensive" spy maneuvers in foreign countries. In July 1941, Roosevelt established the Office of the Coordinator of Information, a new civilian intelligence organization attached to the White House. The following year, the Office of the COI was renamed the Office of Strategic Services, which was the fore-runner of the CIA.

So this is where the CIA began—with J. Edgar Hoover telling the British to go to hell, and the British not appreciating it.

This was also when the British began making friendly advances toward Elizebeth Friedman.

The British already had a mature radio intelligence agency, the Radio Security Service (RSS), that excelled at the art of wire-less interception. But due to sheer geography, the British listen-ing posts couldn't hear signals from some parts of the globe. The men of British Security Co-ordination wanted access to any in-tercepted and solved messages that America happened to have. And they realized that when it came to radio intelligence and hard-core codebreaking, the place to be in America was the coast guard. Unlike the FBI, Elizebeth's unit had access to intercepts from its own listening stations, and its cryptanalytic section "was incomparably better than that of the FBI," in the British view, because the coast guard's codebreakers had spent the last decade testing their skills on smugglers, whose networks happened to look a lot like Nazi spy networks. "The whole system" of rum-running had the air of a German spy network in miniature," BSC historians later wrote. "Hence, on the outbreak of war, the Coast Guard was already experienced in the tricks of the illicit wireless operator."

BSC sent a few men to meet with Elizebeth in Washington and chat about the problem of Nazi spies in the Western Hemisphere, and they all hit it off right away. The men had considerable ex-pertise and experience in radio intelligence, particularly a husky, apple-cheeked colonel named F. J. M. Stratton, who had taught astronomy before the war, specializing in studies of supernovas, distant exploding stars that registered as sudden and perplexing balls of light on Stratton's photographic plates. Before that he served in the radio corps of the British army in the First World War, developing a reputation as the happiest man in the trenches despite sleeping only four hours a night. His fellow soldiers called him "Chubby" on account of his bulk and his jollity. Elizebeth thought he looked like Santa Claus.

As they got to talking that first time, Stratton and Elizebeth, they realized that if they combined their resources and their knowledge, they'd have a better chance against the Nazi spies than if they were working alone. The British operated radio posts across Europe staffed by 1,500 secret listeners, many of them volunteer hobbyists, and the intercepts from those stations would fill gaps in the intercepts from the coast guard and the FCC, and vice versa. When the British couldn't hear something, the coast guard could hear it, and when the coast guard couldn't hear it, the British could.

Aside from that, Stratton enjoyed deep connections to Bletchley Park and the already massive codebreaking operation there, where some analysts had been focusing specifically on Nazi spy codes. It might make sense to share knowledge.

By now Elizebeth and her coast guard codebreakers had also begun working directly with the FBI at the request of J. Edgar Hoover. He wanted assistance with several different unknown code systems. Elizebeth obliged. She found that some of the spies who interested the FBI were using book ciphers, and others relied on "turning grilles" much like the grilles that the Friedmans drew one year in their family Christmas card. The spies wrote letters in holes punched through a piece of paper of certain dimensions according to certain rules, and Elizebeth had to make five or six separate deductive leaps to figure out those rules to determine the exact shape of the piece of paper using only clues derived from the messages themselves.

Not only did Elizebeth break the codes for the FBI, she made special devices and tools for the G-men so that they could easily solve future intercepts on their own. For instance, when she solved a book cipher, she gave the FBI the name and description of the book, and when she solved a grille system, she made grilles for the FBI Technical Laboratory. In other words, when the FBI was able to solve its own messages, this was only because Elizebeth had given the Technical Laboratory the means of solution—the laboratory run by the G-man she herself had trained in 1940.

While Elizebeth solved these individual puzzles as fast as she could, immersing herself in the gritty details, the larger goal behind it all, preventing a fascist takeover of South America, remained an obsession at the highest levels of U.S. government. On December 29, 1940, in a "Fireside Chat" radio speech from the Diplomatic Room of the White House, FDR argued that it was time for America to rethink its role in the world. It was futile to hope that fascism would leave America alone if America returned the favor. Instead, the nation must become an "arsenal of democracy," a force to defend and spread freedom abroad. During the 36 minutes and 56 seconds of the speech, he mentioned South America twice and used the word "hemisphere" 10 times. He said, "Any South American country, in Nazi hands, would always constitute a jumping-off place for German attack on any one of the other republics of this hemisphere." Without getting into specifics, Roosevelt referred to "secret emissaries" of the Axis, fascist spies like the ones Elizebeth was tracking. "The evil forces which have crushed and undermined and corrupted so many others are already within our own gates. Your Government knows much about them and every day is ferreting them out."

Hitler responded that England would soon be destroyed along with all other "democratic war criminals" and promised a rapid Nazi victory within the first few months of 1941. On New Year's Eve, Londoners climbed into the blacked-out streets, over the charred remains of buildings, and sang "Auld Lang Syne."

Elizebeth heard the news four days later, on January 4, 1941. She rushed to Walter Reed General Hospital in northern Washington.

The main building was majestic, meant to be a comforting sight to wounded warriors: three stories of red brick, with soaring white columns in front. Staff directed Elizebeth to the Neuropsychiatric Section, a separate structure connected to the hospital by an underground tunnel. She found William there, confined to a large, noisy room. It was three and a half months after his team's breakthrough on the Purple code. Elizebeth counted between sixteen and twenty other psychiatric patients in the room, all men, including some who appeared very disturbed. She was scared and could see that William was scared, too.

Walter Reed was the nation's flagship military hospital, and for soldiers or officers suffering from physical injuries or infectious diseases, it was about as good as could be. During the Great War, men returning from the trenches, many with amputated limbs, would sit on the wide porch in wheelchairs, covered in blankets, looking out at the manicured grounds and the fountain whose bowl was ringed by four stone penguins standing atop concrete pedestals. Psychiatry, however, had never been a priority at Walter Reed or in the army as a whole, and in 1940 and early 1941, the energies of army psychiatrists were almost entirely geared toward keeping the mentally ill *out* of the army, not treating them once they got in.

Walter Reed's chief psychiatrist, Colonel William C. Porter, saw the job of the hospital's Neuropsychiatric Section as one of evaluation and processing rather than healing. The section did offer a range of treatments standard for the time, including chemical sedatives like Amytal, group therapy, and electroshock therapy, but it was too small to provide long-term care, so it functioned instead as a way station, a purgatory. When the section admitted a new patient, doctors and nurses examined him, studied his military records, and observed him for a period of weeks or months before deciding whether he should be discharged from military service. Depending on the decision, the patient was sent back to the army, or home to his family, or in many instances, to a mental asylum.

Sometimes, instead of discharging a patient from the army altogether, the section's doctors recommended he be transferred to a desk job, presumed to be less stressful. The idea that desk work itself might be a cause of debilitating stress—that the army now employed puzzle solvers, cryptologists, who bashed their brains against the stone of codes and bore the heavy burden of secrets—never occurred to the doctors of Walter Reed.

They didn't know what to do with William Friedman when he presented himself. William told the doctors he had collapsed several days earlier and believed he was having a nervous break-down. A psychiatrist asked a battery of questions about his job, his family, and his career. Without mentioning the Purple project, William said his work had been demanding lately. He felt a constant tension that interfered with his ability to function, and sleep provided little relief when he could manage to sleep at all.

The doctors assigned the cryptologist to one of the section's five mental wards. There were three wards for men and two for women, with a maximum capacity of 104 patients. Security guards patrolled the wards. William spent the next two and a half months here, inside the redbrick building, unable to leave until the staff completed their evaluation.

Elizebeth came to visit most days, taking the train to the 116-acre hospital campus and walking briskly past the main building with its cupola and fountain on her way to the Neuro-psychiatric Section. She always wanted to talk to her husband in private during these visits, to see how he was doing, to kiss him and say she loved him, but the setup made it nearly impossible, because the patients were forced to spend their days in the group room, and they all had to share the same psychiatrist, who consulted with each patient within earshot of the others. "In other words, the patient was isolated except for his fellow-patients," Elizebeth later told William's biographer, "who could discuss and consult with each other if they felt inclined to do so."

The patient. Seeing him there was horrible for Elizebeth, and the hospitalization represented such an obvious threat to his livelihood in the army that she had to find ways to distance them both from what was happening. She refused to admit that her husband might have a serious mental illness. She thought the word "depression" was "too strong a term" and preferred "mood swings" or "downswings." At home she answered his personal mail, explaining that William was ill and would get back to people when he could.

Meanwhile, at the Munitions Building, his team of cryptologists continued to harvest the fruit of Purple and plant seeds in new places. Having already built one replica of the Japanese machine, the SIS workers built several more, and in January, two of William's deputies, Abe Sinkov and Leo Rosen, sailed across the Atlantic with two Purple machines, delivering them to grateful British codebreakers at Bletchley Park. Now the British could make their own translations of Japanese messages. It was an important exchange of cryptologic knowledge between America and Britain, one of the first of many during the Second World War, although the British didn't return the favor yet—they weren't ready to share what they knew about German Enigma systems.

In March 1941, the staff of Walter Reed finally made their decision. William Friedman, they believed, should return to army duty. His nervous collapse was an "anxiety reaction" sparked by "prolonged overwork on a top secret project." The hospital discharged him on March 22 into Elizebeth's arms. He went back to work at the army on April 1.

He wasn't quite the same, and never would be. The breakdown and the hospitalization had changed his universe in ways it would take years to measure and understand. For one thing, the ordeal had planted doubt in the military bureaucracy that William Friedman was fit for service. It created a trail of medical documents that would chase him for years, popping up and causing havoc at the oddest times. Three weeks after he left Walter Reed, William received a letter from the army notifying him that he had been honorably discharged "by reason of physical disqualification"—no hearing, no chance at a defense. William made a vigorous protest, pointing out that the hospital had pronounced him fit, but the army forced him to retire and continue as a civilian; eventually he would need to sue to get his old rank and pay reinstated. Later, in 1946, checking his personnel file, William discovered that the government had him classified as a *temporary employee.* It was probably a paperwork snafu, but it struck him as a bizarre indignity—his twenty-five straight years of service to America had hardly been temporary—and his friends were so horrified on his behalf that they threw him a big surprise party at an officers' club and staged a mock court-martial as a send-up of the ridiculousness. The judges recorded their votes on a cipher machine, pronounced him guilty, and presented William with an aluminum medallion that read, "To Wm. F. Friedman for making the intelligible unintelligible and vice versa 1921–1946. Presented by those he has led astray."

William's illness also disrupted the balance of the Friedmans' marriage. William and Elizebeth had always acted as fierce equals. Modesties and flatteries aside, they lived as if neither was smarter than the other, or stronger, which was the truth. From here on, though, Elizebeth often had to be the stronger one, out of pure necessity. She had to care for William during his depressions and keep her job. They needed two incomes to pay the mortgage and their kids' private-school tuition. The Friedmans, like so many middle-class Americans who hurt for money and pinch pennies, were determined that their children receive the same educational opportunities as the "sons of capitalists," Elizebeth once wrote. And throughout the spring and summer of 1941, as William recovered, Elizebeth's job was only getting harder. The Invisible War was intensifying. The documents produced by her team now bore its mark. On the coast guard decrypts, in the lower left corner, beneath the letters of the plaintext, the same two words appeared, over and over, on page after page.

"German Clandestine."

CHAPTER 3

The Hauptsturmführer *and the* Funkmeister

Johannes Siegfried Becker, the most prolific and effective
Nazi spy in the Western Hemisphere during the Second World War.

No code is ever completely solved, you know.
—ELIZEBETH S. FRIEDMAN

The fact that Johannes Siegfried Becker is an obscure figure today, a man without a Wikipedia page, his name producing a few stray Google hits, is a testament to his skill as a spy and also the skill of the woman who became his nemesis, Elizebeth Smith Friedman. They were two cloaked particles meeting across a void at the speed of light and partially annihilating each other, leaving jets of alphabets, a spray of letters falling to the ground.

According to the FBI, which was slower than Elizebeth to

understand his significance, Johannes Siegfried Becker was "one of the most active as well as the most capable of German agents operating in this hemisphere during this war," a spy of rare vision and resourcefulness, directing endless funds and resources with a "deft Teutonic hand." He spoke German, Spanish, Portuguese, and English. He held the rank of SS-*Hauptsturmführer* in the elite Nazi security service, equivalent to a captain, and wore a gold ring carved with the SS's death's head symbol, "a sign of our loyalty to the Führer," Heinrich Himmler wrote in a letter of praise to Becker, and "a warning to be ready at any time to sacrifice our lives as individuals for the life of the whole." He had some forty-seven aliases and several false passports and moved freely across South America, recruiting spies in seven nations, organizing political plots and military coups with Nazi sympathizers, and building clandestine radio stations. In mid-1944 the FBI concluded that the activities of 250 Nazi agents in South America and twenty-nine radio stations could be traced back to Becker by direct or indirect steps.

Yet Becker did not appear on the FBI's radar during 1938, 1939, 1940, 1941, 1942, or 1943, and when the FBI finally began to hunt him, they were too late. He was the Invisible War's invisible shadow. He escaped every trap, slipped every net—except, at last, the one set by Elizebeth. And even Elizebeth would be astonished by Becker's ability to vanish.

For all his talents, Becker started out as something of a screwup, his first missions in South America marked by mediocre results and sexual improprieties. Between 1936 and 1939, while spying for the SS in Brazil and Argentina before the war broke out, he left a trail of irritated German expatriates, men who "disliked his manner" and found him vain. There appeared to be nothing exceptional about Becker. His Nazi Party number was 359,966, marking him as a relatively early convert to the cause of National Socialism but hardly one of its pioneers. He stood five foot ten, with wavy blond hair and a slight paunch, and his face was not overly handsome, leaving his acquaintances wondering how it was that Becker always seemed to have a girlfriend. For a time he worked for an Argentine firm as an importer of German children's toys and doll eyes, claiming to be a woodworking expert, when in truth he spent most of his days watching British ships come and go in the harbor and his nights prowling the bars and dance floors of the city, writing the phone numbers of prostitutes and fascist sympathizers in a personal address book. He caused a scandal in Rio de Janeiro by impregnating the wife of a Brazilian cabinet minister. The Nazi ambassador complained to Berlin that Becker was risking an international incident. No one in South America had anything good to say about Becker's personal habits, and all who met him were struck by his grotesquely long fingernails, which curled down like the talons of a predatory bird.

Still, Becker had one essential quality that set him apart from almost everyone else in his corner of the Nazi universe: he was *adaptable.*

Becker worked for a wide-reaching SS office that placed spies all around the world and communicated with them from a four-story building in Berlin that had once been a Jewish retirement home. Called AMT VI, the SS office employed five hundred people in Berlin and managed another five hundred spies in foreign countries. A minority of the spies were actual SS officers like Becker and the rest were considered "V-men" (*vertrauensmann* is German for "informer"), usually German expatriates and local fascists who wanted to help the cause. There was also a separate German agency, the Abwehr, that sent spies to foreign countries, but the Abwehr predated the Nazi movement, and SS leaders thought the Abwehr was insufficiently ruthless and possibly disloyal. They promoted their own AMT VI as the true Nazi foreign intelligence service.

Not just any Nazi could be selected as an SS intelligence officer, according to an SS handbook. He had to be the purest of Nazis, a man of "absolute loyalty and obedience to the Führer . . . Like the knights of the Holy Grail intelligence officers have the most noble task to protect the most valuable possession and its future realization: the blood of the Germanic race, the National Socialist ideology."

In practice, however, this meant that Becker's organization was riddled with amateurs promoted for their zeal instead of their knowledge. The leader of the South America section of AMT VI, Theodor Paeffgen, was a thirty-one-year-old bureaucrat with "no

qualifications whatever for intelligence work," an American interrogator would later conclude. Paeffgen's previous job with the
SS had involved "combating partisans" in Russia, a euphemism
for killing Jews. Paeffgen's deputy was a former Gestapo thug
named Kurt Gross, who badgered his spies in South America to
send him packages of cognac, coffee, and silk stockings, and often made lewd comments to the buoyant, brown-haired young
woman who managed the section's files, Hedwig Sommer, who
had been forced into working for the SS against her will. (After
the war, Sommer gladly told U.S. interrogators everything she
knew about the section.)

These men cared mainly about ideology, not competence, and
even when they made a rare exception, they were overruled by
other fanatical organs of the Nazi state. One of the section's most
talented spies was a Jewish man from Holland named Weinheimer
who was working for the SS in the hopes of saving his family from
the concentration camps. He had smuggled himself into Chile,
posing as an immigrant, and according to Sommer he sent back a
number of "highly regarded" and "very accurate" reports about
political and economic trends in the Western Hemisphere. Then
Weinheimer learned that the Gestapo had shipped his mother-in-
law to the Bergen-Belsen camp. Kurt Gross asked the Gestapo to
make an exception for the spy's kin, but Gross was unsuccessful,
and the spy stopped sending reports. The Nazis lost one of their
best agents because they wouldn't spare his loved ones from the
death camps.

When it came to the nuts and bolts of intelligence, the SS bosses
in Berlin didn't really know what they were doing—and neither
did Siegfried Becker at first. But unlike his superiors, he was flex-
ible enough to learn from his mistakes. He was a loyal Nazi but
didn't concern himself with the intricacies of Nazi dogma. Hedwig Sommer liked him. "He was an intelligent person," she said.
"He was sincerely desirous of doing a good job. Added to these
attributes was the fact that he was something of an adventurer."

After the Nazi invasion of Poland in September 1939, Becker
had left South America and sailed to Berlin, guessing that his
bosses would want to modify his mission. He was correct. In
meetings at the home office, the SS leaders told Becker that he was

now their top agent in South America, and he needed to go back to the continent and recruit a team of spies.

Berlin gave Becker a trunkful of explosives for blowing up British ships in the harbor. Becker arrived in Buenos Aires with the trunk in December 1940 and was intercepted at the German embassy, where the ambassador opened the trunk, saw the bombs, imagined the diplomatic headaches they would cause, and ordered Becker to dump the bombs in the river. At this point, he abandoned the sabotage mission and began building his new spy network in earnest, traveling across the continent, from Argentina to Brazil to Bolivia to Paraguay, trying to convince German colonists to spy for the Führer.

Many of the would-be V-men in South America proved hopelessly ineffectual—one was a petty crook who seemed to do nothing but wander along the waterfronts carrying a revolver and scaring passersby—but in Rio de Janeiro, Becker soon met and cultivated a formidable spy named Albrecht Engels, a broad-shouldered German businessman with a thick mustache. Engels was already spying for the Abwehr, which meant that he and Becker were not supposed to work together, but neither man minded. Becker thought Engels was a perfect collaborator: married to a Brazilian woman, owner of a thriving firm in Rio, well liked by all Germans in the community.

And Engels, who went by the code name "Alfredo," was impressed by Becker. Ever since Engels started working with the Abwehr, he felt he had been dealing with imbeciles. His Abwehr colleague in São Paulo was a jittery mechanical engineer of Polish ancestry, Josef Starziczny, who went by the code name "Lucas." Starziczny was an elfin man with large ears who lived with his Brazilian mistress and talked too much. He observed the harbor, radioed reports to Germany with his own transmitter, and didn't listen to advice. He made Engels nervous. Becker was different, another caliber of spy: "the only real professional" in South America, Engels later told an FBI interrogator.

Engels's arrangement with Becker was strictly *improvisado*—making things work. Until this point, Engels's duties for the Abwehr had consisted mainly of scouring English newspapers and magazines (*Time, Collier's, Reader's Digest*) for information about U.S. politics. Becker single-handedly turned this press-clipping service into an actual spy network. He built a courier system to exchange information with Germany, convincing employees of the Condor and LATI airlines to carry spy messages in pouches on their flights to Germany and deposit the pouches at a firm owned by an SS man. He taught Engels how to use book ciphers and codes based on pencil-and-paper grids and turning grilles.

And in the spirit of *improvisado*, when Allied pressure shut down Condor and LATI flights to Germany in summer 1941, destroying Becker's courier service, he found ways to communicate wirelessly with Berlin. At first Becker paid a V-man to set up a small shortwave transmitter on the patio of a German expatriate's home. When the signal proved too weak, Becker coaxed the captain of a Swiss ship docked in the Rio harbor, the SS *Windhuk,* to allow the spies to borrow the ship's radio.

Becker signed his wireless messages with one of several code names. The main alias was "Sargo." Engels sent messages under his "Alfredo" alias.

It was difficult to get a reliable signal, and Becker, for all his ability, lacked the technical expertise. He asked the SS to send him a *Funkmeister,* a radio operator, and in September 1941 the SS dispatched Gustav Utzinger to Rio.

Utzinger was the opposite of Becker in many ways: a man of education, a trained chemist. He went by the code name "Luna." Clean-cut and athletic, with brown eyes and close-cropped brown hair, he had served in the 1930s as a *Funkmeister* in the German navy before joining the SS. Later, speaking to an American interrogator, Utzinger claimed that he acted out of "natural patriotic efforts for my Fatherland" and not "the most detestible tendencies of Nazi ideology." The interrogator didn't buy this. Still, after speaking with Utzinger for hours, the interrogator concluded that he was essentially an honest and even somewhat idealistic person: "an extremely able and personable young man who was the product of his era and who acted according to his own lights."

Becker arranged to meet Utzinger in a café outside of Rio to discuss the urgent need for reliable clandestine radio stations. Utzinger's first impression of Becker was unfavorable. He saw Becker as a man "with very little education and few moral scruples in pursuit of his ends." But Utzinger would come to respect Becker over the next several years as the two men worked to spread fascism across South America. In their separate realms of expertise—Becker in espionage, Utzinger in radio—they approached their jobs with the pride of craftsmen. Becker had the contacts and the vision. Utzinger had the technical skill. Soon the *Hauptsturmführer* and the *Funkmeister* would prove to be the most dangerous Nazis in the West.

In the beginning Elizebeth knew the *Hauptsturmführer* **and the** *Funkmeister* only by their aliases, "Sargo" and "Luna."

She first encountered these names in the late spring of 1941, at the tail end of William's hospitalization in the mental ward at Walter Reed. This is when the listening stations of the coast guard and the FCC provided Elizebeth with the first of thousands of intercepts from clandestine transmitters in South America, and she started doing what she had always done: smash the codes, recover the plaintexts, translate them into English, type the translations on fresh sheets of paper (decrypts) ready for study and dissemination, sift the decrypts for clues about the secret identities of the spies, keep immaculate records; build an archive, a library of enemy words.

The original messages had been written in German, Spanish, and Portuguese, and to recover the plaintexts and translate them, Elizebeth worked closely with her lead coast guard linguist, thirty-two-year-old Vladimir Bezdek, a handsome Czechoslovak army veteran with black hair and high cheekbones. Born in Czechoslovakia, Bezdek had escaped to America when the war broke out by sneaking onto a ship. He spoke eight languages fluently: Czech, German, English, French, Polish, Latin, Italian, Russian. He read dictionaries in his free time, for fun, so of course he and Elizebeth got along, checking in with each other throughout the day, puzzling out bits of language together.

It appeared that the Nazis had at least three separate clandestine radio stations up and running in South America. Two were in Brazil, on the eastern coast of the continent, and one was in Chile, on the western coast. The Brazilian stations were in Rio de

Janeiro and a suburb of São Paulo, about two hundred miles south of Rio. All three stations exchanged wireless messages with either Berlin or Hamburg.

Elizabeth gave each radio circuit an alphanumeric label to keep them straight, like 2-B or 3-A. The label was typed on the top of every decrypt from that circuit, beneath the word "S E C R E T," along with the date and time the message was sent, the original language (German, Portuguese), the radio frequency in kilo-cycles, and sometimes the first few groups of the message's raw ciphertext. Below the header came the plaintext message itself, in English, followed by three lines at the bottom identifying the de-crypt as a coast guard product: "CG Decryption," "CG Transla-tion," "CG Typed," the date it was typed, a serial number unique to the message, and the phrase "German Clandestine."

The code names of the suspected Nazi agents were always typed in capital letters, to make them stand out and help every-one on the team get familiar with this strange cast of characters scurrying across the continent next door. You had to get to know your adversary, to see into men's hearts and predict their behavior from a running conversation of potentially enormous stakes that no one else in the world was watching except you. If Elizebeth picked up an inch-thick stack of decrypts and flipped through them quickly with her thumb, as if shuffling a deck of cards, she could see the names of the Nazi agents flick past, a blur of SARGO SARGO SARGO LUNA UTZ ALFREDO LORENZ LUNA ALFREDO LUCAS ALFREDO LUCAS SARGO SARGO SARGO.

The number of times a certain name appeared in the messages was a rough indicator of that person's importance. SARGO ap-peared again and again on the decrypts. He also seemed to call himself SARGENTO, or JOSE, or JUAN. Elizebeth guessed that he was a Nazi spy chief of some kind. The individual known as LUNA tended to speak about technical issues, the details of radio transmitters; Elizebeth pegged him as a radio expert. He went by UTZ in addition to LUNA.

There was a third man in the messages, ALFREDO, who of-ten mentioned his dealings with the other two—ALFREDO, a trusted colleague of SARGO and LUNA—as well as some other names, like HUMBERTO. For Elizebeth, seeing a name like HUMBERTO was a piece of luck, because it was longer and contained some less-frequent letters, like *M* and *B,* and it repeated across multiple messages as a predictable signature. It was a "crib," a piece of repeating text that gives the codebreaker a foothold. A British colleague of hers once said, "When you get a man with a nice long name with about twelve syllables, it can be of the greatest help to us." If Elizebeth could solve for HUMBERTO, she was well on her way to breaking the rest of the code.

At first the spies in South America were using book ciphers. Elizebeth solved them. She watched these men talk and plot and share information: reports of Allied ships in the Rio harbor, political developments in the United States, information about shipments of ores and weapons and beef, the health of crops, the number of planes being built in American factories. In September 1941, the agents switched to a grille-like cipher, and Elizebeth penetrated that, too. After she solved a message and the clerks typed the decrypt, Elizebeth and the other codebreakers and translators would perform a preliminary level of intelligence analysis, lightly marking up the decrypt with colored pencils, calling attention to proper names and places with check marks and sometimes stapling a handwritten note explaining who the speakers were, what function they served in the network, and what they seemed to be discussing. Then the decrypts had to be transmitted to other agencies: army intelligence (G-2), navy intelligence (OP-20-G), the State Department, the British. Another line was added at the bottom of the decrypt, sometimes in pencil, indicating its destination.

Regardless of a message's content, the coast guard provided copies of every solution from the South American circuits to FBI headquarters, at the request of J. Edgar Hoover. The bureau's newly created Special Intelligence Service then circulated the coast guard decrypts throughout the hemisphere, sending them to SIS agents on the ground in South America, giving the agents a leg up on their quarry.

Throughout 1940 and the first half of 1941, the coast guard was pumping solved puzzles to the FBI on a steady basis, dozens per week, hundreds of messages on each clandestine network and ultimately thousands taken all together. Yet this relationship between the coast guard and the FBI only went in one direction. SIS agents in South America never sent useful information or evidence to the coast guard codebreakers. Worse, the FBI systematically obscured all traces of the coast guard's deep involvement in the spy hunt. When Elizebeth sent them a decrypt, the FBI placed it in their own SIS filing system, with a new four-digit identifying number, and the FBI invented new names for the radio networks that Elizebeth had already named.

This is how the history of the Invisible War would become distorted; these are the small decisions that erased Elizebeth from the record and later allowed J. Edgar Hoover to take credit for her achievements. "A considerable amount of the investigation conducted relative to these espionage groups was based on information obtained from the messages transmitted to and received by the clandestine stations," the FBI wrote after the war in a three-volume history of the SIS. "The technical facilities of the Bureau were used to monitor the several German transmitters, and by analysis and coordination of information obtained from the decodes of the messages, furnished by the Technical Laboratory, and the intensive investigation by SIS representatives, the persons referred to in the messages were identified, their cover names ascertained, and their associates were established."

This is highly misleading. The decrypts were indeed "furnished" to agents in the field by the FBI Technical Laboratory, *after the coast guard had furnished the solutions to the Technical Laboratory.* The evidence is on the original documents themselves. Before the coast guard sent the FBI a decrypt, the coast guard clerks typed "SIS Dupe" at the bottom of the sheet, beneath the line that said "CG Translation" and "CG Decryption." These once-secret files, located in the National Archives and finally declassified in 2000, prove that the coast guard, not the FBI, solved these Nazi radio circuits.

Hoover's stinginess on these South American matters was difficult for the coast guard to understand, especially since the coast guard was simultaneously assisting the FBI with a massive spy investigation inside the United States. It centered on a South African man living in New York City, Frederick Joubert Duquesne, a big-game hunter with dark, floppy hair and a grudge against the British dating back to the Great War, when he was arrested carrying a file of newspaper clippings about bomb explosions on ships.

Bureau personnel called the case "the Ducase," riffing off the pronunciation of Duquesne. Several times a week during the spring of 1941, Duquesne went to an office on Ninety-second Street in Manhattan and met with another German spy, William Sebold, exchanging sensitive information about U.S. military capabilities and discussing the activities of more than thirty confederates who had been recruited as spies by the two men. Sebold used a clandestine radio station in Long Island to transmit the information to Hamburg, the messages encrypted with the book cipher based on *All This and Heaven Too,* which Elizebeth had already broken. What Duquesne didn't know was that Sebold was secretly working for the FBI as a double agent. Video cameras in the walls of the office were capturing him on tape, and the radio transmitter in Long Island was controlled by the FBI, which altered the information before sending it to Hamburg.

The bureau reached out to Elizebeth and the coast guard when the FBI radioman in Long Island received an unexpected request from Hamburg: Could he use the Long Island station to relay messages from Nazi spies in Mexico? The clandestine station there wasn't powerful enough to transmit all the way to Hamburg. The FBI agreed, but when the messages for relay started to come, they were in an unknown code.

Elizebeth broke it. The spies in Mexico turned out to be MAX and GLENN, the same agents she had tracked a year earlier.

She gave these plaintexts to the FBI and kept solving new messages sent from Long Island to Hamburg. By the summer of 1941 her team had decrypted hundreds of notes exchanged by Duquesne and the other members of the ring. These messages not only provided hard evidence against the spies that could be used in court; they also revealed links between the spies in New York and Nazi agents in South America and Mexico, pointing the FBI to suspects they hadn't known about before. The coast guard's patient codebreaking, combined with the FBI's surveillance footage and the cooperation of the double agent William Sebold, led to what J. Edgar Hoover called "the greatest spy roundup" in U.S.

history, a series of raids in June 1941 conducted by ninety-three FBI agents and sweeping up Duquesne and thirty-two members of his ring. Nineteen pleaded guilty to espionage charges and the remaining fourteen, including Duquesne, were put on trial three months later in Brooklyn. President Roosevelt followed the trial closely; if the time came for America to declare war, he needed to know there wasn't an enemy spy network on U.S. soil, able to perform sabotage. After six weeks of sensational testimony by FBI agents and Duquesne himself, all defendants were convicted, and the thirty-three spies were sentenced to three hundred years collectively.

The wild success of the "Ducase" had two large and lasting effects on America. The first was that it discouraged future Nazi attempts at spying within the borders of the United States. The second was that it made J. Edgar Hoover a legend. Hollywood later filmed a movie about the Duquesne spies, *The House on 92nd Street*, in close cooperation with Hoover himself. The Ducase "gave birth to the popular cultural belief that the Bureau was the nation's first line of defense against foreign and domestic espionage," writes the former FBI counterintelligence agent Raymond J. Batvinis. "It launched the popular myth of Hoover as the guardian of 'the American way of life.'"

Elizebeth, who received no credit for her contributions to the Ducase, wasn't nearly as impressed with the FBI's performance. It bothered her that FBI agents had described the spies' cryptographic practices in detail at the trial: "The FBI exposed the secret messages and methods without as much as asking a by-your-leave from the Treasury Department where the solutions and systems were achieved," she wrote later.

It seemed obvious to her that the FBI was too cavalier about publicity, and just as obvious that she couldn't do anything about it. The bureau was more powerful than the coast guard. What Hoover wanted, Hoover got, and that fall, as the Nazis marched on Moscow and the U.S. government shifted toward a war footing, Hoover continued to demand the coast guard's decryptions of spy messages, and Elizebeth's team continued to provide them.

During the final weeks before Pearl Harbor, October and November 1941, she could feel herself losing control of her codebreaking team. The military was starting to take over civilian functions. On November 1, a day after a Nazi U-boat destroyed an American ship off the coast of Ireland, killing more than one hundred sailors, Roosevelt signed an executive order declaring that the coast guard was no longer a Treasury agency. Instead, effective immediately, the coast guard was part of the U.S. Navy, and all coast guard personnel were subject to the authority of Navy Secretary Frank Knox. Basically, with a stroke of the presidential pen, Elizabeth and all her colleagues had been drafted into the navy.

She had no objection to working for the navy per se, but she was convinced that moving the team out of Treasury would disrupt their work and harm their effectiveness. Elizabeth complained to a Treasury undersecretary, Herbert Gaston, and Gaston relayed her objection at a Treasury staff meeting on November 5, 1941, in Secretary Henry Morgenthau's office.

The group that day consisted of thirteen men and Morgenthau's secretary, Henrietta Klotz. They gathered around his desk. The Oval Office in the White House was visible through a nearby window. Morgenthau could sometimes see the silhouettes of FDR and visitors moving around, and flashbulbs popping.

At 10:45 A.M., the men began to talk about income tax rates, agriculture legislation, the price of automobile tires, and the recent federal conviction of Nucky Johnson, the criminal boss of Atlantic City, of tax fraud. When it seemed like every issue had been addressed, every nugget of gossip shared, Morgenthau said, "Anything else?"

"One matter that doesn't much belong to me," Gaston said, "but since I took it up—I mentioned it to you on the phone, so I will mention it again. That is the matter of Mrs. Friedman."

Gaston, a birdlike man with wire-rimmed spectacles and a fine suit, explained to the group that Elizabeth wanted to stay at Treasury.

"She is very discontented about the prospect of having to work for the navy. She is very rebellious and gloomy."

No one was sure what to do. Should Morgenthau call up the secretary of the navy, Frank Knox, and ask for Elizabeth back? Should Treasury fight to keep her, or let the navy have her?

Gaston said, "She has a good organization. Perhaps it is as

good as there is in the government, as you know, on cryptanalysis." He added that Elizebeth "would have some usefulness" if she stayed in Treasury and left the wartime spy-catching work to the army, navy, and FBI. She could help the T-men investigate bank accounts controlled by the Axis powers, for instance.

Morgenthau said that whatever happened, he did not want to answer angry questions from a very gloomy and rebellious woman. "I just don't want it to appear that I am taking the initiative."

Then Harry Dexter White spoke up. Forty-nine years old and balding, dressed in a three-piece suit and owlish spectacles, White looked every bit the economics professor he used to be, a man of charts and formulae. This impression may have been carefully cultivated. Documents unearthed in the 1950s would link White to Russia's top intelligence agent in Washington. Some historians believe that during the 1930s and '40s, White was spying for the Kremlin from within Treasury's innermost sanctum.

In the meeting, White gave his opinion on the matter of Mrs. Friedman. "I don't like to butt into this, Mr. Secretary, but I understood that she is one of the best in the country, is that correct?"

Gaston replied, "She says her husband is very much better than she is, but I think she is very good."

Morgenthau ended the meeting with an intention to talk to Elizebeth about the situation. He never did, though, because a month later, on December 7, 1941, the Japanese attacked the U.S. naval base at Pearl Harbor.

When news of the bombing reached the Friedman home, William started pacing and stammering under his breath that he did not understand. Elizebeth heard him say, "But they knew, they knew, they knew," over and over.

He left immediately for the army's cryptologic bunker inside the Munitions Building. Personnel started to stream in and mill around. The colonels wore a variety of expressions, some red-eyed and worn, others strenuously poker-faced. They heard that two thousand Americans had been killed, maybe more, including

1,177 crewmen aboard the battleship USS *Arizona*, incinerated by an armor-piercing bomb that had burrowed its way into the forward ammunition hold. Twenty-one ships sunk, almost two hundred planes destroyed. A good portion of the Pacific Fleet lay at the bottom of the ocean.

Over the next few days, more than one codebreaker wrote his will. They witnessed the documents for each other.

The codebreakers had known for days, if not weeks, that a large Japanese attack was coming. William and the rest of his team had seen the MAGIC intercepts. It was obvious from MAGIC that Japan had been poised to strike; the only mystery was where. What surprised William on December 7 was not the attack itself but the location. He thought it would happen in Manila, not Pearl Harbor.

In the years that followed, William would become obsessed with the question of what went wrong. He analyzed thousands of pages of Pearl Harbor documents and wrote a three-volume report that boiled down to this: MAGIC had strongly indicated an attack on December 7, but the decrypts had gotten bottled up through a series of farcical missteps in the dissemination stage of the process, and U.S. leaders weren't alerted to the danger in time to take action. It was nuanced: The crucial MAGIC decrypts had been slow to arrive in Pearl Harbor partly because the military hadn't given the Pearl Harbor commanders a Purple machine of their own, a direct tap into the MAGIC fire hose. This decision had been made out of a reasonable desire to limit the distribution of Purple machines in order to minimize the chances of the Japanese learning about the MAGIC secret.

It was a prime example of the brutal choices that codebreakers must live with. Do you take risks to keep a secret that may save hundreds of thousands of future lives, or do you expose the secret to save a small number of lives right now? William once referred to this broad dilemma as "cryptologic schizophrenia," adding, "What to do? Thus far, no real psychiatric or psychoanalytic cure has been found for the illness."

Cryptologic schizophrenia may have explained an unusual personal interaction that the Friedmans had on the day of the attack, December 7. That evening, after work, Elizebeth and

William were at home when they heard the clack-ack-ack of their elephant door knocker. They found a red-faced British man on their doorstep: Captain Eddie Hastings, the BSC officer who had pushed America to create a new spy agency, the Office of the Coordinator of Information. Elizebeth had gotten to know Hastings, and so had William.

According to a declassified NSA report of a postwar interview with Elizebeth, what Captain Hastings did next would become "one of the most vivid recollections of her life." Hastings wobbled into the Friedman home and sat down. He mentioned Pearl Harbor. Then he started to laugh. The attack had just been announced a few hours earlier on the radio. Elizebeth looked at him, baffled, as he kept laughing. "Mrs. Friedman was shocked and offended," reads the NSA report. "Apparently Hastings found the surprise element of the attack amusing. Nevertheless their friendship continued."

Maybe Hastings was giddy from the stress of the day. Maybe he really did think it was darkly funny that MAGIC, for all its power, couldn't save the American sailors and pilots at Pearl Harbor, that the Americans could almost literally read the minds of Japanese leaders and yet fail to prevent a huge Japanese attack. Elizebeth would never understand the British man's laughter. It was one of those mysteries in the intelligence profession that leaves you to dangle, that you think about years afterward, that comes back to you in calm moments on a plane or in your bed at night, making you realize that as much as intelligence seems to be about knowing things, about gaining power through knowledge, it is just as much about not knowing them, or getting them wrong, or seeing other people get them wrong, and having to go on living with the uncertainty, with the not knowing, and thinking about what might have been.

"Yesterday, December 7th, 1941—a date which will live in infamy— the United States of America was suddenly and deliberately attacked by naval and air forces of the Empire of Japan."

Less than twenty-four hours after the bombings, President Franklin Delano Roosevelt stood at a podium in Congress, speaking into a bouquet of microphones, asking for a declaration of war. His son, James, stood next to him in his marine uniform. Earlier in the day, James and his father's aides had fitted his paralytic legs with the three metal braces that were required to support the president when he could not be seen in his wheelchair. Sixty-two million Americans listened to the speech by radio. It lasted seven minutes. Within an hour, Congress had authorized war with Japan. Three days later, Germany declared war on America.

Before a single regiment of U.S. soldiers set foot on European soil, the war changed American culture. It was a stomach that ingested a large diverse nation and started breaking it down into widgets. Hollywood movies and Disney cartoons were about the war now. Business was about the war. Work was about the war, and school was about the war. It was the only time before or since when Americans became emotionally invested in the idea of self-deprivation and frugality. Third graders roamed their neighborhoods in packs, gathering scrap materials, tires, and paper and cooking fat and old sneakers whose soles could be sacrificed for the rubber. The Big Three automakers stopped making cars and started making planes. Factory workers took secrecy oaths. Everybody had a secret now. The government issued ration stamps for eggs, milk, bread, gasoline, contained in ration books, manila-colored pamphlets. Elizebeth's ration book listed her height as five foot three and her weight as 120 pounds.

After Pearl Harbor all government matters were urgent and the military didn't want civilians in charge of sensitive functions. The coast guard decided to appoint a new chief of the Cryptanalytic Unit, Leonard T. Jones, a young lieutenant who had taken an army training course in cryptanalysis. Just like that, Elizebeth was demoted from Cryptanalyst-in-Charge to mere Cryptanalyst. She was no longer the leader of the team she had invented, staffed, trained, and nurtured.

This upset her, but she remained the unit's civilian commander and thought Jones showed promise as a codebreaker, so she didn't complain, and anyway, it wasn't like she had a choice. Men told her what to do, and her services were in high demand. Every few

days someone was calling up Henry Morgenthau, wanting to borrow Mrs. Friedman for various cryptologic tasks.

The battle for her attention rose to the highest levels of Washington. She got to know James Roosevelt, FDR's son, and William Donovan, a tall, irascible former army colonel with a manic
personality, whose soldiers used to call him "Wild Bill." FDR had
asked his son to help Donovan launch the Office of the Coordinator of Information, the spy organization that would become the
OSS and later the CIA.

Donovan was starting from zero, in borrowed office space.
One of the first things a spy agency needs is a way to communicate securely with its people in the field. It needs codes and
ciphers, and mechanical aids to generate them, and clerks to
write them, and training for the clerks. Donovan didn't have any
of this, and didn't know the first thing about codes or ciphers,
so James Roosevelt approached Elizebeth to lend her expertise,
and Donovan reinforced the demand by sending a letter directly
to Henry Morgenthau that requested Elizebeth by name, citing
an "urgent need for her services pending the establishment of
our permanent code section" (Morgenthau grumbled at a staff
meeting, "He wants Mrs. Friedman"). This became her first mission after Pearl Harbor. Detailed to Donovan's office on a temporary basis, she spent three and a half weeks creating the first
permanent cryptographic section for the proto-OSS and proto-
proto-CIA.

She built it from scratch, making alphabet strips and other aids
to generate ciphers, obtaining hard-to-find cipher devices through
navy channels, installing the machines, and customizing them
according to the new agency's needs. She interviewed potential
cryptographic staffers and made recommendations to Donovan,
which he ignored, treating her the whole time like a servant and
failing to appreciate basic principles of communications security.
When the job was complete, Elizebeth wrote him a seethingly
polite letter that conveyed her feelings between the lines, her horror that an important national function was going to be directed
by a man who struck her as foolish and cavalier (Donovan's OSS
would be defined by recklessness). She sent the letter via James

Roosevelt to make sure Roosevelt was aware of Donovan's short-
comings as a guardian of information:

> *My experience and observations during my temporary duty*
> *with your organization, lead me to make the following*
> *recommendations:*
> *—That the representatives going to the field in every case be*
> *required to spend sufficient time to become thoroughly* drilled
> *in the systems of communication provided for them. This drill*
> *and resulting mastery cannot be accomplished in a few hours.*
> *It should extend for a few hours daily over a minimum of five*
> *days, and with certain types of mind a longer time will be*
> *required.*
> *—That a general indoctrination in and discussion of and*
> *handling of classified information be undertaken throughout*
> *your organization. . . . This matter of indoctrination is a long*
> *and difficult process. . . .*

She signed the letter "Dr. Elizebeth Smith Friedman" to under-
score her credentials. (Her alma mater, Hillsdale College, had
awarded her an honorary L.L.D. in 1938.) Then she returned to
her own office, to her trusty desk and her fine coast guard colleagues, relieved to be back "home."

Fresh piles of intercepts from South America awaited her there,
and once again she dug in, eavesdropping on the latest activities of
the spies. "Sargo" and "Alfredo" still appeared to be in charge of the
network there, synthesizing information from their agents across
Brazil and Chile and sending reports to Germany over the radio,
but there was a new strain of malevolence in their messages. After
Pearl Harbor, Brazil had declared solidarity with America, and the
Nazis responded by going after Brazil, firing torpedoes at Brazilian
ships for the first time. The positions of the ships were provided by
"Sargo," "Alfredo," and their men. Outraged Brazilian authorities
moved against German businesses. "Measures against members of
the Axis are assuming drastic form," one spy in Brazil radioed to
Germany. "Bank deposits already blocked. We are destroying all
compromising documents, maintaining radio operation as long as

possible. Heil Hitler." In January 1942 Hitler launched Operation Drumbeat, a coordinated U-boat assault on American and British merchant ships carrying war supplies, and "Sargo" and "Alfredo" helped with this effort, too. In three months the ruthless U-boats sent one million tons of material to the bottom of the sea and by summer 1942 the U-boat captains had murdered five thousand Allied seamen. "All along the Atlantic coast," writes the historian John Bryden, "Americans could look out and see plumes of smoke by day and red fires by night." The messages Elizebeth solved were dense with detail about Allied vessels coming and going in South American waters:

> MARCH 14, 1942 AT 0038
> Departed Montevideo: 4th (American SS) F.Q. *Barstow* to Curacao and (American SS) *Western Sword* to USA. Departed Rio de Janeiro: 11th (American SS) *Ruth* to Baltimore; 12th (American SS) *Lammot du Pont* to Buenos Aires. Arrived Rio de Janeiro: 12th (American SS) *Delmar* from New Orleans; and 13th (British MS) *Devis* from Glasgow.

Elizebeth passed these decrypts along the chain as quickly as she could, knowing that Nazi U-boats might already be hunting any of these U.S. or British ships and hoping that the Allied captains could be warned.

In the first weeks of March, Elizebeth also solved a sinister series of intercepts given to her by an FCC listening station on the coast of Rhode Island. The messages suggested the Nazis were preparing to destroy a troopship, the RMS *Queen Mary,* that was carrying 8,398 American servicemen:

> MARCH 7, 1942
> On board Queen Mary, Indians, Americans, Englishmen, tanks, disassembled airplanes. Came from Dutch Indies via South America.

> MARCH 8
> Queen Mary departed on March 8 1800 local time.

MARCH 12
The Queen Mary on the 11th at 1800 MEZ was reported by the ship Campeiro on the seas (near) Recife.

MARCH 13
The Queen Mary on the 12th at 1500 MEZ was reported near the coast at Ceara in the direction of Belem through Piratiny.

MARCH 14
Concerning Queen Mary, the troops of young people of white race number seven to eight thousand men.

As it turned out, Hitler had placed a bounty on the *Queen Mary:* any U-boat captain that destroyed her would win the Iron Cross with Oak Leaves and one million Reichsmarks. Elizebeth's decryptions (and similar decryptions provided by other Allied codebreaking units) were quickly shared with the *Queen Mary* captain, who was able to take evasive maneuvers, sneaking past a U-boat that was lurking in wait and saving the lives of more than eight thousand U.S. troops and his crew. It was a good example of why these clandestine radio circuits were important: as long as Elizebeth kept solving the messages, she could see danger coming, and America had an edge.

Which is why she grew increasingly confused in February and March when the spies started talking about being chased by police. Spies in Chile reported that their "hiding place has been searched three times." The men in São Paulo told Berlin that the temperature was 31 degrees Celsius and "getting worse"—in other words, they were feeling heat from police—then their transmitter went eerily quiet. In Rio, the Abwehr spy chief Albrecht Engels radioed, "Throughout country sharp police action against Germans." On March 17, 1942, he told Berlin that the docked Swiss ship whose radio they sometimes borrowed, the SS *Windhuk,* had been raided by Brazilian police, the second officer drowned in a struggle and the rest of the crew imprisoned.

Elizebeth, noticing this sharp uptick of panic in the messages she solved, guessed that authorities in Brazil and possibly Chile were conducting some kind of spy roundup, arresting Nazi agents and seizing their radio equipment. She couldn't tell if the FBI was leading the effort, local police, or a combination.

Whatever the case, it wasn't good. She and everyone else at the coast guard felt strongly that now was not the right time to move in and make arrests. The codebreakers were learning so much about the larger structure of the Nazi networks, more and more each day, and if the spies figured out that they were being chased because their codes had been broken, they would surely switch to new codes, perhaps stronger ones. And until Elizebeth managed to break the new codes, which could take weeks or months, depending on the level of difficulty, the Allies would be blind to Nazi activities across the continent. All of South America would suddenly go dark. If the spies started targeting another U.S. troopship like the *Queen Mary*, officials might not be able to warn the ship before a torpedo ripped into its hull.

A veteran codebreaker like Elizebeth understood these things. But the FBI, new to this line of work, did not, and before Elizebeth could figure out what was going on in South America with the police action and stop it from happening, FBI agents there grabbed the golden goose and cut off its head.

"Don't! You'll blow the house up!"

Josef Starziczny, a.k.a. "Lucas," the Abwehr agent in São Paulo, told the Brazilian detective to drop the suitcase. The detective set it down gently.

It was March 15, 1942. Elpido Reali had come to this house in the suburbs of São Paulo armed with a search warrant, intending to arrest a man he had been told was a Nazi spy. He knocked, entered, and immediately saw a spy camera, telephoto lenses, a darkroom, and a radio receiver. Starziczny's mistress was here, wearing a dressing gown, a confused look on her face.

"That suitcase," Reali said. "You said it will blow the house up." Starziczny shook his head. No, there was no bomb inside.

Reali popped the latch and saw a portable radio transmitter.

A Kriegsmarine code book tumbled out. Starziczny had been using the codes to send the coordinates of Allied ships to German U-boats. Seeing the code book, the spy reached for a revolver on a nearby shelf, apparently planning to kill himself—"The Gestapo will never forgive me"—then thought better of it and allowed Reali to take him to the police station.

Soon after the arrest, in Rio, Albrecht Engels, a.k.a. "Alfredo," phoned Starziczny's house in São Paulo. Someone picked up the phone. Engels didn't recognize the person's voice. He hung up.

Engels assumed that Starziczny had been arrested and that he would not hold up under police interrogation (he was correct on both counts), and now Engels activated his emergency plan to protect the spy network he had spent months building under the guidance of Johannes Siegfried Becker, the brilliant SS captain. He arranged to move the radio transmitter in Rio to a new location, gave his code book and $89,000 in cash to a confederate, and fired off a string of wireless messages to Berlin warning them that the network was in danger, the last of these on March 18:

> MEYER CLASEN *in Porto Alegre arrested denounced* LEO *and* ARNOLD *thereupon* ARNOLD *arrested in Sao Paulo and transferred to Porto Alegre. I fear that* MEYER *(has) also given away (denounced) the radio procedure therefore I shall lie low until further notice.*

That day Engels was arrested by the Delegacia de Ordem Politica e Social (DOPS), the Brazilian federal police. They took him to one of their prisons, threw him into a dark cell with no toilet, told him to confess, and kept him there for weeks, punishing him with frequent interruptions to his sleep.

The DOPS took almost ninety members of the spy ring into custody over the next two months, at the insistence of the FBI's Jack West, the bureau's top man in Brazil, the head of its SIS operations across the country, and its legal attaché in Rio. West believed that the intensifying torpedo attacks on Allied merchant ships in the Atlantic meant it was time to take action against the spies who had been providing coordinates to the U-boat captains, and the bolder the action, the better. While the DOPS rounded up the spies, a young FCC employee named Robert Linx drove around Rio in an automobile full of direction-finding equipment, telling police where to find the clandestine radio transmitters, which were then seized and impounded.

According to *The Shadow War,* a 1986 account by the historians Leslie Rout and John Bratzel, "Jack West's conclusion was that piecemeal action was useless; a hard, sweeping blow" was the only way to take the Nazi radio stations off the air at once, to put the known spies in prison "and keep them off the airwaves."

But while the FBI's motive was sound, its tactics were questionable. The spies resisted their Brazilian interrogators. A month passed with few confessions. West grew impatient, suspecting that right-wing elements within the Brazilian police were working against him. And that's when he made the fateful decision to go above their heads. He took copies of the messages that Elizebeth and the coast guard had solved—verbatim copies of the spies' intercepted and decrypted radio messages—and showed *hundreds* of these messages to the president of Brazil, the foreign minister, and the air force minister. J. Edgar Hoover later confirmed that Allied agents "delivered the complete information" about the radios and the messages to the Brazilian government.

The gambit had the desired effect: Brazilian police started to get tougher on the prisoners. The men were stripped naked and questioned. Some police officers showed them pages of decrypts (Elizebeth's decrypts) and demanded that the prisoners fill in a handful of missing words. At least two men were beaten until unconscious. One had his fingers dislocated. One was repeatedly kneed in the scrotum while naked, and burned with cigarettes. One was interrogated in a six-by-three-foot cell with no bed. Police in São Paulo questioned one naked suspect for two days straight, pouring cold water on his skin and blasting a high-speed fan in his face until he lost control of his mental faculties. A few prisoners resisted. The Abwehr spy Friedrich Kempter went on a hunger strike after being given a plate of food full of rocks; the FBI grew concerned that Kempter would become too weak to talk and finally arranged a meal of steak and french fries. This was a rare occasion when the FBI intervened to stop brutality or torture; the rest of the time they either participated or looked the other way, in a grim foreshadowing of the bureau's future misadventures in Latin America during the 1970s and '80s.

The FBI's plan didn't work. The roundup failed. To deliver a deathblow to the Nazi network, to keep the spies off the airwaves, the FBI needed to get all the spies at once. But they didn't. Becker, the SS captain, remained at large—the FBI knew little about him anyway. Gustav Utzinger, the radio expert, also got away. He air-mailed one of the radio transmitters to Paraguay and boarded a Brazilian ship on a phony passport, picking up the transmitter in the Paraguayan capital of Asunción.

Even worse for the Allies, Albrecht Engels, the Rio businessman-turned-spy, was able to smuggle three long letters out of prison by passing them to the visiting wife of a colleague. In these letters he described the brutality of the police and alerted Berlin that the spies' codes had been broken and must be changed.

Berlin sounded the alarm across the clandestine network, telling all stations to interrupt communications with Brazil. "Warning," they radioed the spies in Chile. "Alfredo arrested. Take all precautionary measures, above all, separate."

Engels, trapped in his Rio prison cell, found comfort in the thought that the ones who escaped could now rebuild the network and make it more secure than ever, with new codes. He knew that Utzinger was still out there, somewhere. He wasn't sure about Becker. Engels asked a fellow prisoner who was about to be freed to go looking for Becker and send Engels a pack of cigarettes if Becker was safe. Soon afterward, a pack of cigarettes arrived at the prison for Engels. He was glad. As long as Becker and Utzinger, the *Hauptsturmführer* and the *Funkmeister,* were free, there was hope.

CHAPTER 4

Circuit 3-N

```
                        S E C R E T
  From:   Argentine              HDZ    8000 K/cs
  To:     Berlin                 HDZ    8100 K/cs

  January 29, 1943               TOI 1-29-43/0033/Z on ckt. 3-N

  German                         T-5

  #567                           HAHAO SQOAL .....


       For "LEIT":  There follows the frequencies provided
  for the station "G" (GUSTAV) with identification letters
  and calls:

       "GE"  -   7560    ZPU
       "WB"  -   8000    HDZ
       "BA"  -  10800    CEN
       "GH"  -  11345    COB
       "GP"  -  12000    JUR (A)
       "GI"  -  14465    PVR3
       "BD"  -  15120    CTW
       "WO"  -   8300    TFA (V)

       Time of schedule with LEIT:  all even days at 0220 on "WB".

       We have transmitted on all wave-lengths we have, but one
  operates on the antenna which has a length of 130 meters.
  Between 1400 and 1600 tell if you hear us very well at Radio
  Berlin on 17 meters.  We are going to test this wave-length
  also.



  German Clandestine             CG Decryption (FCC& RSS)
                                 CG Translation (GEB)
  Serial CG3-877                 CG Typed 5-18-43.

                        S E C R E T
```

A cipher message from Circuit 3-N, the Nazi clandestine radio link between Argentina and Berlin, solved by Elizebeth's coast guard unit.

William Friedman's depression returned in December 1941, the month of Pearl Harbor. He had trouble sleeping and was besieged by doubts and morbid thoughts. "Flight, fight, or neurosis," he wrote on a loose sheet of paper years later during a similar period of depression, trying to describe the feeling. " 'Floating anxiety' which attaches itself to anything and everything. Fear that E. despises me for being such a weakling." It was scary for a man who prided himself on precision and rationality to feel like he was not in control of his mind or his body. He sometimes referred to this unpleasant condition as the "heebeegeebees," which he abbreviated as "hbgbs" in private notes to himself.

He did not seek help this time, did not go to a psychiatrist or check himself into a mental hospital—after his experience at the understaffed and punitive mental ward of Walter Reed in January 1941 he was not about to repeat this mistake unless completely desperate—and so, as always, the Friedmans concealed the seriousness of his condition to friends and family, and they continued to work, except for three consecutive days in the spring of 1942 when they stayed home to celebrate their twenty-fifth wedding anniversary. That was the celebration, sleeping in. It was amazing. They were so tired. Elizabeth went to the store and bought a whole chicken and some strawberries and figured she would cook their usual anniversary dinner, a simple feast of roast chicken and strawberry shortcake.

They hadn't told their friends about the twenty-five-year milestone but somehow the secret leaked, and that evening, to their delight, colleagues and friends knocked on their door, offering silver-anniversary gifts. Fred and Claire Barkley brought a sterling silver round sandwich tray; Jean Chase Ramsay wore a stunning silver dinner gown; Stub and Enid Perkins appeared with an array of flowers in a glass bowl, yellow and blue and white irises, blue delphinium, flame-colored columbine, white gypsophila. To these Elizabeth added pink and yellow roses she thought to pluck from her own rosebushes, and some white and yellow honeysuckle, too, and by the time the next-door neighbor brought two huge armfuls of his own scarlet roses, the house was dizzy with fragrance.

All day long, telegrams of congratulations arrived from friends near and far. Two of the telegrams were jokes written by William, notifying Elizabeth that she had been awarded an honorary A.B. degree from the Sorbonne, "Artiste de Boudoir," and also a D.S.M. from Harvard, "Doctor of Successful Marriage." In the second telegram he made light of his mental struggles and acknowledged his wife's patience and kindness during his periods of illness, though of course he did not use those words:

> *WHEREAS ELIZEBETH SMITH FRIEDMAN*
> *HAS CONDUCTED IMPORTANT SPECIALIZED*
> *RESEARCH EXTENDING OVER A PERIOD OF*
> *TWENTY FIVE YEARS IN THE VAGARIES AND*
> *IDIOSYNCRASIES OF ERRANT HUSBANDS; AND*
> *WHEREAS DURING THE CONDUCT OF SUCH*
> *RESEARCH SHE HAS BEEN SUBJECTED TO MANY*
> *HAZARDS INVOLVING CONSIDERABLE MENTAL*
> *ANGUISH, PERSONAL CHAGRIN, DAYS OF*
> *ANXIETY, AND NIGHTS OF SLEEPLESSNESS; AND*
> *WHEREAS SAID RESEARCH HAS RESULTED IN*
> *THE DEVELOPMENT OF ADEQUATE METHODS*
> *AND INSTRUMENTALITIES FOR THE CONTROL*
> *OF ONE HUSBAND, TO WIT, WILLIAM FREDERICK*
> *FRIEDMAN, AND HAS MADE HIM LIVABLE*
> *WITH . . .*

The children weren't there to celebrate with their parents. John Ramsay was finishing his sophomore year at prep school in central Pennsylvania, and upon graduation he planned to join the Army Air Corps and head straight to flight school. Barbara was between semesters of college and living in New York City, in an apartment on West Fifty-sixth Street, getting involved in left-ist political causes and dating an activist named Hank. "Hank is beautiful," she wrote to William, "but we're so utterly different. He lived in the slums and led a gang (because he was the tallest and the biggest) and hated cops and swam in the East River. . . . And now we go to bars and stand at the rail with the workmen and talk about Leninism."

William had no interest in Leninism but told his daughter she had a good heart. "I hope you will let nothing interfere with your enthusiasm for helping where help is needed, but don't let the slow, snail's-pace progress upward and onward get you down," he wrote. "Remember always that the dawn of man's conscience is only 3 or 3½ thousand years behind us."

He had always found this a comforting thought, that the age of barbarism was not long past, that if humans failed to be kind it was because they were still children, historically speaking, and the idea rang true to him as he read and disseminated MAGIC intercepts through the spring of 1942, learning secrets about Japanese war strategy in the Pacific and helping to guide the American response. In June 1942, with the two opposing navies speeding toward a fatal clash at the Battle of Midway, William and his codebreakers moved from the Munitions Building to a new location, Arlington Hall, a former private school for girls located on the outskirts of the city. The army had taken over the hundred-acre campus to provide room for an expanded codebreaking operation. Meanwhile, the navy started transferring intelligence personnel to a similar facility, on Nebraska Avenue, also a former private girls' school, anchored by a five-story building dubbed the Naval Communications Annex.

These two campuses soon evolved into an American version of Bletchley Park, deeply secret compounds where workers solved puzzles behind barbed wire and never spoke about what they did. Many were women. It's where the machine era of cryptology began, the era of brute force, women operating machines the size of rooms, American bombes and some of the first IBM punch-card computers.

The women of Arlington Hall and the Naval Annex were mostly WACs and WAVES, members of the army and navy auxiliary programs designed to patch the wartime shortage of male labor. They lived together in barracks and apartments. Hundreds had been trained in secret cryptology courses offered at the Seven Sisters colleges, the likes of Bryn Mawr and Vassar and Mount Holyoke, the professors relying on exercises and concepts first pioneered by William and Elizebeth Friedman. Inside the high-security buildings encircled by barbed wire and guarded by U.S. Marines some of the women sat at long rows of desks, smoking and drinking coffee, and identifying cribs to feed into the bombes, while others operated the bombes that ticked and whirred as they explored the keyspaces of distant Enigma machines. The buildings were hot and unventilated. An Arlington Hall codebreaker named Martha Waller recalled that in the summer, it was often 90

degrees indoors at 8 A.M., and because of the wartime nylon shortage the women couldn't wear nylon stockings, so "we rejoiced in
going bare. . . . Sitting quietly at a desk, one could feel drops of
sweat rolling down one's legs."

Less than a year from now the navy would force Elizebeth and
the Coast Guard Cryptanalytic Unit to move to the Naval Annex,
but even then she would never work inside the large, hot rooms
with the rows of young women at desks. The unit remained separate, a small elite team doing its own thing. From time to time
she and her colleagues would take advantage of the Naval Annex's
technology to make progress on clandestine circuits, relying on
IBM punch-card machines to perform statistical analyses that
saved time in solving certain ciphers, but the initial assaults on the
puzzles emerged from their brains alone. Elizebeth was among the
last of the paper-and-pencil heroes. And in the summer of 1942, as
the U.S. and Japanese navies clashed in the Pacific and the Nazis
ordered French Jews to wear the Star of David, she was taking on
the most fiendish challenge of her career, for the biggest stakes.

Exactly as she had feared, the Nazi spies had changed their
codes after the March police raids in Brazil. As a direct result of
the FBI's roundup, "Germany was unmistakably informed that
the systems had been solved," wrote the uniformed commander
of her unit, Lieutenant Jones, and "the inevitable consequence was
that systems on all clandestine circuits were almost immediately
thereafter completely changed." Within two or three weeks the
Nazis were back online in multiple locations across South America, and from there the radio network expanded, adding nodes in
new cities and countries.

Elizebeth, Lieutenant Jones, and their coast guard teammates
watched in frustration as new circuits lit up throughout the summer and fall of 1942—two, then five, then fifteen—each using a
different and yet-unbroken code. It was as if the FBI had tried to
destroy an approaching asteroid with a single huge bomb but instead just blasted the rock into dozens of sentient fragments able
to regenerate and spread wreckage over a wider swath of earth.

The codebreakers were hardly the only Americans troubled by
the FBI's actions in South America. Intelligence chiefs at the army
and navy couldn't believe it either. "Unfortunately, the matter got

out of hand, and it became public knowledge that the ciphers used by the espionage agents in that territory were being read by our government," wrote Joseph Wenger, head of the navy's OP-20-G, in an internal memo. "It might be much more valuable to the military services to obtain the information flowing through clandestine stations than to close them up." The British were also taken aback—they had never trusted J. Edgar Hoover in the first place—and as more newspaper stories appeared in Brazil about the court cases of the arrested Nazi spies, local British officials had the stories translated and exchanged secret telegrams about the unfortunateness of it all. "You may care to read the attached rough translation of the story as it appeared in the Brazilian Press," one diplomat in Brazil wrote to an official at MI5, Britain's counterintelligence agency. "Rather shattering, I'm afraid."

The whole fiasco triggered four months of jurisdictional squabbles between the different intelligence services, conducted by memo and conference. The army, navy, and British complained about the FBI; the FBI pushed back. These fights resulted in a series of awkward compromises. The army and navy forced the State Department to promise that no further clandestine stations would be seized without their approval, and all parties recognized that the coast guard had the authority and the expertise to monitor clandestine circuits in the Western Hemisphere. But no one could stop the FBI from doing counterespionage in South America. Hoover had clear authority from the president.

And so, unable to remove his power on paper, the other agencies simply started to freeze him out in secret, routing information so that it flowed *around* the FBI as much as possible. British relations with the coast guard "grew steadily closer and more informal," according to the BSC history. "It was understood by both sides that no information received from the coast guard was to be divulged to the FBI." People literally whispered secrets to one another whenever an FBI representative was in the room; the British observed that "valuable items of intelligence were imparted, hurriedly and sotto voce, either before the meetings began or after they had adjourned."

In April, the navy ordered the coast guard to stop disseminating clandestine decrypts itself and provide the decrypts to

OP-20-G instead, for tighter control. That month, representatives of the army, navy, coast guard, British Security Co-ordination, and Canadian intelligence met at a weeklong conference in Washington to talk shop about radio and spies. It was chaired by Commander Wenger of OP-20-G. The FBI was not invited. Elizebeth was. On the day she explained the coast guard's approach, April 8, there were seven male naval officers in the room, three male army officers, four male Canadians, five male British, three male coast guard personnel, and her—listed at the top of her cohort in the meeting minutes:

U.S. Coast Guard
Mrs. Friedman
Lt. Comdr. Polio
Lt. Comdr. Peterson
Mr. Bishop

Her Cryptanalytic Unit had grown since Pearl Harbor, but it was still fairly small, with fewer than twenty cryptanalysts, translators, and clerks. They now worked as a team to fix the mess the FBI had caused: to break the new codes on the multiplying circuits and wrangle the chaos into some kind of order.

Until the arrests in March 1942, most spies had been using book ciphers or single transposition systems (a common, Scrabble-type cipher) that were similar in style and relatively easy to break. After March, the spies switched to weirder, harder stuff: running-key systems, double-transposition systems, poly-alphabetic substitution with columnar transposition. They started hopping on the radio at different times of the day in an attempt to avoid interception. They mixed and matched cipher methods in unpredictable ways. One of the new procedures relied on "rail-fencing"—a more sophisticated application of the same principle that Elizebeth and William once used for writing love notes, only now, instead of the plaintext being JE T'ADORE MON MAR or I LOVE YOU VERY MUCH, it was a snippet of German from a Nazi spy in France: HERZLICHE WEIHNACHTSGRÜßE UND WÜNSCHE ZUM NEUE JAHR: "Warm Christmas greetings and wishes for the new year . . ."

As soon as Elizebeth and Lieutenant Jones could get a handle on one circuit and break the codes, a new one came online. It was a cryptanalyst's nightmare. Here is a partial list of the circuits the unit was monitoring by the end of 1942:

3-G	Hamburg—Valparaíso (Chile)
3-J	Hamburg—South America
4-C	Lisbon (Portugal)—Lourenco Marques (Mozambique)
4-D	Madrid—West Africa
4-F	Hamburg—Lisbon (Portugal)
4-G	Stuttgart—Libya
4-H	Hamburg—Unknown
4-I	Hamburg—Bordeaux (France)
4-L	Hamburg—Gijon (northwest Spain)
4-M	Hamburg—Spain
4-N	Hamburg—Unknown
4-O	Berlin—Madrid
4-P	Hamburg—Madrid
4-Q	Hamburg—Tangier (Morocco)
4-R	Hamburg—Vigo (northwest Spain)
4-S	Berlin—Tetuan (Morocco)
4-T	Berlin—Teheran (Iran)
5-D	Hamburg—The Crimea (USSR)

This was a planet's worth of radio signals, a one-of-a-kind view of the earth as it convulsed with war, borders blurring, power shifting hands. The Nazis had invaded Crimea, in the Soviet Union. Iran was partially occupied by British and Soviet troops. Morocco was controlled by the Nazi-collaborationist Vichy government of France. Portugal was neutral and fiercely contested by both sides. Libya had been conquered, for the time being, by Italian and German troops. The Nazis had spies and saboteurs in all these places, sending back information over clandestine radio transmitters, and Nazi diplomats and even military officers sometimes borrowed the transmitters to speak with Hamburg and Berlin in times of transition and stress. The clandestine network, though built for espionage, was really just another communications channel, a way for Nazis of all sorts to share information in a fluid situation. Nikola Tesla predicted in 1926 that "when wireless is perfectly applied the whole earth will be converted into a huge brain." The clandestine network was the Nazi brain, fragmentary but already encircling the earth, and adding new synapses at a fearsome clip.

The information flickering through the brain wasn't necessarily accurate—some circuits seemed to contain little but "a miscellany of military information partly factual and partly distorted," wrote Lieutenant Jones in a 1944 memo, "evidently pieced together from barroom conversations with merchant seamen"—but even the garbage circuits were useful to monitor. "In all cases," Jones went on, "the reading of these circuits provided assurance that if any serious leak of vital information did occur we would learn of it almost before it became known in Germany and would be in a position to provide immediate safeguards." In other words, the Nazi brain served as a kind of early-warning system for the Allies. Beyond that, the solved messages provided a wealth of information about the Nazi grasp of American military capabilities, because when Germany asked the spies for information about U.S. ballistics or antiaircraft guns, they were revealing that this was information they lacked.

Elizebeth raced to stay on top of the shifting codes, the proliferating patterns. Her worksheets grew weird, beautiful. She filled the grid squares with letters and numbers that made different geometric shapes when you stepped back and looked at the worksheet from a distance. Some of the shapes were parallelograms, some looked like stairs, others like labyrinths. She pulled mischievous letters from the sky and sorted them on the page. The invisible world was all out of whack, misaligned, and she had this set of tricks to knock it back into order.

Once Elizebeth broke the code on a circuit, she solved every message that she came across, no matter how trivial or personal. Sinister messages, personal messages, sober messages that spoke of bombs and guns and ships and submarines, messages that just seemed bizarre. (Berlin to South America: "Can you procure details about the process of making explosives from cacao?" Cacao is the main ingredient of chocolate.) She learned intimate details of the spies' lives. Berlin informed a spy in Iceland that his wife, Erika, had given birth to a healthy baby girl, Jutta. Several times a family member of a spy was allowed to transmit a personal message from Berlin. One of the wives shared Elizebeth's first name, although she spelled it in the traditional way. "My dear JOHNY: My heartiest congratulations on your birthday and thousands of loving greetings and kisses from your Elizabeth."

Elizebeth Friedman solved and disseminated these messages like all the others. Such were her weapons against fascism: pencils, puzzles, circuits, names, dates, places, check marks, handwritten notes affixed to typed pages with staplers, stacks of solved messages rising with the hours and days and weeks.

And she stepped lightly as she worked. Her name did not appear on the documents of the Cryptanalytic Unit. She wasn't the top commander anymore, so she didn't write official memos to other parts of the intelligence community (Lieutenant Jones did that), and she didn't meet directly with FBI officials (Jones did that, too). Jones's name was the name typed on memos about clandestine radio traffic that circulated within the navy's OP-20-G. And although Elizebeth was not shy behind closed doors, sometimes quarreling with Jones about the direction of their work, disagreeing about which puzzles were more urgent or less urgent to tackle (she found his judgment clouded sometimes by careerism, a hunger for promotion that was irritating), she didn't mind being anonymous on the page. Her experiences as a cryptologic celebrity in the 1930s had convinced her that in this secret world, attention was a kind of poison. She said after the war that she considered herself "one of the workers" in the Cryptanalytic Unit—she was certainly being paid like a worker, not a leader, earning $4,200 a year, equal to $63,000 today, as a P-5 civil servant, a middle classification of government employee—and this combination of Elizebeth's tendency to minimize her own contributions and Jones's role as commander is one reason that her role in the war would go undiscovered for so many years.

Still, unavoidably, she left fingerprints in the records, little traces of her labor. Her initials, ESF, were typed at the bottom of some coast guard decrypts. Her handwriting appeared on many notes stapled to the decrypts, preceded by a code name, "GI-A," that meant "analyst" and was shared by others in the unit, and sometimes she wrote directly on the decrypts themselves, a characteristic burst of red or blue colored pencil next to a particularly crucial name or phrase, scribbled in the distinctive slant of Elizebeth's hand that had not changed since she was a college student writing in her diary.

At the end of the day, the unit was hers, not Jones's. She was the one who first proposed its creation in 1930. She staffed it. She trained the cryptanalysts. She guided its work over the years, building it into a powerhouse. This gave her an informal authority. People who had been paying attention knew that she was the beating heart of coast guard codebreaking. In December 1942, two British intelligence liasons met with Elizebeth in Washington to talk about closer coordination between Bletchley Park and the Coast Guard Cryptanalytic Unit. One of the liasons, Major G. G. Stevens, described the meeting in a "Most Secret" cable to his superiors in Britain and suggested that they arrange for a coast guard representative to visit Bletchley. "For this one would like to see Mrs. Friedman go," the British official wrote, "but probably at this end"—i.e., the Washington end—"it would be considered more suitable to send Lieutenant-Commander Jones, the official head of the section."

Given the tension of her job, the pressures of extreme secrecy, and the bleary afternoons stacking alphabets, Elizebeth was glad for her friendship with F. J. M. "Chubby" Stratton, the British astronomer and radio expert who looked like Santa Claus. He was a calming presence. He had a habit of appearing at Elizebeth's desk at unexpected times. She would be at her desk, nose buried in some problem, and look up to see him there, smiling. Without ever revealing anything about himself, he had a way of making you feel like you had known him forever, and that everything would be OK. And, of course, he was brilliant—a bit of a tinkerer. He loved the challenge of locating hidden radio stations. He had invented a device he called a "snifter," a little piece of electronics that fit in the pocket and allowed a person on foot to covertly pinpoint the exact location of a pirate radio transmitter within a building, accurate to the floor. He managed a staff of direction-finding experts who cooperated with the FCC to narrow down the location of pirate signals.

Through terrific effort, teamwork, and long hours, the coast guard codebreakers and their partners finally regained mastery over the clandestine circuits after an unpleasant period of blindness. By winter 1942, Elizebeth's eyes had adjusted. In the dark, lights out, she was watching them now, one letter of the alphabet at a time. She had the Nazi brain in a jar on her desk, alive and glistening, electrodes running out to her pencil. The unit had conquered every new circuit, every new code that came online that year.

Except one.

The circuit exchanged its first messages on October 10, 1942. The messages seemed to resist solution. She wondered if it might be an Enigma circuit, the messages encrypted by an Enigma machine of some kind.

She called it Circuit 3-N.

Presumably the messages on Circuit 3-N were sensitive enough to require a stronger-than-usual code, so she guessed that the messages must be important. She was more correct than she knew: in the end, the fate of the Invisible War would turn on the sinister frequencies of Circuit 3-N.

For now the anxiety of an unsolved puzzle was more than enough to motivate her. Here were some messages she couldn't read and she wanted to read them. The code simply had to be smashed.

At weekly radio intelligence meetings with her British and FCC colleagues, she talked about Circuit 3-N, and all the agencies chipped in with clues. The FCC determined that one of the stations was located in Europe and the other in South America, and the British confirmed the FCC's bearings.

New intercepts from Circuit 3-N arrived in Elizebeth's office each week. By December 1942 she had accumulated twenty-eight encrypted messages. A cursory analysis showed telltale signatures of an Enigma machine.

Elizebeth and the coast guard had already solved one Enigma, back in 1940, a commercial Enigma whose wiring scheme was already known. Now they were up against an Enigma that would turn out to be a machine of unknown wiring, never before solved. It was the real thing in the wild. She had to find a way in.

The Argentine sun felt good on his face and warmed his scalp through his thin layer of brown hair. Gustav Utzinger liked living in Buenos Aires. It was a beautiful old city—clean, musical, full of friendly people who did not ask too many questions. At a small shop in the heart of the city, 1511 Calle Donado, not far from the water, he built and repaired radios for legitimate paying clients.

He was twenty-eight. It did not seem crazy to imagine that he might survive the war, get married, and have children, either here or back in Germany. He had people in Berlin who adored him and worried about him. Utzinger's girlfriend worked at AMT VI, which gave her access to the radio transmitter, and she often sent him personal messages from her and from Utzinger's loved ones. She signed the messages "Blue Eye" and called him "Dark Eye." Sometimes "Blue Eye" transmitted a message written by Utzinger's grandmother in Berlin, who called herself "the Ahnfrau." He tried to take care of his grandmother from afar, sending packages of food from South America, tinned meat and coffee. She told him in radio messages that the packages were greatly appreciated and boosted her spirits. "I am sitting with . . . BLUE EYE and with a bottle of wine, celebrating your birthday," she wrote once. "On the one hand, I should like to have you here, on the other hand, I am proud of your accomplishments . . . received 2 packages this year. Was very pleased and thankful . . . I am well. Most cordial greetings. THE AHNFRAU." She asked him to remember to brush his teeth.

If it had been up to Gustav Utzinger, he may have spent the rest of the war running his little radio business there in Buenos Aires, making good money, placing advertisements in newspapers, doing everything on the up-and-up. But he was in the SS, the elite vanguard of the Nazi state, and the SS did not care about his personal dreams and ambitions, which is why, in the shop's basement, during the final months of 1942, he resumed his "natural patriotic efforts for my Fatherland," and made himself busy assembling and testing a new generation of clandestine radio sets, at the urgent request of his Nazi superiors.

It had been eight months since the FBI tried and failed to smash the Nazi network in Brazil. After the arrests, Utzinger had floated around for a while, trying to earn money to support himself. Escaping to Asunción, Paraguay, he got a job as a radio technician for the Paraguayan air force, which was commanded by an energetic fascist named Pablo Stagni. Under Stagni's wing, Utzinger built transmitters for the military and taught Paraguayan officers the fine points of radio. Sometimes he would see a few of his former Nazi colleagues kicking around Asunción, but all they did was talk—except for one day when they burned the reels of Charlie Chaplin's *The Great Dictator* in the public square.

It wasn't until Utzinger traveled to Buenos Aires that he got pulled back into the clandestine radio game.

He went there with Stagni, who was trying to convince the Nazi naval attaché, Dietrich Niebuhr, to sell him a gun sight for some ancient Krupp cannon that the Paraguayans had seized from the Bolivians during a previous war. Dietrich Niebuhr was far more interested in Utzinger than the cannon and started to bend the radio wizard's ear. Niebuhr said he was under extreme pressure from Berlin to build a clandestine radio transmitter here, and he needed Utzinger's help.

At this point in the war, Argentina was a far more amenable climate for clandestine radio than Brazil. The Brazilians had leaned toward the Allies as the war developed; Argentinians headed in the opposite direction. An American Jewish novelist named Waldo Frank toured Argentina in the summer of 1942 and noticed "a spawn of little nazi and nationalist papers," he wrote in an account of his journey. Frank traveled through cities and small towns delivering speeches about the value of democracy, and in every town a pro-Nazi newspaper attacked. He sensed that Argentina's conservative government, controlled by corrupt landowners, had "a very uncertain grip upon the country" and that fascists had infiltrated the police. By the end of Frank's tour, the government had declared him persona non grata, and before he could escape, five cops stormed into his hotel and beat him on the skull with truncheons.

Naturally the Nazis wanted to take advantage of this receptive atmosphere, especially since they were losing friends around the world at a rapid rate. Only two nations in the Western Hemisphere now maintained formal relations with the Reich—Chile and Argentina—and in early 1943, Chile would sever the link, making Argentina the lone holdout. This is why Berlin and Dietrich Niebuhr were desperate for a wireless link. Argentina was one of Germany's only listening posts in the hemisphere, one of the last places where it was still possible to obtain reliable intelligence about anything happening in the West, including the United States. They had to keep a line open at all costs.

Niebuhr gave Utzinger a powerful Seimens transmitter he had been keeping at the embassy, and in the fall of 1942, Utzinger took it to a small farm outside the city and began making tests. But as it turned out, Niebuhr wasn't the only Nazi in Argentina who needed radio assistance. Utzinger was also approached by an Abwehr spy named Hans Harnisch, code name "Boss," an employee of a German steel firm in Buenos Aires. He also wanted a wireless link to Berlin.

Then, in January 1943, the most mysterious spy of all suddenly appeared in Buenos Aires: *Hauptsturmführer* Siegfried Becker, a.k.a. "Sargo." He had stowed away on a Spanish ship, paying off the crew to smuggle him through customs.

And in his luggage, all the way from Germany, he had brought an Enigma cipher machine.

Becker looked the same as always—blond hair, nice clothes, a dirty mustache, bizarrely long fingernails—but Utzinger thought he detected a new gleam in Becker's eye. He was vibrating with ambition. Becker said that in the months ahead he would need Utzinger to transmit very sensitive information to Berlin. The "embassy crowd" must not be able to read the messages. Whatever Becker had in store for his next chapter, he intended to keep it close.

Utzinger now found himself in a nearly impossible position. Three men from three rival agencies had asked him for radio assistance: Niebuhr with the German embassy, Harnisch with the Abwehr, and Becker with the SS. In theory, he could create three separate radio stations, but this wasn't practical given the limitations of his own equipment. One powerful station was preferable to three weak ones. Instead of building three stations, then,

Utzinger decided to *trick Berlin into thinking that he had*. He designed a mythical radio organization that only existed on paper, a "Potemkin network," so that each agency in Germany would feel it controlled its own station in Argentina. He called this network "Bolivar" and told Berlin that it consisted of three sections, *Rot, Gruen,* and *Blau*—Red, Green, and Blue.

Red was Becker, Utzinger, and their SS collaborators.

Green was Hans Harnisch and the Abwehr.

Blue was the "embassy crowd."

The ruse gave Utzinger the freedom to run his operation as he saw fit without needing to explain his every technical choice to Berlin. He was determined to avoid the mistakes that had gotten the men arrested in Brazil, and with Becker's help he got to work. Together they recruited a new team of spies from Buenos Aires and surrounding towns, "42 loyal and seasoned collaborators," including German immigrants and working-class Argentines who believed in fascism. Utzinger selected new radiomen and drilled them in good security practices. Limit transmissions to short bursts; send decoy messages full of garbage text on prearranged frequencies; transmit at different times each week; never repeat the same message twice.

"I am teaching my boys tough wireless discipline," he radioed to Berlin. "The Yankees are copying every dot of our transmission."

In addition to the new radio link, the spies carved out another useful channel to Berlin, building a courier system that relied on Spanish sailors to smuggle packages and luggage on Spanish vessels. These men, paid in pesos and known as "wolves," allowed Becker and Utzinger to receive shipments of money, pharmaceuticals to sell for cash, and radio parts—items necessary to keep the network afloat. And through the wolf system and other sources, Utzinger also obtained crypto machines. He now had two Enigmas at his disposal, plus a pocket watch–size Kryha device called a Liliput, small and light and easily concealed.

By the end of February 1943, Utzinger and Becker had everything in place: a new radio system; crypto machines, including Enigmas, the best available, each message a perfect tiny fortress;

a team of collaborators. They were ready to transmit reports to Germany in volume. On February 28, Becker hopped on the radio and sent Berlin a cheerful progress update:

> *New organization established with LUNA. LUNA has assembled in splendid fashion a circle of co-workers. Spread and prepared so that I am able to start immediately with work according to plan. SARGENTO.*

Berlin replied with glee. "Old boy, now we are off," they radioed. "Test message. Cordial greetings to all. We are awaiting your blind traffic there on Monday, Wednesday and Saturday at 0200 and 0400."

The spies' SS leaders back home, the officers of AMT VI, could not have been happier to hear that a door was opening in Argentina. The winter months of late 1942 and 1943 had been grim ones in Germany. The Red Army was crushing the Wehrmacht in Stalingrad, and although Nazi censors blocked reports of losses, rumors leaked. In the snow-dusted cities of Germany, it was difficult to find toothbrushes, belts, bicycle tires, and toilet paper. Restaurants complained of patrons stealing glasses. On February 18, at a rally of twenty thousand in Berlin, Minister of Propaganda Joseph Goebbels finally admitted that the Battle of Stalingrad was lost and called for "a war more total and radical than anything that we can even imagine today," an apocalyptic death struggle against the Allies and against Jewry.

All in the homeland that season was darkness, danger, frost—and yet in Argentina, Becker and Utzinger reported, it was the height of summer, and they were making interesting friends.

Becker, for instance, was cultivating a relationship with Juan Domingo Perón, the future three-time president of Argentina, now just a young army colonel with a taste for moral larceny. (He lived with a fourteen-year-old girlfriend whom he called "The Piranha.") Perón belonged to a secret lodge of military officers, the United Officers' Group (GOU), that aimed to overthrow the Argentine president, and he was already thinking beyond his own nation. Inspired by Hitler's domination of Europe, Perón imagined himself at the head of a nationalist movement sweeping across all of South America.

His ambition exceeded his reach. He didn't yet have the contacts that a wider revolution would require. But Siegfried Becker did. During his decade with the SS, working in Germany, Brazil, and Argentina, Becker had gotten to know all kinds of influential South Americans: among them lieutenants, generals, diplomats, and police captains. He carried a small notebook with their phone numbers. These contacts—Becker's little black book of nationalists and fascists—made him an alluring figure to Perón and his friends, and soon the Argentines hashed out an informal deal with the Nazi spies.

Each side would get something it wanted. The Argentines would share secrets about the United States and protect the spies from the FBI and other Allied law agencies. In exchange, the spies would operate behind the scenes to extend Argentina's influence across South America, connecting revolutionaries in one country to like-minded men in another. They would plot coups, overthrow governments, and install fascist-friendly regimes.

The ultimate goal was to assemble a bloc of nations aligned against the United States. "Hitler's struggle in war and in peace will be our guide," Perón and his GOU plotters wrote in a covert manifesto. "With Argentina, Paraguay, Bolivia, and Chile, it will be easy to pressure Uruguay. Then the five united nations will easily draw in Brazil because of its type of government and its large nucleus of Germans."

With Brazil fallen, the "continent will be ours."

At her coast guard desk, Elizebeth reached for a fresh sheet of grid paper. Circuit 3-N. Argentina to Berlin. The unknown Enigma machine.

Twenty-eight unsolved messages from Circuit 3-N now sat in a pile on her desk. She wrote the twenty-eight ciphertexts on the worksheet in pencil, one on top of another, assembling a stack of text so she could solve the messages in depth, like she had done in 1940 to solve the commercial Enigma machine.

The twenty-eight messages all appeared to use the same key—a huge gift to the codebreakers from their Nazi adversaries. It made things easier and allowed Elizebeth to begin solving the individual messages.

She made a frequency count of the letters in the columns. The cipher letter *H* appeared seven times in column no. 2, four times in column no. 3, and so on. This was enough to start guessing at the plaintext letters. Each column was its own MASC, a monoalphabetic cipher, and she hopped around, penciling plain letters in the columns and guessing at German words across the rows, like *bericht* (report) and *wir hoeren* (we hear).

She was following the path she had blazed in 1940 with the commercial Enigma, using her experience with that old machine to discover "an entering wedge" with this new one, and then hammering the wedge until the damn thing split. The team was still relying on the geometry of patterns. Step one: Line up the messages one on top of another. Step two: Solve the plaintexts in depth. Step three: Use the resulting alphabets to deduce the wheel wiring. Step four: Exploit the wiring knowledge to reveal the new keys whenever the adversary changes them.

This time, the coast guard had some competition. Across the sea, at Bletchley Park, the secret British codebreaking campus, a unit called Intelligence Service Knox (ISK) was attacking the same Enigma, trying to solve it independently of the coast guard. Bletchley had been breaking spy ciphers since the start of the war. Now the two Allied teams each solved this new Enigma at about the same time, in December 1942, by different methods that ultimately got them to the same place.

The machine turned out to be a G-model Enigma designed for the Abwehr, similar to the commercial Enigma but with wheels that stepped less regularly. Within the high-security universe of Enigmas, it was a medium-security model, more puzzling than the commercial machine but less so than other Enigmas. Decades later, Elizebeth called the G-model a "less superior Enigma that was used by Germany and her confidential agents—her spies." At the moment, however, solving the machine felt like victory. "There was much celebration," an NSA historian reported after the war, in an interview with Elizebeth.

Now the coast guard could read the messages backward and forward—the old messages already intercepted and the new ones incoming. Looking at the plaintexts, the codebreakers confirmed that the South American end of the circuit was in Buenos Aires, Argentina.

Also, the names of three colors popped out in the messages: Green, Red, and Blue. The spies were tagging their notes by these color names, and each color seemed to represent a different spy leader using a different code.

It was as if multiple groups of Nazi agents had converged around Buenos Aires and were sharing this one circuit, pooling resources to make a kind of last stand.

Having broken into the Green messages, the coast guard codebreakers turned their gaze to the Red traffic, which they already knew was encrypted with a device that the Germans called Lily. "Following messages all enciphered with LILY," Berlin had radioed in February 1943. It seemed clear to Elizebeth and her coworkers that Lily was short for Liliput, as in a Kryha Liliput, a miniature version of the German cipher device.

The Liliput was somewhat more complex than earlier Kryha models, but as a rule, Kryhas were less secure than Enigmas. Elizebeth walked to her shelf, picked up a Kryha, and examined it with her colleagues. It was an older model, but the principle was the same. Two concentric alphabet wheels stepped against each other, the stepping regulated by a control wheel set to a certain starting position. After a month of work, the codebreakers recovered the key with the help of punch cards, the IBM crunching and tabulating the frequencies of certain juxtapositions of cipher letters that helped them understand how the letters were distributed and offered clues about possible keys that were then tested to see if they produced sensible plaintext.

Now the coast guard was able to read the bulk of the circuit's traffic, the Red messages as well as the Green. (They would never solve the Blue messages, but it didn't matter.)

And what two names popped out in the plaintexts?

"Sargo" and "Luna," Elizebeth's old friends.

She recognized the aliases of the spy and the radio wizard. She had first encountered them on the Brazil circuits in 1942, and now, apparently, the duo had reunited in Argentina.

This alone meant that Circuit 3-N was important—the presence of these two important individuals, "Sargo" and "Luna." But there was something else that concerned her. The men seemed to be building a completely new wireless organization, bigger than before. They were hiring radio operators and obsessively testing new radio equipment.

"We have antenna 100 meters long beamed on Berlin," the spies tapped out to Berlin one day. "Hope you like it."

It was clear to Elizebeth that the Nazi network in the West had shifted to Argentina, and that the men were preparing the ground for a significant espionage effort. Something big was about to happen.

Unfortunately for Elizebeth, right as she was starting to figure this out, the navy forced her to interrupt her work and pack up her office.

In March 1943, the coast guard codebreakers were ordered to move from their longtime home at the Treasury to the former girls' school now known as the Naval Communications Annex, the temporary wartime facility on Nebraska Avenue. Elizebeth had to pause her assault on Circuit 3-N while she got settled in the new location. She claimed an office on the second floor and spoke little to anyone outside of her own team. A coast guard employee who worked two doors down from her in the Annex later recalled, "Our work was compartmented. We went as a group for lunch, but the conversation was never about work, but concerned current events—the way the war was going, etc." Elizebeth worked a little with a few of the SPARS, members of the coast guard women's auxiliary, and she sometimes interacted with the young WAVES and WACs on a purely social basis, outside of the Naval Annex, inviting them to her home for tea and asking how they were getting along in Washington. One evening, Martha Waller, the codebreaker at Arlington Hall, returned home to her D.C. apartment to find Elizebeth and William Friedman sitting at dinner with five of her roommates. One of the roommates' father apparently knew William and had set up the dinner. Waller was dumbstruck. The Friedmans, she had heard, were legends. "I think I was mesmerized, and I know I said next to nothing,"

Waller recalls. Here is what she remembers about Elizebeth: "She looked, sounded, and behaved like the professor of English she might well have become."

The month after Elizebeth relocated to the Naval Annex, her husband left the country on his first big mission since his mental breakdown two years earlier. William Friedman traveled to Bletchley Park at the request of the army, which wanted him to negotiate an agreement to share information, expertise, and blueprints for building bombes. He succeeded, but his depression came back while he was overseas, manifesting as insomnia. He swallowed pink Amytal pills to knock himself out and in a personal diary of the trip he repeatedly mentioned sleep problems:

> TUESDAY APRIL 27 . . . *Bed at 10:30 but too tired for good sleeping.*
> SUNDAY MAY 2 . . . *Poor sleeping for some reason or other, maybe tetanus shot still working.*
> MONDAY, MAY 24TH . . . *Got up at about 1 a.m. and took two small pills from Washington cache but didn't do much good. Awoke early & not at all refreshed. Guess this work is very exhausting mentally & I hope to get through with it soon.*
> MONDAY, MAY 24TH . . . *I've noticed that on days when I am "tense" & have "heebeegeebees" I sleep well in night but when don't have them, sleep not so good. —Haven't had hbgbs for many days now. Wish I could solve this mystery of myself.*

Elizebeth didn't see these diary entries but imagined that William must be under great stress, and she worried about his mental state. Throughout April and May, she wrote him constantly while he was gone, sending at least fourteen letters to the military attaché at the American embassy in London for delivery to his secret location. In the letters she focused on small details of life in Washington: their lawn that was cracked and dried from the spring sun; a party with some family friends where they all drank too many old-fashioneds and stayed up until 2 A.M. singing songs around the piano. "I will create pictures for you of the scenes and people present and so seem like a momentary return for you." She mentioned a mutual friend, Colonel John McGrail of the signal corps, one of the intelligence professionals who had written critical annotations in William's copy of *The American Black Chamber*. Like William, McGrail was brilliant, prone to depression, and kind; he sent Elizebeth a corsage of violets on Easter, "the dear sweet thing," Elizebeth wrote to William, adding that McGrail "seemed more depressed than ever." She said that William's Telechron machine in the library, an electric clock, seemed to be ticking at faster-than-normal speed: "Even your telechron misses you. It is running crazily."

She mentioned that she had been skipping breakfast to stay slim and was smoking cigarettes in the evenings, a rare admission that she herself was stressed and overwhelmed by the mental strain of her job. One summer day, the temperature inside the Naval Annex reached 110 degrees, with no air conditioning. She was dripping sweat onto her worksheets. Everyone was. The navy commanders declined to send people home, saying that a war was on.

There was a real danger that if Elizebeth relaxed, even for a week, her Cryptanalytic Unit might fall behind and never catch up. Throughout the summer of 1943, the Nazis rolled out extensive reforms of their crypto systems to improve security. The orders were issued by a respected Nazi cryptographer, Fritz Menzer, who led a staff of twenty-five cipher experts in the High Command of the Wehrmacht. Spies who were still relying on hand ciphers discarded their old methods for *Oberinspektor* Menzer's new procedures, and the coast guard codebreakers had to adapt. They worked furiously to solve "Procedure 62," a double transposition system based on a thirty-one-letter key phrase. In this one, the letters of the message were scrambled once and then scrambled again, and only unscrambled by means of a key phrase that was itself scrambled based on the number of the month.

Clandestine circuits in Spanish-speaking countries began to use Procedure 62. Another new crypto system, "Procedure 40," adopted by spies in Madrid and elsewhere, was a substitution method combined with double transposition, both steps sharing the same key phrase. In the substitution step, the phrase was written in a 5-by-5 square of letters. One such key phrase that Elizebeth's team unearthed was the Spanish proverb *donde menos se piensa salta la liebra*. Written in the 5-by-5 square, skipping all repeated letters, it looked like this:

```
D  O  N  E  M
S  P  I  A  L
T  B  R  C  F
G  H  K  Q  U
V  W  X  Y  Z
```

Literally the proverb means "Where least expected, the hare jumps," though a better translation might be "Opportunity knocks where it is least expected."

And whatever else demanded her attention during the day, Elizebeth kept going back to Circuit 3-N, the link between Argentina and Berlin.

She was more convinced than ever that Circuit 3-N was the most important circuit of all. The volume of traffic abruptly rose in April and continued to expand every week. Some days they sent as many as fifteen different messages to Berlin and just as many replies traveled in the opposite direction—an explosion of new leads for Elizebeth and her team to chase down, a deepening abyss of Nazi text.

Germany seemed as hungry as ever for information about the United States, asking questions about U.S. weapons capabilities and political figures. ("Are there differences of opinion between Roosevelt and his Jewish advisers, above all Roseman, Morgenthau and Frankfurter?") Berlin often sent requests to the spies in long numbered lists:

1. The Fisher Co. in Detroit reportedly constructed a new anti-aircraft gun of about 12-centimeter caliber, which is fired by remote control. Urgent question: Construction, mode of action, performances.
2. What is manufactured in [Henry] Ford's shop in Iron Mountain? Size of the plant? Since when? Monthly production?

3. The USA armor bombs: What caliber? Kind of the material? Cross section plan, explosive charge, quantity of (5 letters garbled) and detonator. Is "Explosive D" used?
4. Details on new development and production of armor-piercing arms and in this, air bombs of USA and England in particular.
5. Details and particular on development and introduction of rocket weapons . . .

Such messages were concerning, of course, but they were also straightforward and familiar. Elizebeth had seen hundreds like them on other circuits. What was completely new and sinister about Circuit 3-N was the *political* intrigue shining through the plaintexts—a whole other level of conspiracy and malice. "Sargo" and "Luna" weren't just two guys transmitting the shipping news anymore. They had friends across the continent, poised to act in the name of revolution. They were building a secret army.

For one, it was clear to Elizebeth that the spies had forged an intimate working relationship with powerful Argentine figures. On June 4, 1943, a group of generals had occupied the *presidente*'s mansion in Buenos Aires, the Casa Rosada, deposing the old regime and installing a new *presidente,* Pedro Pablo Ramírez. The messages contained references to Ramírez (the Nazis called him "Godes") and other coup figures, including Juan Perón and Captain Eduardo Aumann, code name "Moreno," now a high official in the Argentine foreign ministry. Elizebeth noticed that a Nazi agent named "Boss" (Abwehr leader Hans Harnisch) was regularly meeting in secret with these men. "Another important conference with [*Presidente*] GODES revealed his willingness for energetic collaboration in the interests of the Axis powers," "Boss" radioed to Berlin on July 24, 1943, adding later that Aumann and others were "ready in every respect to promote mutual interests," and that "the USA is considered greatest enemy."

It wasn't shocking for Elizebeth to learn that Argentina, a supposedly neutral country, was cooperating with Germany behind closed doors. But the *scope* of the cooperation was surprisingly extensive. She was seeing glimmers of incredible clandestine missions. "Boss" wrote in one message, "Through our efforts Argentine Government has established close contact with nationalist groups in Chile, Bolivia and Paraguay; even with Brazil through V-men residing here." Argentina alone was not enough for the Nazis: They were conspiring to overthrow the governments of Bolivia, Chile, Paraguay, and Brazil. They were trying to run the table, to turn the continent fascist.

"Sargo" appeared to be the man behind the scenes, the invisible agent traveling from country to country, meeting with revolutionaries, bringing money and information, linking them to one another. He told Berlin he had secured the cooperation of the Paraguayan air force chief, Stagni, who "is completely in our camp" and was glad to fly him around the continent on Paraguayan planes. "Sargo" said he was optimistic about the prospects for a military coup in Bolivia, where he had formed a cell of conspirators with the Bolivian minister of mines and an attaché named Elias Belmonte. And as "Sargo" schemed and plotted in the dark, working on the gritty details of revolution, his colleague "Boss" continued to meet with the Argentines and talk about the big picture. In late August "Boss" attended "a secret session of high officers and officials" of the Argentine government and radioed the following to Berlin:

> Final objective is said to be formation of a bloc of South American countries, which would itself protect its interests, without tutelage of others who pretend to do this. Bolivia must not only be freed from USA influence, but also establish social justice. Argentine can carry out this together with or even in spite of the Bolivian government. Great progress was achieved in negotiations with Chile and further improvement is to be expected. The days of the Rios government with its ambiguous leftist policy are numbered. Chilean military circles had prepared everything in order to follow Argentina's example [i.e., to launch a coup] . . . haste is necessary . . . there is dissatisfaction with [Paraguayan president] Moringo. It was intimated that change in government there is not excluded . . .

Perhaps boldest of all, the spies seemed to be arranging a secret weapons deal between Argentina and Nazi Germany. They were trying to figure out a way to get guns and bombs from Berlin to Buenos Aires without the Allies knowing.

Elizebeth was able to follow every twist of the weapons deal in the plaintexts. The details, she discovered, would be negotiated in Berlin by an Argentine envoy, a local man who would sail from Buenos Aires under diplomatic cover. He had been promised meetings with Himmler and Hitler. "An agent will depart from Argentina to Germany," one of the spies informed Berlin in July. "Name, rank, mission to follow."

In a subsequent message, Elizebeth learned the man's name: Osmar Hellmuth.

Donde menos se piensa salta la liebra. She heard a knock.

Osmar Hellmuth had never felt so important before. He had never done anything quite this exciting. One minute you are at the German Club of Buenos Aires, having a nice conversation with some nice German fellows in riding boots, and the next minute you are back at your flat on the Calle Esmerelda with some different German fellows, discussing an international weapons deal, and you are introduced to a man with unusually long and curling fingernails and told that "this gentleman" will make all arrangements for you to sail across the world for a private audience with the Führer.

Hellmuth—forty, not too bright, heavyset, with a red mustache and red hair combed straight back—was "easy prey" for the Nazi spy ring, British officials would later conclude. A low-ranking naval officer, he handled minor diplomatic duties for the Argentine government. His portfolio was not enough to make him powerful. But it was enough to afford the kinds of diplomatic protections that brought him to the notice of Siegfried Becker.

It was the summer of 1943, and Becker's ambition was growing with his power. He now lived in an upscale neighborhood of Buenos Aires, not far from the palatial home of Juan Perón, whose star had been rising after the coup.

The two men spoke often, in secret, which is how Becker came to understand that the Argentines had a pressing desire for weapons. Fearing an invasion from Brazil, their enemy and main rival on the continent, they wanted Becker to help procure weapons from Berlin. This was a complicated request. At this stage of the war, weapons were hard to come by—Germany could not easily spare them—and Becker could not simply ask Berlin for weapons because the foreign ministry was controlled by a bureaucratic rival of SS foreign intelligence. The only way to get the weapons, then, was in secret, without "the embassy crowd" finding out.

This is when Becker approached Osmar Hellmuth, the former insurance salesman, and made him an offer.

Becker explained to the naive Argentine that the sale of weapons had to be negotiated in person. An envoy was needed, an intermediary. If Hellmuth agreed to accept the mission, Becker explained, he would board a ship in Buenos Aires, the *Cabo de Hornos*. The ship would sail to the port of Trinidad on the northern shore of the continent, where British officials searched all vessels bound for Europe, and then depart for Spain. Upon Hellmuth's arrival in Bilbao, Spain, he should check into the Hotel Carlton and wait for an SS agent to approach and speak the words "Greetings from Siegfried Becker." Hellmuth was to reply, "Ah! The *Hauptstürmfuhrer*!" The SS man in Bilbao would then arrange for Hellmuth to meet with Nazi leaders, including Himmler and Hitler, and after the deal was arranged, Hellmuth would be rewarded with a cushy consular job in Barcelona.

"I had a marvellous opportunity to go to Europe," Hellmuth explained later, "free of expense with a good salary, on an extremely interesting mission, and with good prospects." He didn't understand the risks. Becker did. Becker just assumed that if the mission failed, or if the Allies found out about it, no one would be able to trace its genesis back to him. In a brilliant espionage career, this may have been his one fatal mistake.

Becker wasn't alone in his miscalculations that summer. Gustav Utzinger, the ring's radio expert, was also failing to grasp the danger of his position. The Americans—Elizebeth Friedman and her team—had broken his cipher machines and were now reading his every transmission, but for Utzinger, the prospect of a Yankee breaking an Enigma machine was beyond his comprehension.

This wasn't to say that he slept soundly at night; like any good radio expert, Utzinger lived in a fog of professional paranoia. He simply assumed that any breaches of his network must be the fault of his incompetent counterparts in Berlin, a consequence of their "dilettantism and lack of imagination," in his words. They were always making stupid mistakes. They stayed on the airwaves for too long and repeated messages. One evening in July, the radio operator in Berlin transmitted the same message, "OK HELLO," for fifteen straight minutes over the same frequency. Utzinger scolded Berlin: "The enemy has such an easy time!"

Gradually, Utzinger's mood improved. By November 1943, the network seemed poised for a string of terrific successes. Becker assured him that it was almost time for the plotters in Bolivia and elsewhere to activate, the coups to be attempted. That month, Becker befriended a Chilean gunnery sergeant who had just returned from the United States after completing a one-year weapons course offered by the U.S. Navy. The Chilean's descriptions of U.S. naval capabilities and tactics, relayed to Berlin over the radio, amounted to a detailed American game plan against Nazi ships and U-boats:

In day fighting, the heavy artillery is said to use the following manner of ranging fire: salvos every 7 seconds. The first salvo 300 yards over the distance, measured by radar. The second salvo the radar distance. The fifth and sixth salvos 200 yards shorter or longer than the fourth salvo. In night fighting, the ranging is carried out by ladder and radar . . . Each shell contains aniline compound which produces intensive coloring of the water, so that the location of the fire can be observed . . . Depth charges [against U-boats] resemble English Vickers [depth charges]. Models with 300 and 600 lb. of TNT charge. The fuze contains 3.25 lbs. of granular TNT . . . exterior depth setting for 30, 50, 75, 100, 150, 200, 250, 300 feet . . .

The spies wrote this message and the others with their Enigma, ALCSA JYFMK JFNVH KYOIM, transmitting the letters in Morse, dot-dash (A), dot-dash-dot-dot (L), dash-dot-dash-dot (C), not suspecting that in Washington, an American woman was sitting at a loom, spinning these ugly loops of letters into a sensible fabric of plaintext.

Becker and Utzinger thought they had everything wrapped up tightly. As far as they could see, there was only one major loose thread in their system, one element beyond their control: Osmar Hellmuth, the former insurance salesman, about to become a diplomat abroad.

During the final days of September, Becker handed Hellmuth a letter detailing some precision radio instruments they wanted him to purchase and bring back. He also gave Hellmuth several trunks containing sixty kilograms of gifts for friends in Germany. He radioed to Berlin that Hellmuth was on his way. "HELLMUTH enjoys the absolute confidence of the Argentine Government," Becker wrote. "He is going to bring you lists of the government's wishes . . . The Argentine Government demands strictest secrecy about the mission." One last time, in Buenos Aires, the *Hauptsturmführer* wished his man luck, and on October 2, 1943, Osmar Hellmuth sailed into a faultless blue sea.

Elizebeth and her teammates solved the message about the Chil-ean sergeant who infiltrated the navy's gunnery school and escaped with its secrets. They watched in real time as the confidential agents of Germany and Argentina conspired to flip the chessboard of global politics together. Thanks to the Cryptanalytic Unit's successful assault on Circuit 3-N, Elizebeth understood the structure of Nazi espionage in South America. The money, the actors, the codes, the connections—she had the map now.

She forwarded her decrypts along the chain as fast as possible, often annotating them with brief written notes describing the named agents and explaining what they seemed to be doing.

Luna is probably Gustav Utzinger, the radio expert of the spy ring.

This is the latest in a series of messages dealing with the attempt of the German espionage ring, members of the Argentine general staff and the Brazilian integralists to form an anti-US bloc in South America.

These were ULTRA messages, stamped TOP SECRET ULTRA at the top.

By now, thanks to Elizebeth's ULTRA decrypts and those of British codebreakers, everyone in the English-speaking spy world knew that Argentina was conspiring with the Nazis. The ULTRA messages were clear.

But they were also forbidden fruit. The Allies would have liked to show the decrypts to the Argentine government and demand that they stop. But then, of course, the Argentines would know that the Allies had broken the Nazi codes, and then the Argentines would tell the Nazis, and the lifeblood of ULTRA would instantly stop flowing through the world of Allied intelligence—a catastrophe.

There seemed to be no way out of this bind until October 1943, when Elizebeth solved messages from Circuit 3-N describing plans to dispatch an envoy named Osmar Hellmuth to negotiate a weapons deal between Germany and Argentina. This seemed to provide the Allies with an unprecedented opportunity. A relatively obscure Argentine man, this Hellmuth, was working very closely with Nazis, and he was about to get on a ship and sail to Bilbao, Spain. Perhaps he could be intercepted en route and forced to divulge his Nazi contacts. The Allies could say they got the information from the confession, not the Enigma messages that Elizebeth had solved. Then British and U.S. officials could finally take steps to disrupt the Nazi network in Argentina without exposing the ULTRA secret.

This plan required a bold and possibly illegal act by the British. Hellmuth was a diplomat. He had protections. He could not simply be kidnapped. Could he?

The British yanked Osmar Hellmuth off his ship in the middle of the night. They ignored his diplomatic privileges. They did not listen to his protests and did not allow him to make a phone call to his embassy. The *Cabo de Hornos* had been docked in the port of Trinidad before departing for Spain. The British placed Hellmuth on a different ship, bound for England. It sailed east across the Atlantic and arrived in Portsmouth, England, on November 12, 1943. From there the confused and indignant Hellmuth was carried to a secret interrogation facility in a mansion in southwest London called Camp 020, part of a network of nine interrogation centers operated during the war by MI5, British counterintelligence.

Argentine officials made frantic calls to British diplomats and asked where their citizen was. The British diplomats claimed they didn't know.

He was placed in solitary confinement at Camp 020 for the first two weeks. The guards confiscated his seven trunks of gifts for German officials and the letter with the details of the precision instruments he was supposed to purchase. Then the prisoner was brought to see the commandant of the facility, Colonel Robin Stephens, a broad-shouldered man who wore the tan wool jacket of a British military commander, medals pinned above the left breast. A monocle affixed to his right eye made him look like a broken owl. His men called him "Tin Eye." His motto regarding the art of interrogation was "truth in the shortest possible time." He always said he did not believe in torture—he claimed he was so skilled at eliciting confessions that he did not need to use it—but after the war former Camp 020 prisoners told credible stories of being beaten by Stephens's men, whipped, subjected to mock executions, deprived of sleep for long periods, made to stand in excruciating "stress positions," and starved.

"I am speaking with authority," the commandant told Hellmuth, "and full authority of Great Britain in war. My observations do not invite any replies from you and I shall regard any interruption as an incipient indiscipline."

Stephens started to berate the prisoner. He said that the British had confiscated his letter asking him to buy precision instruments for the spies of the Reich, proving that he was a Nazi spy. He said Hellmuth was a Nazi stooge, in way over his head, playing a game he did not understand. Stephens made fun of the gifts in Hellmuth's trunks: "There are about seven trunks of gifts. Food for the pot-bellied masters of Germany. Silk stockings for their clod-hopping women. Chocolates for their sniveling children." Stephens said that Hellmuth's only chance for escaping this facility was to tell the full truth and reveal everything he knew about the Nazi spy network in South America. "When you lie, we shall know instantly."

Hellmuth stalled for a few days as the interrogators of Camp 020 demanded information. Initially the British found him to be "possessed, almost arrogant." Hellmuth claimed he did not know much about SS intelligence activities in Argentina; he was familiar with one or two Germans from social circles and that was all. The interrogators asked about the man they knew only as "Sargo"—the man they suspected was "the head of Himmler's faction in Buenos Aires" and "the prime mover of the secret mission." The power behind the scenes, the Nazi mastermind in South America. Who is "Sargo"? What is his real name? What is his rank? Hellmuth said he did not know. He began so many sentences with the word *probablemente,* "probably," that the interrogators became infuriated and banned him from saying it ever again.

Eventually, the interrogators said that if he didn't start talking, his treatment in the camp "must necessarily deteriorate," a hint he would be tortured.

Hellmuth softened. He told the British about how Argentina had worked with the SS to depose the Bolivian government and install a Nazi-friendly dictator. He told them about the clandestine radio stations and the radio technician who ran them, Gustav Utzinger, code name "Luna."

Most important of all, Osmar Hellmuth divulged the secret that his British interrogators had been burning to know. Hellmuth's confession, made possible by Elizebeth's decryptions, would soon spark a fantastic chain of global events, forever turning the tide of the Invisible War in the Allies' favor.

Who is "Sargo"?

Poor and unlucky Osmar Hellmuth, cold and alone, had been wondering for weeks if his personal sense of honor was worth this suffering, and now he decided it was not. "Sargo," he told his captors, was Siegfried Becker, an SS captain, age thirty-two, five foot ten with a strong build and blond hair.

CHAPTER 5

The Doll Lady

While the Nazi spies waited for news of Hellmuth's mission, not knowing that he had been kidnapped, they upgraded their security procedures in Argentina.

Since arriving here, Becker and Utzinger had encrypted hundreds of messages by typing them on Lily, their miniature Kryha. These were Red messages, meant for their SS superiors in Berlin. The springs and holes of the Kryha were now wearing out from overuse, and Utzinger asked Berlin to smuggle them a new cipher device through Becker's network of wolves.

Instead of a Kryha, Berlin sent a new Enigma machine.

"Enigma arrived via RED," Utzinger reported to Berlin on November 4, 1943. "Thank you very much." He typed this message on his older Enigma machine, the Green machine. He went on, "From our message 150 we shall encipher with the new Enigma . . . LUNA."

"It is a birthday surprise for LUNA," Berlin replied.

Utzinger now possessed three Enigmas. Throughout November and December 1943 he flashed Enigma messages to Berlin, newly confident in the security of his codes and increasingly upbeat about the prospects of his spy organization.

Momentum in South America seemed to be shifting in their favor. Right-wing movements continued to surge, and Becker and Perón were making progress in flipping governments. On December 20, 1943, a right-wing Bolivian general named Gualberto

Villarroel occupied the presidential palace in La Paz with his troops, assuming power in a successful coup. Becker and his Bolivian conspirators had set the coup into motion. The spy was elated. He arranged a meeting between Juan Perón and a Brazilian Integralist leader; Perón told the Brazilian that the "first fruit" of the continent-wide revolution had been plucked and would soon spread to "Chile, Paraguay, Peru and even Uruguay." The bloc of Hitler-inspired regimes, the Argentine revolution, the transplantation of fascist ideology to the soil of South America—these were no longer dreams but projects already begun, and any project begun has a chance of being completed.

Eight days after the coup in Bolivia, on December 28, 1943, Becker and Utzinger radioed warm greetings to Berlin, signing the message, as ever, with their code names. "We extend hearty wishes for a happy, successful New Year to all our loved ones and comrades in the war-torn homeland," wrote the SS spies. "Our thoughts are always with you and our Führer. SARGO, LUNA, and all collaborators."

Elizebeth never experienced a catharsis like soldiers do on a battle-field, a decisive moment when she got to stand over her fallen enemy with a sword and plunge a killing stroke into his heart. Rather, all through the war, she dissected fascists in the dark. If you were her adversary you never felt the blade go in. You bled slowly, painlessly, for months or for years, from tiny internal wounds, and then sometimes there was a terrible morning when you woke up groggy and confused, and your kidney was sitting in a bowl of ice on the counter.

She knew about the new Enigma machine sent to Argentina in November 1943—the Red Enigma—because the spies had discussed its delivery in Green messages and she had been reading those for a while.

The Red Enigma posed a new challenge for the codebreakers. It turned out that Berlin had forgotten to include keys in the shipment. This forced them to send a new key to Argentina over the radio. To protect that key, to keep it extra safe, Berlin decided to double the crypto. They sent Argentina a series of twenty-seven messages that were encrypted twice: first with a Kryha machine (using a new Kryha key), then with the new, Red Enigma. In other words, the plaintexts were typed on the Kryha, generating ciphertexts, and then those ciphertext letters were typed on the Enigma, generating a second set of ciphertext letters.

It wasn't possible for the coast guard to read the messages in depth because the crypto was doubled. The messages were like gnarly logs of wood covered with two distinct layers of bark. All the same, with hints from three sources—their own prior solutions, the navy's IBM machine, and the Germans themselves—the codebreakers found a way to strip off the bark and saw the wood into neat two-by-fours of plaintext. In an earlier message, Berlin had said that the new Kryha key incorporated part of the old key, which the coast guard already possessed. This reduced the number of possibilities for the new key, allowing the codebreakers to write a punch-card program that sorted the ciphertexts and aligned them in proper depth. Now the codebreakers could solve the plaintexts as usual and work backward toward the wiring.

During December 1943 and January 1944, as the coast guard worked toward a complete solution of the Red Enigma, wheel wiring and all, the team's prior solutions started to pay off on an international scale. Solving messages on Circuit 3-N had given Allied officials a priceless view of Nazi espionage in South America, and now, with Osmar Hellmuth's confession in hand, American and British diplomats were able to hammer Argentina for its cozy relations with Nazi spies. The pressure proved too great, and on January 26, 1944, the Argentine government announced that it was severing all relations with Germany and Japan. Argentina had been the Nazis' last friend in the West, the "last neutral bulwark," and now that bulwark was destroyed.

The following month, the coast guard solved the wiring of Red, their third Enigma of the war.

On February 19, 1944, Elizebeth's commander, Lieutenant Jones, sent a secret cable to Bletchley Park informing the British of the team's achievement:

CG have solved . . . red. Details later.

Five days later, Jones transmitted the wiring details for all three wheels:

Following is wiring for new . . . red machine . . .
Outside wheel
P R Y B G A U T E V M K C Q D S J W L O F Z I X H N . . .

The British codebreakers, always competitive, sent a cable back, notifying the coast guard that they had just solved Red themselves:

Many thanks. As this has just been solved here, details not
required.

The SS in Berlin heard that Argentina had severed relations with Germany the same way everyone else did. They saw it in the news: a Reuters report of January 26, 1944. That day Berlin sent a panicked radio message to its spies in Argentina, begging for information on what happened. "We urgently need reports, whether it is true, and reports on the backgrounds and purpose."

The South America section of AMT VI had recently gained a new commander: Kurt Gross, the corrupt former Gestapo agent who nagged his agents to send him chocolates, and he now made a series of catastrophic assumptions. Unable to imagine that Elizebeth or any other adversary had been able to break their codes, Gross guessed that one of Becker's "wolves," the Spanish couriers, had betrayed him. "We cannot ward off the impression that there is a leak in your courier organization," he radioed to Argentina. "We suggest urgently once more that you scrutinize the men most critically."

Gross told the spies in Argentina to redouble their efforts. "Operation concerning USA and South America must now even more go on in full revolutions," Gross radioed. "Himmler's motto for 1944 is: 'We shall fight as long and no matter where, until the damned enemy gives up.'"

Throughout the early months of 1944, Allied planes bombed

Berlin. The headquarters of AMT VI suffered a direct hit. The spies' family members in Berlin used AMT VI's wireless to report that they were still alive. "Dear DARK EYE," began one message from Utzinger's girlfriend, "Blue Eye." "Air raids of Tommy cannot shake us . . . With the old zest and with 'Heil Hitler.' Your BLUE EYE." Utzinger's grandmother, "the Ahnfrau," wrote, "Here, life goes on, in spite of everything. With scornful and triumphant laughter, amid the ruins of homes, we take up our work immediately, with suppressed fury, in cardboard and wooden compartments."

Gross needed to hear good news from Argentina. His requests to the spies took on a newly apocalyptic tone. In one message, he asked Becker and Utzinger to investigate U.S. chemical weapons stocks and U.S. vulnerability to chemical warfare attacks:

Chemical warfare materials: What types of fluids or solid substances are on hand or in production? Their appearance? Effect on the body, eyes, respiratory system, clothes, metals. . . . What are the enemy's means of protection against gas?

The same week that Argentina announced its break with Germany, in late January 1944, a sharp-eyed reporter for the London *Sunday Express* noticed U.S. warships massing off the wharves of Montevideo, Uruguay, almost within sight of the Buenos Aires waterfront. He made some phone calls and learned that something big and strange had just happened in the clandestine world of spies and counterspies. He wrote a story.

BRITAIN SMASHES SOUTH AMERICA SPY RING
Argentine H.Q. of Hitler's best agents

MONTEVIDEO (URUGUAY): The hulls of American men-o'-war were spotted in the sparkling waters of La Plata estuary. The warships were a sign that not much longer would secret German radio stations flash sailing dates and expected routes of Allied troopships and merchantmen to waiting U-boats. They were a sign that the Nazi dream of forming a solid block of Fascist dictatorships across South America is ended now that the Argentine has at last severed relations with Berlin and Tokyo . . . the coup that ended the Argentine's neutrality, fostered for years by Axis diplomacy and money, was the arrest of Osmar Alberto Hellmuth . . .

Within a few more weeks the "Hellmuth Affair," as it came to be known, was splashed across newspapers around the world. It had all the elements of a movie: Nazi masterminds, a pawn doing their dirty work, a kidnapping, a secret interrogation. It was already being spun into legend. Songs were being sung. Young Ziegfield, a popular calypso singer in Trinidad, performed a "Security Calypso" about the hapless Argentine named Osmar Hellmuth:

> *He was on his way to cause a lot of trouble*
> *Osmar Hellmuth that Argentine Consul*
> *Caught with sufficient incriminating documents*
> *Offer no alibi to establish innocence . . .*
> *Just for the matter of a few paltry cents*
> *He sold his people and lost their confidence*
> *But in the end they will shoot the scamp*
> *Because he is safely locked up in one of Britain's Internment*
>   *Camps.*

This publicity created immediate problems for two key institutions. One was the Argentine government. The Hellmuth Affair showed that Argentina was colluding at the highest levels with the Nazi state. This was very dangerous for Argentine politicians; they knew that if the Nazis lost the war, the Allies would surely punish them as collaborators. The public break in relations with the Reich wasn't going to be enough. Argentina needed to do something else, something bigger, more dramatic, to show that it wasn't a tool of Berlin.

The second institution that was unhappy about the Hellmuth Affair was the FBI. Here was the juiciest spy tale in the world, playing out within the bureau's jurisdiction, yet the bureau had been the last to know. The coast guard and the British had withheld their decrypts, fearing that the FBI would leak them. On December 16, 1943, an FBI assistant director sent a peeved memo to Hoover: "It is certain that the information was actually obtained from decodes of the extremely active clandestine radio traffic between Argentina and Europe, which decodes we have been endeavoring to obtain for a long period."

If there was anything J. Edgar Hoover hated, it was being out of the loop. But he was about to get the bureau back into the picture.

Argentina needed an "out." It needed a big public display of neutrality, a story to tell about its independence from Nazi influence, and the more spectacular, the better. Well, the FBI was good at telling stories. Perhaps the bureau could provide one.

The story, the "out" chosen by the FBI, was the story of Siegfried Becker.

Becker: a character out of a novel. A Nazi spy with long curling fingernails. A man who carried explosives in trunks and Enigma machines in his luggage. A seducer of the wives of Brazilian politicians. A stowaway on ocean-crossing ships. An SS-*Hauptsturmführer* who wore the ring of the death's head, a gift from Himmler. A friend of secret fascists in high places. A plotter of coups. A Johnny Appleseed of pirate radio, the human link between clandestine radio stations that spanned an entire continent.

The FBI had obtained some of this information from earlier coast guard decryptions and much of it from their own interrogations of the spies they arrested in Brazil in 1942. They reasoned that they could share the information with Argentina without exposing the ULTRA secret. The bare facts of Becker's career were already lurid, and in the FBI's hands, they could be arranged into a tale of a master spy, a kind of Nazi James Bond.

Francis Crosby, the FBI's legal attaché in Buenos Aires, made the case in a February 15, 1944, letter to Hoover, accompanied by a memorandum on Becker. "The memorandum was written with a view to furnishing the Argentine government with the material necessary for a spectacular spy story," Crosby wrote. "Becker would make excellent copy and perhaps provide the Argentines with the 'out' they are seeking in the Hellmuth case. The rational[e] which occurred to us is about as follows. Becker had an extremely successful career as an espionage agent in several other neighboring republics.... Details about the cases in which he figures would make excellent copy. However, this 'master spy' did not fall into the hands of the law until he ran afoul of the extremely astute police of the splendid Republic of Argentina, who immediately upon learning of his activities in Argentina, terminated his brilliant career." Then came Crosby's three-page memorandum for the ambassador, a pulpy, magazine-ready narrative:

> To the best of our knowledge, the deft, Teutonic hand of Siegfried Becker first appeared on this hemisphere when Becker organized clandestine radio station CEL in Rio de Janeiro.... It is possible to follow Becker into clandestine radio station CIT, also in Brazil . . . in Ecuador . . . clandestine radio station PYL in Chile, and a group in Mexico. . . . The hand of Becker is in some instances heavy and immediate, in others light and remote. However, it is possible to demonstrate the connection at all times. . . .

While the FBI circulated the legend of Siegfried Becker through diplomatic channels, they were going after him and his men on the ground in South America. The bureau now had two dozen SIS agents across the continent, covering all the major cities, finally acclimated to their local posts, cozy with local police, and able to get intelligence from confidential informants. Hoover ordered all of the FBI attachés across Latin America to search their files for any information on Becker, "one of the most important German agents in Latin America, who is presently a fugitive and believed to be hiding somewhere in Argentina." At the same time, SIS agents like Crosby sprang into action on the ground, scoping out bars and restaurants where Becker was said to hang out and interviewing prostitutes he had known.

The FBI also began to close in on the pirate radio stations themselves, the ones Elizebeth had been listening to all along, remotely, from her office in Washington. American agents went climbing over Chilean and Paraguayan mountains with handheld direction-finders, and FCC technicians drove through Buenos

Aires and the surrounding countryside with direction-finding automobiles.

Stations were seized, one by one, the radio equipment impounded. This was fun for the FBI agents, this classic police work. They were good at it. It was what they did best. They weren't codebreakers. They were investigators. They were cops. They built cases from physical evidence and in-person conversations, and they arrested people.

Utzinger wanted to go dark. That was his first instinct. In Argen-tina, when the SS radio expert heard that the Hellmuth mission had failed in the worst possible way—that the envoy had confessed to the British and the Argentine government was now breaking relations—Utzinger thought the spies should stop using the radio entirely. Something was not right. He needed some time to think.

He was overruled. Berlin insisted that the radio transmissions continue and even be increased, and Becker assured him that the break in relations was a "sham." According to Becker's military and political contacts, the Argentine government was just making some noise to appease the Americans. Everything would be fine once the commotion died down.

Having little choice in the matter, Utzinger continued to transmit. But his work with Becker only grew harder throughout the first months of 1944. The two men sensed they were being surveilled, followed. Utzinger took what security precautions he could, moving the transmitter to a friend's farm outside Buenos Aires and hiding it beneath a chicken coop. The authorities found it anyway. One day in February 1944, when Utzinger and Becker were somewhere else, Argentine cops raided the farm and arrested several of their collaborators, taking the V-men to a police station and applying painful shocks with the *picana eléctrica* (electric cattle prod) to extract confessions. A prisoner named Gaucho, who may have been Becker's bodyguard, had his eardrum destroyed. Another, Herbert Jurmann, a Hitler Youth leader, decided to commit suicide rather than confess, throwing himself out of a third-story window. "He fell on February 19," Utzinger radioed to Berlin, "faithful to his oath, for us a model and obligation."

Jurmann's death cast a pall over the work of the spies in Argentina. They were frightened and demoralized. The Reich seemed increasingly distant and German defeat increasingly plausible. No one was sure they wanted to die for National Socialism anymore. The Spanish wolves grew less willing to cooperate; two of the wolves were captured and hanged by the British. Utzinger's grandmother, "the Ahnfrau," sent him an ominous message over the radio, asking where he had placed their ancestral papers, the important documents of their family history. She feared that the papers would be destroyed by bombs.

Utzinger still could not shake the feeling that something was very wrong, that the Allies could read his words. But he assumed that the problem was the incompetence and poor security habits of his counterparts in Berlin. Over and over they had demonstrated themselves to be idiots; he had no reason to think that anything more subtle was amiss. So he tried his best to keep the radio network operational while maintaining a self-protective level of paranoid awareness. When looking at the raw intelligence gathered by his V-men, he paid special attention to news about Allied spy hunts, anything that might give him advance warning of a raid and save his skin.

He picked up an intriguing piece of news that he shared with the home office on March 22. According to the U.S. press, a "Mrs. VALERIE DICKINSON, N.Y." was being charged with treason. (Her first name was actually Velvalee.) She owned a shop in New York that sold dolls and doll clothing. She had sent suspicious letters to an address in Buenos Aires. The letters appeared to concern dolls, but the Americans believed Dickinson was a Japanese spy, communicating in code. The press was calling her the Doll Lady.

Velvalee Dickinson whirled around on the two FBI men and tried to scratch out their eyes. It was January 21, 1944. The agents had staked out the vault at the Bank of New York, waiting for Dickinson to walk in and open her safe-deposit box, and as soon as she did, unlocking a drawer that contained $15,900 in cash, the FBI agents said they had a warrant for her arrest. Dickinson shouted that she didn't know why. She was fifty years old, a widow, a frail-looking ninety-four pounds, with brunette hair. She made such a kicking commotion that the men had to pick her up by the arm-pits and carry her away.

The FBI arrested Dickinson because of five suspicious letters that had been previously intercepted by postal inspectors and forwarded to the bureau. The letters talked about dolls and the condition of dolls, some of which were damaged: "English dolls," "foreign dolls," a "doll hospital," and a "Siamese dancer" doll "tore in middle." The first of the five letters read in part, "You asked me to tell you about my collection. A month ago I had to give a talk to an art club, so I talked about my dolls and figurines. The only new dolls I have are these three lovely new Irish dolls. One of these three dolls is an old Fishermen with a Net over his back. Another is an old woman with wood on her back and the third is a little boy." The letter had been addressed to Señora Inéz Lopez Molinari in Buenos Aires. No such person existed; the letter was returned to the address listed on the envelope, the address of one of Dickinson's customers, a Mrs. Mary E. Wallace in Springfield, Ohio, who was confused to read the letter, as she had not written it.

Dickinson owned a doll shop on Madison Avenue in New York and had developed a reputation for her artistry—she sold dolls for as much as $750 apiece—yet the bureau discovered that she had fallen into debt after the death of her husband, that she was a member of the Japanese-American Society, and she had visited the West Coast in January 1942, immediately after Pearl Harbor. The FBI tested the shapes of ink on the letters against Dickinson's seized typewriter and confirmed a match; the bureau's investigation also revealed social ties between Dickinson and Japanese consular officials.

After the agents arrested Dickinson in January 1944, a federal prosecutor took up the case: Edward C. Wallace, the U.S. attorney for the Southern District of New York. Wallace had worked with Elizabeth in the smuggling days and hoped to get her opinion on the Doll Lady's letters, but first he called the supervisor of the FBI's New York office and asked if the bureau had any objection to showing Elizabeth the letters.

Within the FBI, the prosecutor's request provoked a remark-able exchange of at least eight phone calls, teletype messages, and memos that traveled up the chain from the FBI's New York office to Washington and ultimately to the desk of J. Edgar Hoover. The gist of these communications was that the prosecutor, Edward Wallace, wanted Elizebeth and spoke highly of her—"According to Mr. Wallace," an FBI agent in Washington wrote in a memo to Hoover's deputy, "Mrs. Friedman and her husband, who is a cryptographer for the Army, are recognized as the leading authorities in the country and have written numerous books on the subject"—but FBI agents worried that Elizebeth would siphon publicity from the bureau, stealing its spotlight. The agents seemed reluctant to speak of Elizebeth as an independent analyst separate from her husband. Although no one had ever discussed involving William in the case, the supervisor of the FBI's New York office fretted that the Friedmans, *plural,* "might, in the event of a successful espionage prosecution, attempt to lay claim for any work that they might have performed in this connection."

The New York office sent Hoover a teletype on March 18, 1944: "ADVISE AS TO SUBMISSION QUESTIONED LETTERS TO ELIZEBETH FRIEDMAN FOR EXAMINATION." Hoover responded with a dismissive shrug of a memo: "Concerning the project to submit the documents to Mrs. Friedman. . . . There appears no point is to be gained by multiplying the number of examiners." But he posed no formal objection, so U.S. Attorney Wallace went ahead and sent Elizebeth the Doll Lady's letters, and Elizebeth analyzed them and crystallized her thoughts into a five-page letter before traveling to New York at the feds' expense to discuss the case with Wallace in person.

"My dear Mr. Wallace," Elizebeth began in her letter, "Within the last two days I have spent a few hours examining the Dickinson letters. I am setting forth here some queries and statements which may be accepted for what they are worth, mindful of your statement on the telephone that you hope to obtain 'leads,' and that you understand that the code in the letters is the 'intangible' type of method not susceptible to scientific proof."

After making it clear that this was not the usual sort of cryptanalysis that she did, that this was only her *opinion,* Elizebeth went on to discuss what the Doll Lady was really talking about when she talked about dolls.

The letters, she said, were a textbook example of "open code," a way of communicating secretly out in the open, without necessarily arousing suspicion. "Granddaughter's doll" in one letter might refer to a U.S. ship that had been damaged at Pearl Harbor and was being repaired. "Family" meant the Japanese fleet. "English dolls" meant three classes of English ships, such as a battleship, battle cruiser, or destroyer. Where Dickinson wrote, "One of these three dolls is an old Fishermen with a Net over his back" and "another is an old woman with wood on her back and the third is a little boy," she probably meant, "One of these three warships is a minesweeper, and another is a warship with superstructure, and the third is a small warship." ("Destroyer?" Elizabeth guessed. "Torpedo boat? Auxiliary warship?")

Elizabeth also pointed out that the street number of the address in the five letters—Señora Inéz, O'Higgins Street, Buenos Aires—was given as five different numbers (1414 O'Higgins, 2563 O'Higgins, etc.), suggesting that the messages were never meant to reach their destination and were intended to be intercepted en route, in an airline pouch or a censorship office, by a friendly Axis confederate.

Elizebeth's letter shows her analytical brilliance; it also shows her native cautiousness, her reluctance to say anything that couldn't absolutely be proven. Words in an open code can have multiple meanings. She didn't want to testify in court for this reason. Hoover saw it differently. To him, the vagueness of an open code was an advantage, not a disadvantage, enabling his agents "to give the more extended estimates and alternative possibilities" during cross-examination.

The FBI had gathered other damning evidence against Dickinson, including the unexplained cash and her relations with Japanese officials, and the government charged Dickinson with espionage, as a spy for the imperial Japanese government. The charge carried the death penalty. As far as anyone knew, Dickinson was the first woman to be accused of espionage on American soil since the war's start. "So far," wrote the *Washington Sunday Star,* "on this side of the water, Mrs. Dickinson is the woman spy of this war." During her first court appearance in New York in May 1944, Dickinson looked subdued, wore a black hat pinned with imitation white flowers, and twisted a handkerchief behind her back with black-gloved hands, glancing around the courtroom at the FBI agents and prosecutors and reporters: "Who are all these people?" she said. When the prosecutor spoke, "She even yawned, decorously behind a hand," the *Washington Times-Herald* reported.

Dickinson pleaded guilty. At her sentencing three months later she denied that she was a spy, breaking down in court and swearing that she didn't know a "battleship from any other ship except that it's larger." She ended up with ten years in prison and a $10,000 fine.

All through these proceedings, Elizebeth stayed out of the public eye. When the Doll Lady case was over, the conviction won, the FBI, as always, informed the press of its heroism, feeding the dramatic details of "the War's No. 1 Woman Spy" to reporters. "What made her become a Japanese spy?" asked the *Star*. "One FBI man who questioned her advanced the idea that she was an introvert, embittered by life, the frustration of childlessness." Elizebeth was not mentioned in the coverage. The articles said variously that the code had been cracked by "FBI cryptographers" or "a check with the Navy." Hoover himself wrote about the Doll Lady in *The American Magazine,* calling her "one of the cleverest woman operators I have encountered. Cultured, businesslike, cunning, and, despite her 45 years of age, most attractive, she presented one of the most difficult problems in detection the FBI has tackled in this war."

And while readers learned of the Doll Lady's treachery from Hoover, the woman who analyzed the Doll Lady's letters in her spare time, quietly, as a side project, returned to her primary task of hunting Nazi spies.

Through the rest of 1944, as Elizebeth and her coast guard team continued to decrypt Nazi radio messages, she noticed an uptick in paranoia in the plaintexts, a creeping sense of doom. All along, Elizebeth had been watching the Nazis build a spy network, and now, after invisibly undermining that network, she was watching it die—and stepping on its neck when need be.

She tracked the spies as they tried to escape or hide, decrypting their desperate wireless notes. Becker went into seclusion in

April 1944. "He is hidden in the center of Buenos Aires," Utzinger wrote. "He works at night only. Keep your fingers crossed for him. LUNA."

On August 11, 1944, Utzinger sent one of his final radio messages: "The enemy succeeded in locating two of our stations in 60 days."

Seven days later, on August 18, 1944, Gustav Utzinger, a.k.a. "Luna," was arrested by the Argentine federal police along with forty of his associates.

Utzinger later described the scene to an FBI interrogator. On a table at the jail, police arranged items they had taken from Nazi Party members years earlier, including swastika flags, pictures of Hitler, and hunting weapons, and put those next to a resistance coil from the spool of a film projector. Utzinger had managed to destroy his radio equipment before capture, so the police claimed the coil was a "bobina de tanque de un transmisor potente de los espías nazis"—*tank-spool of a powerful transmitter of the nazi spies*—and photographed the whole cornucopia to release to the press. It was for show; the Argentines were mainly concerned with covering up their own links to the Nazis. Five days after Utzinger's arrest, Juan Perón himself appeared at the jail and told police that "the investigation was to show merely the breaking up of a great German intelligence machine; and that the mention of contact with any political or military personalities or their foreign colleagues must be suppressed."

This wasn't the end of the spy hunt in South America. Siegfried Becker, the wily SS captain, remained at large, the target of a deepening manhunt by FBI agents and police in multiple countries, and Elizebeth would continue decrypting clandestine messages across dozens of Nazi radio circuits for the next thirteen months. But the arrest of Gustav Utzinger in August 1944 marked the beginning of the end of the Invisible War—"the final chapter to any effective espionage activity" carried out by Nazis in the Western Hemisphere, according to Utzinger's colleague, Hedwig Sommer. Nazi spies would never again pose a threat to America.

Elizebeth had not done it alone. It was a team effort, with strong work from Chubby Stratton and other British officers, the FCC, and the FBI. But in her own piece of the war, Elizebeth was a central figure. She had to smash the codes on the page before anyone else could smash the spy networks on the ground. "Technical advantages played a big role in the undercover struggles," Rout and Bratzel concluded in 1986's *The Shadow War.* "Technical brilliance in cryptography and radio interception plus hard work by field agents proved to be the unbeatable combination which made victory possible." The two historians credited the FBI for both the fieldwork and the technical brilliance (the coast guard's files were classified at the time), and authors of more recent books have also praised the bureau for destroying the Nazi networks in South America. But the FBI didn't intercept the messages. It didn't monitor the Nazi circuits. It didn't break the codes. It didn't solve any Enigma machines. The coast guard did this stuff—the little codebreaking team that Elizebeth created from nothing.

During the Second World War, an American woman figured out how to sweep the globe of undercover Nazis. The proof was on paper: four thousand typed decryptions of clandestine Nazi messages that her team shared with the global intelligence community. She had conquered at least forty-eight different clandestine radio circuits and three Enigma machines to get these plaintexts. The pages found their way to the navy and to the army. To FBI headquarters in Washington and bureaus around the world. To Britain. There was no mistaking their origin. Each sheet said "CG Decryption" at the bottom, in black ink. These pieces of paper saved lives. They almost certainly stopped coups. They put fascist spies in prison. They drove wedges between Germany and other nations that were trying to sustain and prolong Nazi terror. By any measure, Elizebeth was a great heroine of the Second World War.

The British knew it. The navy knew it. The FBI knew it. But the American public never did, because Elizebeth wasn't allowed to speak. She and every other codebreaker who worked on ULTRA material was bound by oath to keep the ULTRA secret. Even if she had been free to discuss her triumphs, explaining them to the public would have taken some time.

J. Edgar Hoover did not have these constraints. His power

allowed him to manipulate the press and disclose secrets without consequence. And because his agents were old-school detectives, not technical wizards like Elizebeth, Hoover was able to frame the Invisible War in terms of instantly familiar images: disappearing inks, saboteurs, hidden cameras, police raids on clandestine radio stations, gumshoes in snap-brim hats.

So this was the picture of the spy hunt that the public ended up receiving. They got Hoover's story, not Elizebeth's.

Hoover made sure of it. In the fall of 1944, with the Wehrmacht collapsing across Europe, and the Red Army moving toward Berlin, he launched a publicity blitz to claim credit for winning the Invisible War.

He published a seven-page story in *The American Magazine* titled "How the Nazi Spy Invasion Was Smashed." The sub-headline read, "One of the great undercover victories of the war—the defeat of a vast Axis plan to penetrate South America—is revealed here for the first time by the Director of the FBI." Hoover claimed that his bureau had disrupted seven thousand Axis operations, catching two hundred and fifty spies and seizing twenty-nine radio stations, and that these actions had "stopped Hitler in South America. He needed his men and radio stations to carry out his plans of conquest and sabotage. . . . Without his machine he was lost." Hoover did not mention Elizebeth or the coast guard, but he did thank the policemen of Brazil and Argentina, a significant number of whom were fascists, torturers, or both.

He also starred in a fifteen-minute film that was shown to U.S. troops abroad, *The Battle of the United States,* a highlight reel of the Invisible War.

The film, made with the help of beloved Hollywood director Frank Capra, opens with a blast of patriotic music and a waving American flag that fades to a shot of Hoover at his FBI desk, sitting in front of the Stars and Stripes. Hoover wears a pin-striped gray suit and a garish tie. His hair is neatly combed. His hands are clasped on the desk. "I want to talk to you fighting men and women about the Battle of the United States," he says, looking into the lens.

Cut to a shot of a wooden door marked FBI CONFERENCE ROOM. The door opens and the camera moves inside. Seven FBI agents are gathered around an oversize map of South America. Ominous music. The map expands to fill the entire screen. Animated planes fly above the map and drop bombs on the Panama Canal; radio towers writhe with cartoon electricity; a cartoon Nazi appears, dressed in fatigues, holding a bayonet. He stands on Argentina, facing the United States and growing taller until he lunges and stabs Wyoming.

The film goes on to describe roll-up of the radio networks and the destruction of the Duquesne spy ring as solo feats of FBI tenacity. In the final scene Hoover faces the camera and addresses the intended audience, the troops still fighting overseas in the last months of the war. "The attack against all German and Japanese agents in this country," he says, music swelling, "was as vigorous and as victorious as the attacks you have made against the enemy." He ends with a flourish: "We of the FBI feel that we are part of a team, to make America a great and decent place to live. We are on that team, all of us, together."

In December 1944, around the time U.S. troops were watching Hoover's film, Elizebeth sat down to write the Friedman family Christmas card. It wasn't a clever puzzle or a game like in years past. It didn't contain any pictures of Bill and Elizebeth and the kids, or any secret messages. It was just an old-fashioned letter in plain English. After four years of war, and tens of millions of people killed, writing a letter felt like a good and human thing to do.

She typed

*BULLETIN ** 1944 ** FRIEDMAN*

at the top of a fresh white sheet of paper.

"We keep wondering what has transpired in the lives of our friends, and their families," Elizebeth continued. "Perhaps they too are wondering about us."

The director of the FBI had been boasting about catching spies he did not really catch. Elizebeth, who did catch them, bragged about her family.

"Bill, Will, Billy," she typed, and summed up what was allowed to be said publicly about her husband's professional activities during the war:

> Having been retired from active duty in 1941 for physical disability, he has spent from nine to sixteen hours a day carrying on, doing a terrific war job. (Such things make liars of the Army Medical Service, and who doesn't know of like instances?) In March of 1944 he was the first man to be awarded the highest distinction given by the War Department: the Exceptional Service Award with gold wreath, equivalent to the Distinguished Service Cross awarded to those persons who in the field of combat, "perform exceptional service over and beyond the call of duty."

"P.S.," William added, out of modesty. "I didn't write this.—Bill."

As for Elizebeth's own accomplishments in the year 1944, there was nothing to report. She wrote that she was "just carrying on a routine navy job, in an unglorious fashion, unlike her distinguished husband."

"P.S.," William wrote. "Elizebeth always was, is, and continues to be the most fascinatin' woman I've ever known."

Elizebeth spent the rest of the Christmas letter discussing her clever children. John Ramsay was now manager of the football team at his prep school, chairman of the Dance Committee, and president of the Senior Club. Barbara was learning Spanish at her job with the U.S. Office of Censorship in Panama. "There is no rationing in Panama," the mother reported, "cigarettes are plentiful and eight cents a pack, living expenses are about one third of their cost in the States, and her work, she says, is fascinating." Elizebeth signed off by wishing all of the Friedmans' friends and loved ones "a reunited family in 1945."

It was the spring of incendiary bombs. The Allies lit German cities on fire, one after another, in the first three months of 1945. The quantities of bombs were measured in thousands of tons. The RAF dropped more than thirty thousand tons of bombs on Germany in the month of January alone. On a single night in March, the RAF sent 223 planes above Würzburg, dropping bombs, lighting fires that tore through the beautiful old wooden buildings and sent people fleeing through the blazing streets to the river. A grandmother clutched her grandson to her chest, trying to protect him from the flames. Their bodies were found melted together.

The death marches from Auschwitz began in January 1945 as the Red Army closed in from the east. SS guards led the prisoners away from the camp on desolate roads and forced them to walk until they collapsed, then shot the ones who remained standing. The Allies liberated the Dachau concentration camp on April 29. Hitler killed himself on April 30.

On April 19, 1945, eleven days before Hitler's suicide, Siegfried Becker, "Sargo," the greatest Nazi spy in the West, was arrested in Buenos Aires by the Coordinación Federal, the Argentine state police. He had dyed his hair black and was living with a girlfriend named Teresa. There were twenty-six thousand pesos in his pocket, and police confiscated an address book containing the names of more than one hundred associates in Buenos Aires, Barcelona, Bilbao, Rio, São Paulo, Hamburg, and Berlin.

Police took him to Villa Devoto prison, located in a poor neighborhood in northwest Buenos Aires. Becker was outraged. He started to talk. He gave statements about his links with high Argentine officials all the way up to Perón, who was now preparing to run for president.

The statements were quickly doctored, but Becker had made his point: He could hurt Perón if he opened his mouth. Officials responded by treating him as gently as possible. Perón assigned his personal bodyguard, Major Menendez, to look after Becker's health and well-being, and Becker was allowed to spend much of his day in the warden's office instead of a cell. At Christmastime, Becker sent "a huge basket of delicacies and champagne" to officials of the Coordinación Federal and arranged for a friend to deliver seven stuffed turkeys to the prison.

Gustav Utzinger, "Luna," the radio expert of the spy ring, heard about the champagne and turkeys straight from Becker. They happened to be detained in the same prison, Villa Devoto, at the same time, although Utzinger hardly enjoyed Becker's privileges. Horrified by the corruption and brutality of the prison officials, Utzinger frequently defied them and was punished with solitary confinement and beatings.

In February 1946, Argentine voters elected Perón to his first term as president. He moved into the palace with his second wife, the alluring actress Eva Duarte. He no longer feared what the spies might reveal; he was powerful now, insulated from consequence. So he released them. Utzinger tried to start a radio business but was later re-arrested and deported to postwar Germany. Becker stayed in Buenos Aires and was never heard from again. According to the Argentine journalist Uki Goñi, the *Hauptsturmführer* used his connections to smuggle Nazi war criminals into Argentina, helping them escape prosecution. It is likely that he lived to be an old man in Buenos Aires and died a natural death there.

Some of Becker's friends across the continent did not enjoy his luck. On July 21, 1946, an angry mob invaded the presidential palace in La Paz and fatally shot Gualberto Villarroel, who had ruled Bolivia since Becker's coup. The president's corpse was thrown from a palace balcony and hung from a lamppost in the public square.

The Friedmans had been tired and stressed for years now. They had continued to function in spite of it, had kept going in to work every day, Elizebeth to the Naval Annex, William to Arlington Hall. But at the start of 1945, when it seemed like the enemy was on the run, and it finally felt permissible to relax ever so slightly, their bodies fell apart in unison. William got bronchitis and limped around the house, wheezing. Elizebeth took care of him and slept a lot on weekends. On George Washington's birthday, February 22, 1945, they were invited to a dance party with some army friends. Elizebeth realized that it had been years since she had put on nice clothes to go out and have fun, and she allowed herself to spend an extra-long time getting dressed, dabbing Shalimar perfume on her neck, choosing pearl earrings, finding just the right dress, before heading out the door with her hobbling husband, trailing a cloud of warm, luxuriant scent.

President Roosevelt died on April 12, of a brain hemorrhage. Elizebeth was crushed. She had never liked politicians, but Roosevelt was the exception, a man who seemed both decent and brilliant, who believed in democracy, science, equality, and international cooperation, values she held dear, and Elizebeth feared that in his absence, "evil influences" like the Ku Klux Klan would sweep the country: "Our country *will* go on. But who can say what catastrophic results will come from his going? Or worse, the results, hidden, subversive, that take place with no fanfare, no appearance on the surface, but so quietly, hiddenly evil."

The next month, a family friend died of a sudden illness: Colonel John McGrail, the army intelligence expert who had sent Elizebeth an Easter corsage of violets when William was away at Bletchley Park. The Friedmans felt that McGrail had worked himself to exhaustion in the war. During the colonel's burial at Arlington National Cemetery, Elizebeth stood at the side of his widow, Florence, and held her arm as six white horses carried McGrail's flag-draped casket to the grave.

Elizebeth didn't think there was any pattern to death. It was random and cruel. She only hoped it would not strike her own loved ones. They were still scattered across the planet, subject to the winds of the war. John Ramsay, in flight school in Alabama, told Elizebeth on the phone that he had been ordered to report to an unknown destination. He said, with trepidation, "I will let you know where I am, when I am." William kept murmuring in her ear that the U.S. Army was preparing a final mission for him, an assignment that would take him to Europe for weeks or even months. And Barbara was still in Panama, dating young navy officers about to report to the front and worrying they would be killed. Elizebeth tried to prepare her daughter for that possibility: "Remember, darling, my blessed fatalistic philosophy—when the number is up, he goes; whether he is on a davenport in his own home, or firing broadsides at the enemy." She signed off, "Your getting-to-be-an-old-woman, Mother."

Maybe it was her physical exhaustion, or her grief, or her ongoing fear for the safety of her family, but when Elizebeth heard on May 8 that the Nazis had formally surrendered to the Allied forces, that the war with Germany was over, it did not feel over. "We have difficulty believing it is really true," Elizebeth wrote to her daughter in Panama. "New York, we hear, celebrated. But work went on, we didn't even stop to hear the proclamation."

The picture from the news was fluid and confusing. Japan had not surrendered. Its diplomats spoke with a new humility but the generals vowed to fight on. Stalin's Red Army stood poised to occupy defeated territories in the East. Stalin was rumored to be building an atom bomb. The Americans were rumored to be building an atom bomb. Elizebeth wrote to Barbara, "It's absolutely terrifying."

The heat of the Washington summer grew brutal. In the master bedroom of 3932 Military Road she slept with the windows open in the vain hope that a draft might blow in. John Ramsay sent her a poem he had written in flight school, describing his desire to live the adventurous life, "to grip the rock-bound cliffs / that jealously guard the house of wisdom." His mother, delighted by her son's ambition and proud of his interest in poetry, re-copied the poem in a letter to her daughter.

In early July, Elizebeth learned that William's orders had finally come through. He was being sent to Europe for a ninety-day assignment with the Allied forces there. It would be his final mission of the war.

William told his wife he would be conducting research. He made it seem routine. It was not routine. He had been recruited for a mission called TICOM, a joint U.S. and British effort to seize intelligence secrets from former Nazi territories.

One historian has called TICOM, short for Target Intelligence Committee, "the last great secret of World War II." The aim was to preserve Western dominance in whatever the next war might be: perhaps, it seemed, against the Soviet Union. This meant preventing knowledge and technology from falling into the hands of Joseph Stalin. The Western Front needed to be scoured and picked clean of secrets—any Nazi cryptologic inventions snapped up, any information about MAGIC or ULTRA secured—for the United States and Britain to maintain a codebreaking edge in future battles. And in pursuit of this goal, William Friedman would soon journey into the deepest, most secretive chambers of the

Third Reich. He would find himself riding a tram into thin air to reach the Kehlsteinhaus, or Eagle's Nest: the private mountain lair of Hitler himself.

William and Elizebeth drove to the military air terminal together on July 14, 1945, a warm, clear blue morning in Washington. He wore his dress uniform of a khaki jacket, a khaki cap, khaki pants, and brown boots. She thought he looked handsome and wished the army allowed the men to sew stars on the shoulders. She kissed him goodbye and he boarded a Douglas C-54 transport plane along with twenty-three other passengers: army officers, several scientists, a WAC stewardess. Then Elizebeth went back to the family's car, which was parked on a little hill, and waited until the C-54 took off, watching it rise into the clouds, shrinking into a speck of silver.

CHAPTER 6

Hitler's Lair

William Friedman gripped the sides of a jeep as it shook and rumbled up the twisting incline to the laboratory of the Nazi scientist. The codebreaker was in a small town in Bavaria miles from any battlefield, ascending a mountain called Feuerstein. The jeep continued to climb. Looking backward William could see the intact homes and shops of the town below, growing smaller in the distance. Up ahead, the Laboratorium Feuerstein loomed into view. The impression was of reaching a castle on a mountaintop. The building was enormous. Red Cross signs were painted on the roof, a ruse to pass off the institute as a hospital and prevent RAF pilots from dropping bombs.

Dr. Oskar Vierling, an engineer from a poor family, had run this place. Before the war Vierling specialized in acoustics research, investigating the properties of sound and inventing new kinds of instruments; his "electrochord," an electrical organ, was a favorite of the Nazi minister of propaganda, Joseph Goebbels, who adapted the instrument for party rallies, using it to play forceful blasts of chords at predetermined spots in his speeches. In the late 1930s the Nazis insisted that Vierling focus his activities on war machines, and the Laboratorium Feuerstein, named after the mountain which it crowned, began to fill with scientists and assistants under the Doktor's direction, as many as two hundred people.

The Allies wanted to know what Vierling had invented during

the war, particularly any devices related to intelligence or cryptology, so they had dispatched William Friedman, along with a dozen colleagues from army intelligence, to make an inventory of the laboratory. Earlier the defeated Nazis had ordered Vierling to destroy all of his inventions, but he had hidden his favorites in a locked room in the basement and was now glad to show them to the Americans.

The interior of the lab was cavernous, Gothic. William felt like a character in a murder mystery, about to meet an intricate demise. He analyzed Vierling's prototypes. There was a machine said to encrypt the human voice the way that Enigma encrypted text, and a device for scrambling speech to make it unrecognizable to anyone listening on a wiretap. There was an "acoustic torpedo," a machine that shot bullets of sound; a coating for submarines that made them invisible to radar; and a "speech stretcher," an audio playback device that sped or slowed a recording without changing the pitch. It was hard for William to avoid drawing a comparison between Vierling and George Fabyan, the American robber baron who built a deviant temple to science—this place was like a Nazi Riverbank. Inside the Laboratorium Feuerstein, William ate a brief supper of hot dogs, potatoes, coffee, and crushed peaches, and he stayed late into the evening with his American colleagues, the topics of their conversations growing spookier, starting with Albert Einstein and the theory of relativity before veering off to more occult topics like the possibility of extrasensory perception.

Vierling's laboratory was only one target of TICOM, the mission to lock down the intelligence secrets of the war. The Allies deployed six TICOM teams to Europe starting in April 1945, each containing eight to fifteen intelligence personnel from both the United States and Britain. William's team began its work in late July. The mission carried him for hundreds of miles across southern Germany, France, and Czechoslovakia. He took notes. He had spent the war in office buildings. This was his first look at the landscape of physical battle, and it filled him with "a heavy feeling of sadness" difficult to describe. The German countryside was intact, the plots of wheat and rye and barley ripening, the green forests seemingly untouched, but the people were broken, and the machines were broken. The highways between cities were full of what the army called DPs: displaced persons. Mothers and fathers walked along the road with their children, carrying personal belongings on their backs or hauling small quantities of wood in handcarts, to burn for fuel. Wrecked trucks and tanks had been tipped into ditches, and women wept in the backs of clattering wagons.

He ate a C-ration for the first time, canned meat and beans. It didn't agree with his stomach. He ate Spam. He stayed at an army installation code-named BARN and thought that if it had been up to him he would have named it something less boring, like LEPIDOPTERA, the scientific name for a butterfly, but then he thought about it some more and realized that LEPIDOP-TERA would be more difficult for soldiers to remember and they might end up getting lost. He rode at slow speed through the fire-bombed cities of the Reich, through Frankfurt, Nuremberg, Aschaffenburg, and Würzburg, feeling like an ant crawling across the body of a dead child. "The destruction to be seen in cities such as these should be noted by anybody who believes in war," he wrote, "because it can tell more about what happens in modern warfare than reams of literature." More than once, while scouring an abandoned Nazi garrison, the men on William's team found copies of his own cryptologic publications from years past, trans-lated into German or French by the Nazis. This was hardly a sur-prise, given the ubiquity of William's contributions to the science, and the men got a kick out of it, grinning as they showed him these discoveries—and the bibliophile in William could not resist taking these extremely rare documents as souvenirs and keeping them for his own library—but it was the strangest compliment. Imagine walking into the devil's library and seeing your book on his shelf.

William always believed the war was worth fighting. But he saw it as a grim duty, not a crusade, and his experience of fight-ing the war had permanently destroyed his faith in the way the world was put together. Earlier that year, his daughter asked him in a letter if he believed in Zionism, the project to create a Jewish homeland. He said no. "Zionism is only one of many virulent forms of a detestable disease known as 'nationalism,'" William wrote to Barbara. "The sooner we realize that we are all God's children regardless of color, race, creed, nationality, etc., the better for all nations and the world as a whole." He didn't believe in nations anymore, not even his own. This is what he had tried to tell his daughter. The world is very fragile, more fragile than it is healthy to believe if you want to get out of bed and make it through the day.

William did not sleep well the night before he visited Hitler's alpine lair: Kehlsteinhaus, the Eagle's Nest, a meeting center and getaway built by the party in 1939 and given to the Führer as a gift on his fiftieth birthday. An army friend of William's made steak sandwiches with raw onion at 1 A.M., and the cryptologist woke to the smell and downed the steak with Scotch. At 8 A.M. he rode in the army staff jeep at the front of a fifty-vehicle convoy to the town of Berchtesgaden, overlooking the Austrian Alps. The path to the lair went straight up a mountain for four and a half miles, no guardrails to prevent an errant car from plummeting several thousand feet to the valley. The driver kept the jeep in first gear all the way.

The convoy reached a plateau with a parking area at the base of the lair, and William entered a hundred-foot-long tunnel protected by two enormous, heavily ornamented bronze gates. The tunnel was wide enough for three automobiles and brightly lit with electric lamps. At the far end of the tunnel, an elevator shaft the height of a fifteen-story building rose the rest of the way, to the mountain's pinnacle. The elevator operator was the same German who had worked for Hitler and his deputies all during the war. William talked to him for a bit. "He gave us a little speech in his defense, saying that he was assigned to the job—the Nazis wouldn't let him get away, etc., etc."

The ride to the top took three minutes. Then William and his fellow officers were led through a passageway and into the first chamber of the lair, a small, wood-paneled dining room where Hitler held banquets with visiting world leaders. The Führer didn't visit often—the long automobile climb made him impatient, and the change in atmospheric pressure disagreed with his constitution—but Goering and Ribbentrop spent a lot of time at the Kehlsteinhaus, and Hitler's mistress, Eva Braun, loved to entertain friends here. She lugged her Scottish terriers up the mountain and let them romp in the thin air. She threw a wedding party for her sister and the groom, an SS captain later shot by Hitler for desertion.

Beyond the dining room and down a few stairs was an octagonal room with thick granite walls and a 360-degree view of the Alps. From this height the mountains appeared at eye level, jagged triangles of green and blue, like incisors rising from earth's jaw, "indescribably beautiful," William thought. A third room contained a fireplace of red marble, a gift from Mussolini. Most of the furniture was intact. The Americans milled around like tourists atop the Empire State Building, unsure what to do with themselves after the first minute or two. William took pictures and wished he had brought a film camera.

The group descended in the elevator and drove a mile back down the mountain to a level area where the Nazi party had built Hitler a separate residence, a private house. The front of the house was a twenty-five-foot-wide plate-glass window with no glass in it. The building had been almost completely destroyed by an RAF blockbuster bomb and by subsequent visits from American soldiers, who wrecked what was left of the house before the army stopped them. William thought it was a shame. "I think it is too bad that this whole installation was not left absolutely intact to serve as an everlasting and terrible monument to the folly of a people led to perdition by a madman's lust for power."

The floor of Hitler's house was littered with chunks of rock and marble. William reached down and picked up a piece. When he got home, he decided, he would keep it on his desk, as a reminder. "I shall have it made into a paper-weight."

That day, back in the States, Elizebeth was in Michigan, visiting her sister. She stayed at a hotel in Ann Arbor that brought her a bowl of ice in the afternoon. She looked at this simple object, this bowl of smoking ice, the unthinkable luxury of it, in wonder and awe, and remembered that it was not normal for humans to spend their afternoons trapped inside a 100-degree building in Washington, sweating through their clothes, solving puzzles to save the free world. She remembered she was alive.

One evening she read an issue of *The New Yorker* straight through.

Germany was only the first leg of William's TICOM mission. The second leg took him to Bletchley Park, headquarters of British codebreaking, in late July. He arrived there on July 28, eight days before America dropped the first atomic bomb on Japan.

As he had done on his previous visit to Bletchley, William kept a detailed diary. One entry described a meeting with Alan Turing: "At 1535 a visit with Dr. Turing. He is leaving GC&CS, to my surprise. Says he's going into electronic calculating devices and may come to the U.S. for a visit soon. Invited him to visit us if he comes to Washington." This turned out to be the final encounter of William Friedman and Alan Turing. The two geniuses would never see each other again. In 1952, the British government stripped Turing's security clearance on grounds that he was a homosexual, and officials coerced him into taking estrogen injections. Turing's maid later found him dead of an apparent suicide, a half-eaten apple by the side of his bed, traces of cyanide in his blood. A government witch hunt had destroyed one of the war's greatest heroes.

William's goal in England in July 1945 was to learn how the war had looked to his Nazi adversaries, the code and cipher experts employed by the Reich. He read cryptologic materials seized by British intelligence and observed interrogations of Nazi prisoners. Twice he visited a manor in the country village of Beaconsfield, noting that he could "say no more." This was a POW camp where he encountered at least three high-value German POWs, including two of Nazi Germany's top cryptologists, Dr. Wilhelm Fricke and Erich Hüttenhain. William didn't conduct the interrogations but he did observe and suggest questions. After listening to the POWs and analyzing the documents, William concluded that Germany had never lost faith in the security of the Enigma machine. They thought Enigma was unbreakable all the way to the end. He was proud to learn that Nazi codebreakers had never managed to defeat America's best cipher machine, the SIGABA, which he had invented with Frank Rowlett.

He had a lot of downtime in England. The pace of things in the codebreaking offices had slowed. The buildings seemed to be emptying out. At night he took Amytals and crawled into bed. One evening a friend took him to a burlesque show in London, the famous *Les Folies-Bergère*. They sat in two plush seats in the front row. The women wore g-strings and glittery, spangly tops, and William admired the looks of intense concentration on their faces, their cool self-possession. "The girls devote their complete and absorbed attention to their work—not even a glance or a wink at any member of the audience."

Once or twice in the slack moments of the days, the British asked him to tell stories about crazy old George Fabyan and his merry band of conspiracy theorists. People seemed to love the Riverbank stories, and William loved to tell them. It was funny how he felt more and more generous toward Fabyan by the year. You get older and want to connect to the people who understand. You try to speak with the young and find that something is wrong with your ears. They use their own slang, their own code, and you start to feel nostalgic about your former enemies, who at least shared the same intense moment on earth and spoke words you could understand. Besides, if not for George Fabyan, William would not now be carrying a piece of Adolf Hitler's smashed marble floor in his pocket.

It had all begun in the most bizarre fashion.

William was in London on August 6, 1945, the first day of the nuclear age. He was asleep and dreaming in his room at the Hunt Hotel when the atomic bomb fell on Hiroshima and destroyed it in seconds, its people and its history, like a page torn from a book. A latch opened in the American B-29 at sixteen minutes after midnight, London time, which was 8:16 A.M. Hiroshima time, and the soldiers on the plane became the first humans to see, from a safe distance, what this technology could do to a living city. "Here was a whole damn town nearly as big as Dallas," the radio operator of the *Enola Gay* later recalled, "one minute all in good shape and the next minute disappeared and covered with fires and smoke." Above Japan, the plane peeled away from the mushroom cloud, while in London the cryptologist's eyeballs darted frantically behind closed eyelids. He was having a sex dream about Enid, the wife of his friend Stub. When he woke in the morning, he felt confused. He had never thought about Enid that way. He wrote in his diary that he would have to tell her. She'd think it was funny.

It took a while for the news to reach London, and William was busy with his work, so he didn't hear about Hiroshima until breakfast on the morning of August 7. He went to lunch that day with Eddie Hastings, the Royal Navy captain who had visited the Friedmans' house on the day of Pearl Harbor, and after drinking a few martinis in befuddled silence, William and Hastings spoke about the bomb. They were in agreement. They didn't understand why it was necessary to kill so many civilians merely to demonstrate the bomb's power. Hastings thought "it was [a] serious mistake to drop the first one on a big city—should have stated the case, given warning, dropped 1st one on a vacant area & then make renewed call to surrender," William recorded in his diary. "I think he is right. Early reports indicate over 350,000[*] people wiped out in Hiroshima—perfectly ghastly, no matter who the enemy may be." He added, "We all here agree that this new weapon represents the last call on man to give up war—or else!"

The second atomic bomb fell two days later, on Nagasaki.

At home in Washington, waiting for the war to end and for her family to come home, Elizebeth sat on the porch in the evening and wrote letters to the children and William, taking breaks to go upstairs and listen to radio bulletins. The weather turned cool and rainy. Each time she wrote to William she had to use a different address because he seemed to be moving all over the place. She worried he wasn't getting her letters. His letters reached her after a time lag of seven or eight days, which meant that their letters were crossing midstream.

---

[*]   The true number was closer to 75,000 dead and another 75,000 injured.

On August 7, when the radio carried news of the atomic bombing of Hiroshima, the news didn't bother Elizebeth as much as it did William. She wrote to her husband that day, "Everyone is saying this will end the war P.D.Q."—pretty damn quick. "I wonder! Much too good to be true, say I."

The morning after the Nagasaki bombing, August 10, William got up in his London hotel, shaved, showered, ate a breakfast of bacon and egg, and took a bus to one of the city offices of his British colleagues. It was a bright day and the sun felt good on his face. After lunch he decided to get out and watch some tennis at a nearby public court, a mixed-doubles match, and he was enjoying the high quality of play when at 1 P.M. a U.S. Army lieutenant came running to tell him the Japanese had accepted surrender terms presented to the emperor. Unofficially, the war was over.

Word reached Elizebeth at the Naval Annex that day, and at 7:30 P.M. she started a new letter to her husband, writing at the top, "A day we will remember!" She told him she was getting tired of managing everything at home, fixing things around the house, sopping up after a small flood in the basement, getting the car ready for its annual inspection ("the fenders cost $30, the other work $13"), "all chores and no play with my Sweetheart. But maybe if V.J. comes true, we can both go play for a long vacation."

"By the time this reaches you," William wrote four days later, "the end of the war will be a fact." He said he was sorry she was tired and knew it was hard to be alone. He suggested sweeping the leaves from the sewer inlet on the side of the house to prevent rainwater from backing up and leaking into the basement. William also replied to a warm letter from John Ramsay, who had asked his father which books he should read to educate himself in spare hours at the Army Air Corps barracks. William recommended *So Little Time*, a war novel by J. P. Marquand ("so good"), and ended the ten-page letter by praising his son's vocabulary: "You've improved remarkably in penmanship and format. You do yourself proud, in fact. I found only one or two orthographic irregularities or aberrations (misspellings, to you!). They are of no consequence. Dad."

Japan surrendered unconditionally on August 14. President Truman declared a two-day holiday for federal workers. A crowd of 75,000 gathered at the White House, ringing bells, blowing horns. Elizebeth stayed in and tried to recover from a stomach bug, drinking clear consommé and ginger tea. "Bobbie, darling," she wrote to her daughter, who was scheduled to return from Panama in September, "I sure am counting days—only 23 more and you will be here! Oh, frabjous, frabjous day!"

She watched the lights come on in America. Gas stations resumed normal operations. The newspapers said nylon hose would be available soon. Shoes by Christmas. The military was starting to release large numbers of personnel who were no longer needed. Arlington Hall ordered a 50 percent staff cut by September 31 and another 25 percent cut by December 31. The Naval Annex in Washington had been emptying for weeks and was down to a skeleton force, quieter than Elizebeth had ever seen it. The temporary workers, including most of the WAVES, no longer necessary in peacetime, had been released without so much as a thank-you cake, and permanent employees were escaping the Annex for offices in more modern buildings.

Elizebeth knew she needed to make a decision about her future, about what to do with her postwar life. Her superiors at the coast guard said they wanted her to stay, to keep her code-breaking unit together in peacetime and return to smuggling investigations, but she couldn't see the point. There wasn't a lot of smuggling traffic anymore.

She noticed with detached amusement that American intelligence officials were scrambling to cast themselves and their agencies in a favorable light so that they might keep their jobs in peacetime. "The O.S.S. is starting a deluge of publicity," Elizebeth wrote to William, mentioning the wartime spy agency of Wild Bill Donovan, for whom she had worked in the early months of the war. "Fight against extinction, I suppose."

Elizebeth heard a radio interview with a New York man who taught cryptology classes. He was discussing the importance of codebreaking to the Allied victory. Elizebeth knew him to be a minor figure and wondered, in an offhand way, how he ever got the spotlight.

She turned fifty-three years old on August 26, a cold Sunday. Her coast guard colleague, Lieutenant Jones, came over with his wife, Gertrude, and Elizebeth Friedman, secret hunter of Nazis, cooked a dinner of minced clams in cheese sauce, a tossed French salad, and hot borscht that everyone agreed tasted wonderful in the cold evening. William spaced out his birthday presents to Elizebeth over four days, starting the previous Friday with a cable from Bletchley Park. ABSENCE ON YOUR BIRTHDAY DARLING SADDENS ME BUT HOPE FLORAL AMBASSADOR . . . WILL VOUCHSAFE UNDYING LOVE. On Saturday a mutual friend hand-delivered a box of perfume to her door. Sunday morning, a dozen roses came, along with one of his business cards. The front said only, MR WILLIAM F. FRIEDMAN, and on the back he had written, "I love you! I love you! I love you! Bill." The next day, Monday, a letter arrived from William, written eight days earlier and timed to reach her right now. "I find it hard to tell you how much I miss you and love you—you're the most wonderful person to have for wife, helpmate, lover, and all," he wrote. "Save some special kisses for me when I get back. . . . I miss you."

She realized how tricky it must have been to coordinate all these gifts and messages across the distances of the war, the disrupted postal and cable lines, and land them to her at the exact moment of his choosing, around her special day. The timing alone was a performance of devotion.

"Dearest," Elizebeth wrote, "what a darling you are!"

He was ready to come home "very soon," he told her in a letter from London three days after her birthday. He said he was already thinking about what would come next for him and for the family. He wanted to make the kind of money that would give them the freedom to travel and pursue their dreams. National security concerns had always prevented William from profiting from the cipher machines he invented, but the war was over now. Surely he would be permitted to patent his ideas and commercialize them? When he was finished describing these thoughts to Elizebeth he rotated the sheet of paper and filled the side margin: "I LOVE YOU! I LOVE YOU! I LOVE YOU! *VERY MUCH* (I shall have that printed as a border on my special stationery to you.)"

Four days after that, on September 2, in Tokyo Bay, General

Douglas MacArthur accepted the Japanese surrender aboard the battleship USS *Missouri*. Elizebeth heard Truman say on the radio, "It is our responsibility—ours, the living—to see to it that this victory shall be a monument worthy of the dead who died to win it." She agreed that there was no point in having fought a war to preserve freedom if people used that freedom to start more wars. She wondered if William heard the same Truman broadcast. "You are the dearest and best husband any woman ever had!" she wrote on September 4. "Roses lasted until today. All, all my love, Elsbeth."

Around September 12, in Prestwick, England, William finally boarded an army transport plane. He flew to the Azores, then Bermuda, then New York, each minute of the long flights an agony of anticipation. It was raining when he landed in New York and got on a train for Washington. The sky when he stepped out of Union Station was a slab of gray and the air was violent with fat drops that followed him to his office at Arlington Hall. Arlington Hall was like Bletchley Park had been, emptier than he ever remembered it, big empty rooms and echoing hallways and a handful of people carrying boxes around and packing up files. He could not concentrate on anything because he knew he would see Elizebeth soon. He waited out the day and it was still raining like crazy when he left the heavily guarded military facility and shoved his dripping luggage in a taxi and rode home, to the house at 3932 Military Road.

Elizebeth opened the door. She cried out in joy. His clothes were wet. His mustache was wet. She reached up and threw her arms around him and squeezed as hard as she could.

The months after V-J Day were a period of limbo for U.S. intelli-gence. All the agencies were thinking about how to extend the gains of the war and also justify their own existence in peacetime, when the government would surely contract. The future of cryptology was especially murky.

It was obvious to William and many others that there ought to be a centralized cryptologic function in America, one agency that gathered intelligence from wireless signals and broke the codes that must be broken. As an elder in the cryptologic community, a person who had not only invented many of its tools but also built a successful organization within the army to apply those tools, William was involved in these discussions at the highest levels—discussions that would give birth, in 1952, to the National Security Agency. In the meantime he entered a phase of furious personal documentation, writing technical descriptions of his cipher machines and applying for new patents in hopes of commercializing the inventions.

Elizebeth was documenting, too; not for commerce but rather for teaching and history. At the Naval Annex she sorted through the voluminous files of her coast guard unit, tens of thousands of intercepts, worksheets, memos, translations, and decrypts. Working with Lieutenant Jones and other colleagues, she produced a detailed technical account of their unit's work between 1940 and 1945, a 329-page book that detailed all forty-eight of the Nazi clandestine radio circuits and how the coast guard broke the codes. The book was secret, meant only for other intelligence agencies to use as a reference and perhaps also for historians of codebreaking in the far future. Five copies were printed, with dark green covers, and every page of every copy was stamped TOP SECRET ULTRA.

With the technical history complete, Elizebeth was told to mark a percentage of the unit's documents for preservation and destroy the rest. She decided to keep four thousand decrypts— the typed, solved messages from the forty-eight Nazi radio circuits. These she organized for transport to the classified areas of the National Archives in Washington. The phrase "government tombs" occurred to her. That's what it felt like. She was burying her experiences in Uncle Sam's mausoleum.

When the task was done, Elizebeth prepared to leave the Naval Annex for the last time. The navy forced her and all other departing workers to sign secrecy oaths that demanded their silence unto death. They could never tell anyone what they did in the war, under penalty of prosecution, for as long as they lived. They could not even tell their grandchildren.

At the end of her final workday, Elizebeth walked down the

stairs from the second floor to the first, went out past the turn-stile where the first marine guard stood watch, then past the second marine guard, to the other side of the barbed-wire fences, until she was standing on the sidewalk on Nebraska Avenue. She crossed the street, paused for a few seconds, and looked back at the grubby, flat-roofed building where she had spent her war. She knew in that moment that she would never again return "to that particular form of endeavor"—breaking codes for the coast guard. "I was back in the world-at-large once more," she wrote later. "It was the end of a Period, an Era."

She was still a coast guard employee, and soon Elizebeth found herself back at her old desk in her old prewar office in the Treasury Annex, near the White House. But she had an exit plan. She was only going to stay long enough to complete a single job. At the Naval Annex she had sorted and filed the records of her clandestine war against the Nazis. Now, at Treasury, she needed to do the same for her smuggling cases of the 1920s and '30s. The smuggling records had been gathering dust during the war—"thrilling records in many respects, detective stories of high interest in many cases," Elizebeth recalled. "The past had been rich in accomplishments. I should see that everything was prepared for posterity to comprehend, if posterity should ever choose to examine the archives."

From the late fall of 1945 to summer 1946, Elizebeth conducted her last campaign for the United States: organizing and indexing the paper archive of her cat-and-mouse tussles with rum lords and drug gangs. Because the records were old and contained no national secrets, she was allowed to keep personal copies for her own library. Then, the task complete, she recommended to Treasury that the department abolish her coast guard unit, along with her job, on the grounds that it served no national purpose in peacetime. They obliged. On August 14, 1946, the coast guard notified her that, "In view of the curtailment of cryptanalytic activities previously performed by the U.S. Coast Guard, it has been necessary to effect a reduction in personnel," and she was hereby terminated at the close of September 12, 1946. Her salary at the time, the most she ever earned, was $5,390, or $67,000 in today's dollars.

J. Edgar Hoover used his influence to expand the FBI after the war. Elizebeth used it to get out of the game.

She had never really wanted to be a government employee anyway. It was only the constant requests from "people on my doorstep" that had gotten her into it in the first place. Now, with the war over, her thoughts turned to projects and desires she had put on hold to serve her country. She still wanted to finish her long-in-progress children's book about the history of the alphabet. She wanted to visit Barbara at Radcliffe and see how John Ramsay was living at the Army Air Corps base in Biloxi, Mississippi. And she wanted to reconnect with William and find a way to collaborate with him. The Friedmans had lived for years in an awkward and isolating silence, working in separate but adjacent government bunkers, afraid to speak freely even in their own home. No more! Goodbye to that! They wanted to work together on something again, and they had the perfect idea.

Elizebeth and William had never lost their fascination with the varieties of occult theories they first encountered in their youth at Riverbank; they never stopped wondering why people believed things that weren't true. The previous December, when the war was still on, they had attended a sold-out Washington show by the Amazing Dunninger, the foremost mentalist of the day. A New Yorker with a poof of brown hair and a tuxedo, Dunninger was both debunker and illusionist; he explained onstage how spirit mediums usually worked, showed that he was not using any of those tricks—and then read the minds of audience members anyway. William and twenty-five other intelligence men planted themselves here and there throughout the crowd at Constitution Hall in an attempt to learn his methods and "came away with *theories* as to how it's done, but no proof," Elizebeth wrote in a letter to her daughter. "The mere fact that Dunninger is still going strong is proof that human beings, the credulous dears, *want* to believe in the mysterious and supernatural."

It had not escaped Elizebeth and William that many people continued to believe the theory that the two of them had rejected in their earliest days at Riverbank, way back in 1917: that Francis Bacon placed cipher messages in Shakespeare's plays.

The community of Bacon obsessives was still around, alive and kicking, publishing new articles and arguments. After Mrs. Gallup died in 1934, followed by George Fabyan in 1936, the Baconians lost two of their most famous and energetic proponents, but others picked up the torch. In 1938 the son of Teddy Roosevelt, Theodore Jr., asked the Friedmans for an opinion on a cipher system devised by an economist named Dr. Walter McCook Cunningham. Roosevelt Jr. was vice president of the Doubleday publishing firm and Dr. Cunningham had submitted a manuscript about his cipher. The method was based on anagrams, and the Friedmans quickly recognized it as bunk. To demonstrate the cipher's folly, they applied Cunningham's method to a page from *Julius Caesar* to produce the following message, which they sent to Roosevelt Jr.:

> *Dear Reader: Theodore Roosevelt is the true author of this play but I, Bacon, stole it from him and have the credit. Friedman can prove that this is so by this cock-eyed cypher invented by Doctor C.*

The experience got them thinking that they should lay out their skeptical arguments in a book of their own, explaining once and for all why these ideas about secret messages in Shakespeare were only fantasies. The Friedmans obtained a pittance of a book deal from a British publisher (advance on royalties: 250 pounds) and went to work. For the sake of the project, they decided to sell their beloved house on Military Road and bought a spacious, high-ceilinged house on Capitol Hill within walking distance of two libraries where they needed to do research, the Folger Shakespeare Library and the Library of Congress. Many who live on Capitol Hill are lobbyists. The Friedmans moved there to be close to libraries.

They transported their own precious books and papers to the new house, reassembling their private library in the den of the second floor, and rehung the axe on the wall as a warning to potential book thieves. And together, researching and writing, they galloped back through the past, weighing the arguments of Baconians and cutting them to pieces. In their hands *The Shakespeare*

Ciphers Examined became a story about the drug of self-delusion and the joy of truth. One section analyzed the cipher system of a French general that had revealed the secret phrase IF HE SHALL PUBLISH. The Friedmans showed that the cipher could just as easily have produced the text IN HER DAMP PUBES. George Fabyan received the full brunt of their scrutiny. The Friedmans wrote that while Fabyan possessed "great natural gifts of energy and dynamism," he was a salesman, not a scientist, and suppressed facts he didn't like. As for Mrs. Gallup, "a sincere and honourable woman, and no fraud," she "found in her texts what she wanted to find" and "was therefore at the mercy of the promptings of her expectant mind."

The Friedmans wrote with a ruthless honesty because that's who they were as people. Still, working on the book made them realize how much they owed the misguided mentors of their youth. In the preface they thanked Mrs. Gallup, "whose work on the question of Shakespearean authorship aroused our life-long interest in the subject," and they thanked Fabyan, too—for introducing them to Mrs. Gallup. They were genuinely grateful. Elizebeth said she and her husband had decided to "give the devil his due," and in later years Elizebeth even went as far as admitting, "Vile creature that he was in many ways, George Fabyan really launched two or three things that were of vital importance to this country," which was true. For all his malice and superstition, Fabyan threw enough money at actual scientists to accelerate the discovery of actual knowledge. He funded investigations of Nature with a fortune that other tycoons would have spent on yachts and jewels. He succeeded in creating the first real code-breaking institution in America, Riverbank Laboratories, an idea factory christened by wartime realities. It not only forged a new science of immense power; it also spawned a love affair that spread the science and ultimately sharpened it into an antifascist weapon. The modern-day universe of codes and ciphers began in a cottage on the prairie, with a pair of young lovers smiling at each other across a table and a rich man urging them to be spectacular.

Until she started researching the book in 1946, Elizebeth always insisted that her life in secret writing was an accident, a series of unpredictable chases, mazes, escapes, and detective capers. Now, viewing her life from a distance, she understands there might be order in it after all, a taut line stretching back through the decades and terminating at that mad place on the prairie.

To help herself write vividly about Riverbank, Elizebeth sits in the new house on Capitol Hill. She closes her eyes. She tries to imagine herself thirty years earlier, in the summer of 1916, a young woman at a rich man's estate, unmarried and free, her whole life in front of her.

A fragrance of overripe banana wafts up. William's fruit flies in the windmill.

The fire pit at night. The chemical reek of a mortar bursting near the ordnance lab. Fatty pork on her dinner plate from pigs slaughtered at Fabyan's word.

Silver blade of river, dome of prairie sky.

She remembers riding bicycles with her friend William Friedman, rushing past lawns and flowers thickened with summer rain, a blur of green and pink. She remembers the low Illinois sun streaming through the windows of the Lodge as she works there with Mrs. Gallup, struggling to see what the older woman saw, squinting through a magnifying glass at a page of Shakespeare, trying and failing to free the imprisoned ghost of Francis Bacon.

Mrs. Gallup and Fabyan keep telling her, try harder. The messages are there.

And there comes a day when Elizebeth just thinks: no.

There is nothing wrong with me. What's wrong is with other people.

This is the moment that hurls her out to the rest of her life. The savaging of Nazis, the birth of a science: It begins on the day when a twenty-three-year-old American woman decides to trust her doubt and dig with her own mind.

The room is dark but her pencil is sharp. An envelope of puzzles arrives from Washington, sent by men who have the largest of responsibilities and the tiniest of clues. With William she examines the puzzles. He is game, he looks at her with eyes like little bonfires, he is in love with her. She is not in love yet but she would not be ashamed to fall in love with such a bright and kind person. She stares at the odd blocks of text and starts to flip and stack and rearrange them on a scratch pad, a kindling of letters, a friction of alphabets hot to the touch, and then a flame catches and then catches again, until she understands that she can ignite whenever she wants, that a power is there for the taking, for her and for anyone, and nothing will ever be the same. The ribs of a pattern shine through. Something rises at the nib of her pencil and her heart whomps away. The skeletons of words leap out and make her jump.

Girl Cryptanalyst and All That

The Friedmans in their home library, 1957.

The government came for their books on an otherwise ordinary Tuesday in 1958. Scattered clouds, cool midwinter sun. William and Elizebeth were inside their home on Capitol Hill and heard a knock on the front door. They opened it and saw at least three men from the government. Behind them, on the street, was a rented truck, as if the men planned to remove something large from the house.

The Friedmans let them inside. One of the men, S. Wesley Reynolds, was the NSA's director of security. A second man worked for Reynolds, and a third worked for the U.S. attorney general.

The men asked to see the home library. The Friedmans brought them up to the second floor.

Elizabeth was sixty-six now, William sixty-seven. His health was precarious but the men didn't know that. They said they had orders to remove a list of books and documents that the NSA wished to reclassify according to a Defense Department order of July 8, 1957, Directive 5200.1, which declared that cryptologic documents previously marked "Restricted," a low level of classification, were now upgraded to "Confidential," a higher level. To the horror of the Friedmans, the men started to pull things off the shelves. They removed forty-eight items, including an entire personal safe full of William's documents, several manuals he had written about cryptology, envelopes of his lecture cards and notes, and his own articles from every phase of his career, including Riverbank, forty years ago.

According to a rumor that later spread through the agency, William "went berserk and he was throwing books around and saying, 'Take this, take that.'" The junior NSA employee who went to the house denied this but admitted that both Friedmans appeared "obviously upset by the action being taken." The NSA's Reynolds wrote in a memo three days later, "Mr. Friedman voiced no objections to my taking this material, however, it was quite obvious that he felt deeply hurt and that the material was being taken for reasons other than Security. He stated that this material deals with the history of cryptography and should belong to the American people."

William didn't understand why information about hand ciphers from the First World War needed to be seized. The ciphers were obsolete. Was it really necessary to seize papers from 1917 and 1918? To raid their home, their sanctuary, their archive of knowledge? He told a friend, "The NSA took away from me everything that some nitwit regarded as being of a classified nature."

As the men worked, carrying files out to the truck, Elizebeth looked on in silent rage, barely suppressing her tongue. She considered this a violation of their privacy and worried it was bad for William's health, which had corroded in the thirteen years since the war, darkening with the mood of a city where counterintelligence had become an obsession. Soviet spies had stolen nuclear secrets from the Manhattan Project, and the FBI and the House Un-American Activities Committee went hunting for communist agents. "The mad march of red fascism is a cause for concern in America," J. Edgar Hoover said to HUAC, promising that the bureau would attack and expose "the diabolic machinations of sinister figures." Senator Joe McCarthy destroyed people's careers with no evidence at all.

William's depression had returned in 1947. At first he complained to a doctor of "psychic giddiness" while walking and playing golf; the condition manifested itself as a tendency to walk to the left. The giddiness was followed by increasingly severe bouts of insomnia. Unable to sleep, on January 23, 1949, he checked himself into the psychiatric ward of the Veterans Administration hospital in Washington, where doctors placed William with a group of deeply psychotic patients. He hated it there. He went home and continued to deteriorate. By January 1950, William was unable to work or solve puzzles, his mind and muscles seeming to move at one-third or one-quarter speed, and suffering from acute despair. He had suicidal thoughts. His son found a rope and a noose at the house. A friend noticed a length of rope in the backseat of William's car and asked about it. William replied in a joking tone, "I'm looking for a tree to hang myself."

Desperate for a solution, he sought out a new psychiatrist in March 1950, Dr. Zigmond Lebensohn of George Washington University Hospital, who was an early proponent of electroshock therapy. William agreed to try it. The first course of shocks began on March 31, 1950. The legendary William Friedman was repeatedly electrocuted while awake, possibly without muscle relaxants (they were not widely used at the time), a heavily padded tongue depressor placed in his mouth to prevent him from breaking his jaw by grinding his teeth when the seizure hit. After six courses of shocks, five to fifteen shocks per course, William was sent home on April 11, 1950. Lebensohn observed that the patient "was almost elated when he was discharged and in a characteristically effusive way he kissed the nurses goodbye in a rather avuncular fashion. About a month or so later I saw him and his wife at a Toscanini concert at Constitution Hall."

William's illness took a toll on Elizebeth. Hair graying at the temples, perhaps shrunken by an inch (she considered herself five feet and two inches tall now instead of five three), she was 110 pounds and thinner than she'd been since she was a girl. In the polite phrasing of a girlfriend, "Anxiety kept her figure slim." Retired from government and earning a tiny pension, she spent increasing amounts of time taking care of William. On mornings when he was depressed, she helped him get dressed, drove him to work, walked in with him, placed a pen in his hand, and moved his hand to get the pen moving. She answered his professional mail when he was incapacitated in mental wards. Somehow she still made time for friends and hobbies. She surprised her friends by getting serious about cooking, hosting dinner parties themed around the dishes of India, Mexico, Italy: "I found it an outlet for some hidden creative instincts perhaps." She looked after her neighbors, once appearing on a sick neighbor's doorstep with a tray of roast lamb, roast potatoes, gravy, and a yellow rosebud in a vase. She stayed active in the League of Women Voters, researching the legal status of women, international relations, finance, and the urgent need for D.C. statehood. "At the drop of a hat," she wrote, "I will turn on a spigot labeled SUFFRAGE FOR THE DISTRICT OF COLUMBIA!"

It often seemed that she had forgotten her own career in codebreaking, that she was content to see her identity and history wash away. This wasn't the case. In 1951 she received an invitation to speak about her life in codebreaking to a women's social club in Chicago founded by the first female judge in Illinois. At first Elizebeth urged the group to reconsider: "That part of my life is over, my dear," she wrote the chairwoman. "You are asking a Has-Been to speak! Your audience will feel cheated, I am sure." But then she wrote a speech and traveled to Chicago with a suitcase full of lantern slides and at least fifteen mutilated sheets of paper, typed and cut with scissors and taped back together into a new order while she had agonized about what to say, and as soon as Elizebeth introduced herself to the women of the club, the beautiful hopeful postwar women of Chicago, they were hanging on her every word.

Speaking in a pink ballroom at the Blackstone Hotel, where the women had gathered for a dinner-dance, Elizebeth made it clear she wasn't free to talk about her life during the Second World War, but she was happy to share anything else, to answer any question at all. "Perhaps you may think that the expression 'code and cipher expert' describes a person who must live in a world apart," she said, then explained why this is a misconception. Your child's report card is a code. A is good, F is bad. It's not a world apart. It's just the world.

Elizebeth showed slides of code messages from her famous cases. The *I'm Alone*. The heroin network of the Ezra twins, "SOLVED BY WOMAN." The polite Canadian gangsters of the Consolidated Exporters Corporation. The women of Chicago kept her there, asking questions, transfixed, and afterward, Elizebeth received more speaking invitations, traveling to Detroit and giving her talk to a pair of neighborhood groups in private homes; one of the groups asked her questions for two and a half hours. They seemed to think that the story of Elizebeth Smith Friedman was one of the greatest they had ever heard.

Every once in a while, the urge struck Elizebeth to write it all down in one place. She wondered if history would remember her. One winter she and William traveled to England and attended a luncheon at Cambridge with two of their colleagues from the war, including Elizebeth's cheerful comrade, the astronomer Chubby Stratton. The men at the table got to talking about the war. "As befits a woman in the monastic traditions of Cambridge, I said little," Elizebeth recalled later, "but my own recollections began to boil up from the cauldron of memories."

After the luncheon she took out a sheet of lined yellow paper, wrote "FOREWORD" at the top, then described her feelings after V-J Day in 1945, when she "folded my tent to steal away" from the coast guard after six years of "exciting, round-the-clock adventures as we counter-spied into the minds and activities of the agents attempting to spy into those of the United States." She continued for seven pages, hinting at the dramas and capers of her war without going into specifics, the way an author does at the beginning of a book.

If Elizebeth intended this to be her memoir of the Invisible War, she never wrote the rest. The seven handwritten pages and a typed version of the same are all that exist. She later tucked the typescript into a manila folder marked "foreword to uncompleted work."

President Truman established the National Security Agency on November 4, 1952, at the peak of McCarthy's popularity and two and a half years after William's shock treatments. The NSA fused the signals intelligence units of the army and navy into one organization, including the unit that William founded and nurtured between the wars.

From the start the NSA was the most secret of agencies, basic facts of its existence concealed. William accepted a job there as a counselor and adviser, a role befitting a respected elder. But the agency had less and less use for him as it grew through the 1950s. It hired thousands of young linguists and cryptanalysts who were trained by the textbooks William wrote but who didn't necessarily listen when he spoke. It broke ground on a new campus in Fort Meade, Maryland, where today at least twenty thousand people work inside two large cubes of eavesdropping-resistant blue-black glass, and invested heavily in computers for breaking codes. William thought computers were "mostly nonsensical and completely nitwit gadgets for daily affairs," he wrote in a morose letter to the historian Roberta Wohlstetter. And as the NSA grew larger and stronger, it began to use that strength in ways that made William uncomfortable. It scooped up enormous quantities of signals seemingly because it could, towering haystacks of intelligence that would make it difficult to find the needles, and it continued to conceal and classify more and more kinds of documents that William thought should be publicly available. At other times in his life he had argued for greater secrecy, as when he objected to Herbert Yardley's book in the 1930s; now he muttered darkly to friends about a "secrecy virus" loose in government.

He suffered his first heart attack in April 1955, followed quickly by a second while in the hospital recovering from the first. That fall William retired from the NSA as a full-time employee. The agency gave him a nice ceremony and a consultant contract to keep him around; the director of the NSA at the time, Ralph Canine, admired William. Then a new director replaced Canine, a man with more inflexible views about secrecy and no personal fondness for the great codebreaker, and the agency raided the Friedmans' home library, and William became depressed again. He wanted to criticize the agency in public, to sound the alarm about the secrecy virus, but feared the NSA would withdraw his security clearance, severing him from his community and many of his own writings.

Whether or not the agency was specifically trying to humiliate him or just rigidly following regulations, William *felt* persecuted, and in his mentally delicate position, the ordeal was enough to push him to the edge. "Frightening to be alone [with] suicidal thoughts," he scribbled on a loose sheet of paper. "For fifty years have struggled with this off and on. . . . Repression by secrecy restrictions—fear of punishment chimerical but still there."

As his disillusionment with the NSA intensified into full-blown paranoia, he reconsidered his long intent to donate his papers to the Library of Congress. He couldn't bear to hand over the contents of his private library, his proudest possession, to the same government that had sent men to raid it. After some thought he decided instead to bestow his archive to the George C. Marshall Foundation, a private institution at the Virginia Military Institute in Lexington, Virginia. With Elizebeth's help he began organizing and indexing his vast trove of treasures in preparation for transfer to the Marshall Library: thousands of books, papers, memos, photographs, prototype board games, and other cryptologic curios. For a brief time the project seemed to revive him. "I now have a great desire to live," he wrote, "to bring the Marshall Foundation project to a completely satisfactory conclusion." His body did not cooperate. He suffered more heart attacks. His feet swelled so much he could not climb the stairs at the Folger Shakespeare Library when he went to hear lectures.

Elizebeth cared for him as always, taking notes on his condition in a daybook.

> MARCH 15, 1969: *Bill had fall in night. Confused and loss of memory momentarily.*
> JULY 20: *MAN ON THE MOON. ES & WFF watched on CBS until 3 a.m. when Neil Armstrong and 'Buzz' had finished moon walk and return to the module.*
> SEPTEMBER 24: WFF *birthday. Asked for spare ribs!*

A few minutes after midnight on November 2, 1969, he had his last heart attack and stopped breathing. Elizebeth called the doctor. William could not be revived. The doctor stayed at the house until after 2 A.M. to comfort her while William's body was taken away.

Overwhelmed, she picked up the daybook, out of habit.

> *My beloved died at 12:15.*

She started a brief letter to Barbara, who was traveling in Rome.

> *Dear heart be courageous. Your beloved father died. . . .*
> *Rejoice that he suffered only a very short time.*

More than 750 letters and cards of sympathy arrived at the house over the next weeks. Joseph Mauborgne called William "the greatest brain of the century," a man with an "ever shining place in history." The novelist Herman Wouk wrote to Elizebeth, "His effect on world history was incalculable, greater than that of kings & captains. Yet what a modest man!" Juanita Morris Moody, a codebreaker who got her start at William's Arlington Hall in 1943 and went on to supervise the NSA's Soviet desk, told Elizebeth that her husband was the last of his kind: "Our business now involves many more people and disciplines," Moody wrote. "It has become more abstract and impersonal. There are no more William Friedmans nor will there ever be."

Elizebeth received, from the Board of Management of the Cosmos Club, the men-only social club in Washington to which William had belonged, a "Woman's Privilege Card," granting access to the club's facilities for a period of two years.

She designed his tombstone.

WILLIAM F. FRIEDMAN
LIEUTENANT COLONEL
UNITED STATES ARMY
1891 · · · 1969
KNOWLEDGE IS POWER

Elizebeth decided to embed a secret message in the stone, in Bacon's cipher, in the letters of Bacon's quote. She specified that certain letters be carved with serifs and the rest without. The serifs were the *a*-form, sans-serifs the *b*-form:

KnOwl / edGeI / spOwE
(*a*- & *b*-forms shown as lower & upper case)
babaa / aabab / aabab
W / F / F

WFF: her husband's initials. It was a signature in cipher.

The army buried William with full honors at Arlington National Cemetery, the casket draped with a flag and carried by six black horses along the winding roads of the cemetery to the grave, accompanied by drummers. People from every branch of the military attended the funeral, and so did the antiwar U.S. senator from Minnesota, Eugene McCarthy. Elizebeth and the children were amazed to see him. The kids had worked on his 1968 presidential campaign. It turned out that McCarthy had worked as a codebreaker at Arlington Hall in 1944 under William's command. The family had had no idea.

After the funeral John Ramsay sent an emotional thank-you letter to McCarthy. "Your presence there seemed to make the idea of a military funeral a little more bearable for all of us. . . . I thought you might like to know that my father was a gentle and peaceful man who detested killing and war, secrecy, spying and all the things you and I hate. But he had a mad love affair with the world of secret writing to which he devoted his life and for which he felt many deep pangs of guilt. In spite of all his honors, he was not a happy man."

Elizebeth became William's avenger. Bitter about his treatment over the years by the army and the NSA, and worried that his contributions would be forgotten or erased, she set out to make sure that William received the credit he deserved. She took on this burden at the expense of curating her own legacy, which her grief and her anger now made a secondary concern.

Immediately after his funeral, in the now-empty house, she sat at William's own desk, the one with the 1918 KNOWLEDGE IS POWER photo under the glass, and worked to complete the annotated bibliography of his papers. The task occupied her for eight to ten hours a day. She mourned her husband while writing crisp descriptions of his articles and books on index cards. She did it out of a sense of duty to William, who would have wanted the project completed, and she also hoped that the collection, once open to the public, would entice a first-rate historian to write a biography of William, a book to cement his reputation.

The Marshall Library paid for a typist to help her one to two days a week and it still took months to finish the 3,002 cards for the 3,002 unique items in William's collection. Then she arranged to transport all of the material from Washington to the library, three hours south. Men came to the house one day in 1971 and loaded the boxes into trucks, along with William's desk. She told friends it felt like watching Bill die all over again. She followed the trucks on the highway in her beat-up, ten-year-old Plymouth, engine wheezing all the way to Lexington: "I guess I'm just a little old lady standing in the center of ruin and decay."

At the Marshall Library she worked six-hour days to manage the details of the transfer, making sure the papers were handled just so, out of love and respect for Bill. The archivists were thrilled to have her guidance (she "was entertained like a queen," she said) and got her on tape speaking about the donated materials, the Friedmans' life together, and Elizebeth's own career. And though she kept the focus on Bill, she also told stories about herself and donated thousands of her own personal papers to the

Marshall, separate from her husband's collection. Elizebeth's papers included documents she had preserved from the smuggling era of the 1920s and '30s, personal letters, her unfinished book manuscripts, diaries, and a lot more, but she had not indexed and annotated the collection like she did with William's. The archivists helped organize Elizebeth's files into twenty-two archival boxes, reverently stored behind the metal doors of the vault on the first floor.

In years that followed, researchers journeyed to the Marshall Library and used the Friedman files to write books that wouldn't have been possible before. The author James Bamford relied partly on William's collection to piece together his 1981 book, *The Puzzle Palace,* the first popular history of the NSA, whose publication the agency tried and failed to stop. The NSA sent representatives to the library twice, in 1979 and 1983, each time removing an unknown number of William's items, but the Friedmans had done such a careful job of indexing that a sharp-eyed professor at Virginia Military Institute, Rose Mary Sheldon, noticed that about 200 of the 3,002 index cards were missing. Sheldon submitted a series of Freedom of Information Act requests that eventually prodded the NSA to release 7,000 additional Friedman documents. In the last two decades the agency has gotten more comfortable telling its history—today it holds public cryptologic history conferences and operates a museum—but it took a while, and in the meantime, the Friedmans had created this alternate archive, beyond U.S. government control, where anyone could learn about U.S. codebreaking.

Even so, the attention of researchers fell lopsidedly on one Friedman and not the other. Elizebeth's papers at the library, unindexed and therefore mysterious, largely gathered dust while people explored William's. The world forgot about her and remembered him, which is what she had expected anyway. In 1975 the NSA informed Elizebeth that it planned to name the main auditorium at Fort Meade in William's honor and asked her to inspect and approve a bronze bust of his head. She attended the dedication ceremony. The NSA men's chorus sang "The Testament of Freedom." The following year a biography of William was published, *The Man Who Broke Purple,* which Elizebeth felt was a competent account of her husband's professional life but did not capture "the man I knew and loved."

She struggled in her final years as her savings dried up and her arteries hardened. She missed Bill so much. In her letters she sounded like a battle-hardened version of the girl who set Riverbank aflame, quick as ever but no longer joyful. "There is just one thing in this world I would now advise all unborn babies," Elizebeth typed one morning in a long letter addressed to no one ("I just had to blow off some steam"). She continued, "Either be born Rich or BE BORN POOR. It is we in between who PAY-PAY-Pay-y-y-y." She disliked the direction her field was taking, its increasing reliance on computers. She gave an interview to a *Houston Chronicle* reporter who found her "lounging in a turquoise silk robe from China, a gift from her husband in 1928." She told him computers are a curse. "The problem with machines is that nobody ever gets the thrill of seeing a message come out." She let her children know she wanted her body to be cremated when she died, with no funeral services. "In a few years there will be no place left on earth to bury any one, and before too long, I think, all cemeteries will have to be disposed of," she wrote. "Why add one jot or tittle to the mess already in existence?"

Elizebeth was eighty-eight when her arteries failed. She died on October 31, 1980, in a nursing home in Plainfield, New Jersey, four days before Americans elected Ronald Reagan to his first term as president.

The public response to her death was more muted than it had been for William's eleven years earlier. The *Washington Post* and *New York Times* printed respectful obituaries of Elizebeth. None of the obituarists mentioned her feats of codebreaking in World War II; almost certainly none of the writers were aware.

At Arlington National Cemetery her ashes were scattered atop William's grave and her name carved beneath his:

BELOVED WIFE
ELIZEBETH SMITH FRIEDMAN
1892 • • • 1980

For years, nothing much happened.

It took a while for people to rediscover Elizebeth. Bit by bit, people went looking. Mostly women. They suspected there was more to her story than had been told, and they were right. A historian at the Department of Justice, Barbara Osteika, located records of Elizebeth's old smuggling cases and came to see Elizebeth as a "beacon of hope" for women in federal law enforcement, a trailblazer. An FBI cryptanalyst, Jeanne Anderson, who solves the handwritten code and cipher notes of suspected criminals, found transcripts of Elizebeth's trials from the 1930s and studied them for guidance on speaking to juries. And although Elizebeth had never worked there, she also won fans at the NSA, where female cryptanalysts rose to distinction after the war, including Juanita Morris Moody, who briefed U.S. leaders during the Cuban Missile Crisis, and Ann Caracristi, who became the agency's number-two official.

In the 1990s the NSA renamed its auditorium. The William F. Friedman Memorial Auditorium is now the William F. Friedman and Elizebeth S. Friedman Memorial Auditorium. As of 2014 there is a second auditorium in the Washington area bearing her name, at a Justice Department building, thanks to a campaign launched by Barbara Osteika. Above the doors it reads, ELIZEBETH SMITH FRIEDMAN, PIONEER OF INTELLIGENCE-LED POLICING.

These things happened for two reasons: because women went looking for Elizebeth's ghost, and because her ghost was making noise in the archives. She was there inside the Marshall Library, rattling the doors of the vault, and she was in the "government tombs," the National Archives, where her records from the Invisible War were finally declassified. The ghost also cried out from unexpected places. Three of the index cards in William's collection contain brief, verifiably true comments about how J. Edgar Hoover and the FBI took credit for feats of spycatching actually performed by Elizebeth and the coast guard. These comments were obviously written by Elizebeth—William wasn't in a position to know. Each card is a knife slipped between the ribs of Hoover, Elizebeth's patient revenge.

She intended to use all of these archives to write her own story. She never got around to it. Maybe she lost hope. But the files are exactly where she left them, the fragments of an extraordinary life. The files have a weight to them, a texture. They can't be erased any more than Elizebeth's legacy can be erased, because her legacy is embedded in our lives today, in our smartphones and Web browsers, in the science that powers secure-messaging apps used by billions, in the clandestine procedures of corporations and intelligence agencies and in the mundane software loaded onto the iPhones in our pockets.

Secret communication is still a dance of codemakers and codebreakers, locks and lockpickers. The locks are different now, of course. With computation as an aid, everything has been massively sped up and mathematized beyond anything Elizebeth would have comfortably understood. But the game is still based in patterns. Someone designs a pattern that looks like mere clutter, and someone else tries to rearrange the clutter into a picture. Over and over again, gazing at what seemed random in the world, Elizebeth found a tiny spot of sense, and then she stood on that spot and invented a system to transform the rest of the landscape all the way out to the horizon, and this is still the process today. Codebreaking is work and patience and method and mind. And Elizebeth had more of these qualities than perhaps anyone else in her time.

She always remained a little sphinxy. Up to the end of her life she hesitated to blurt out all her secrets, to answer every question in movie detail, whether out of modesty, habit, fear of prosecution, or an appreciation for mystery.

"There are plenty of mysteries that you can leave dangling," she told the NSA's Virginia Valaki during their discussion in 1976. "Enough to allure a reader, I'm sure."

"I've been trying to put together the pieces," Valaki said. "We'll never make the whole picture . . . at least we'll get some of the perspective straightened out."

Valaki was one of Elizebeth's descendants, part of the next generation of women codebreakers who prospered after the war. She first joined the agency in 1954 as a linguist and now edited the NSA technical journal *Cryptolog*.

"Well, thanks again, Mrs. Friedman," Valaki said.

"Well, don't thank me," Elizebeth said. "It's been interesting."

"Sometime I myself would love to do a profile on you," Valaki added.

"Oh!" Elizebeth said.

"Girl cryptanalyst and all that. I would think it would be extremely interesting for people to read."

"What happened the other day?" Elizebeth said, asking the question to herself. She said she had been out in the city, walking on Capitol Hill, when she realized that a couple of young women nearby had seen her and were talking about her. Elizebeth recognized one of the women. They had crossed paths somewhere years earlier, in a professional capacity, and Elizebeth was tickled by the fact that these women considered her some kind of noteworthy figure. "Oh my!"

Valaki shut off the recorder. She and Elizebeth spoke for an unknown amount of time, possibly about mutual acquaintances at intelligence agencies. Then the recorder started again, and before too long, the conversation wound to a close.

They checked the time.

"You mean to say it's only five minutes after one?" Elizebeth said.

"My heavens!" Valaki said.

It had been so long since Elizebeth had talked about her life smashing codes that a simple conversation felt like an opera.

"I'll bet no two women ever said as many words in [so] short a time," Elizebeth said.

The transcript notes that the women laughed.

ACKNOWLEDGMENTS

I owe a lot to the people who shaped this book:

My editor, Julia Cheiffetz at Dey Street. Julia's passion for Elizabeth's story was always there, even when I didn't exactly know how to tell it, and our conversations enriched the book immeasurably. I'm grateful for her sharp eye, her instincts, and her belief. Thanks also to Sean Newcott, Lynn Grady, and the rest of the team at Dey Street: Tom Pitoniak, Kendra Newton, Heidi Richter, Dale Rohrbaugh, Paula Szafranski, and Owen Corrigan.

My agent, Larry Weissman. I am so glad to have the benefit of Larry's counsel and his sensibility for narrative nonfiction. I feel the same about his unflappable partner, Sascha Alper. I can't imagine writing books without their guidance and friendship.

Librarians and archivists: This book would not exist without the archivists who preserved, indexed, and annotated the Friedmans' files with such care. Paul Barron and Jeffrey Kozak at the George C. Marshall Foundation are wonderful humans, and their library is just one of the great American places. I was amazed by Rose Mary Sheldon, the Virginia Military Institute classics professor who spent years assembling her epic "The Friedman Collection: An Analytical Guide." She did it as a labor of love—didn't earn a cent—and was generous with her time and wisdom. The NSA historian Betsy Rohaly Smoot and NSA librarian Rene

Stein shared expertise and files with me. Hannah Walters at the Fabyan Villa Museum showed me around what remains of George Fabyan's Riverbank and answered numerous questions about Riverbank in its prime. Thanks also to Thomas Larson at the New York Public Library's Manuscripts and Archives Division; Jessica Strube at the Geneva History Museum; JoEllen Dickie at the Newberry Library; and the staff of the National Archives at College Park, Maryland.

Kari Walgran has been a friend and sounding board for years. Some of my favorite parts of the book grew from her questions and comments on drafts. Malcolm Burnley and Kirsten Hancock were capable research assistants who found and flagged important files. Phil Tomaselli turned up materials about the Nazi spy hunts in the UK National Archives in Kew. Eduardo Geraque in São Paulo sent documents from police archives there. Linda D. Ostman is a hero for discovering the transcript of the 1933 Consolidated Exporters case in a Texas court repository. I also appreciate research performed by Beth Robertson and Lisette Lacroix in Canada.

Thank you to the American women who spoke to me about their cryptologic experiences in World War II: Judy Parsons, Martha Waller, Pat Leopold, and Helen Nibouar.

I appreciate the historians, cryptologic obsessives, and technology enthusiasts who shared their time and wisdom. Philip Marks, the British expert in machine ciphers, was extremely patient in explaining Enigma systems and reviewing technical passages. Craig Bauer's engaging books about cryptology helped me navigate the subject, and conversations with Craig were always clarifying. The historian Richard McGaha helped me chart a path through the crazy waters of espionage and counterespionage in Argentina. The renegade Canadian author John Bryden pointed me toward the coast guard's clandestine decrypts in the National Archives. Jason Vanderhill in Vancouver knows everything there is to know about Canadian rum syndicates. James Somers is the kind of friend you want to have if you're writing about technology, a terrific writer who is also a programmer. I enjoyed meeting and talking with Barbara Osteika at ATF, a relentless researcher, and William Sherman, the Renaissance scholar who told me about the Riverbank cipher collection at the New York Public Library. Any cryptologic or historical errors in the text are mine.

Thank you to friends who provided advice, encouragement, leads, etc.: Carrie Frye, Sasha Issenberg, Eileen Clancy, Christi Bender, John Whittier-Ferguson, Nathalia Holt, Elonka Dunin, Josh Dean, Jason Leopold, Roy Kesey, Ann Daciuk, Sheila Liming, Puneet Batra, Chris McDougall, Stephen Rodrick, Steve Volk, Samantha Newell, Rob Morlino, Neel Master, Elon Green, and my excellent magazine colleagues—Greg Veis and Rachel Morris at the Huffington Post Highline, and Kristen Hinman and Michael Schaeffer at *Washingtonian*.

I'm indebted to the University of Michigan and the Knight-Wallace Fellowship program for inviting me and my family to Ann Arbor in 2014 and 2015. In a lot of ways, this book is a direct result of the rare alchemy of that program. Thank you so, so much to Charles and Julia Eisendrath for one of the best years of my life, Birgit Rieck and the fellow fellows, John DeCicco, and Carl Simon and the Center for the Study of Complex Systems. And I will always be grateful to Matthew Power for encouraging me to apply in the first place.

Thank you to Duchess Goldblatt for allowing me to borrow one of her lovely sentences.

Finally, thank you to my family: Frank, Sharyn, and Lauren Fagone; Gloria Jewell; Lynn and Rich Bauer; and the Howell clan. Most of all, thank you to the bright, adventurous women in my life, Dana Bauer, and our daughter, Mia Fagone. Dana and Mia inspired the book and kept telling me they wanted to read it. To the two of you:

```
O V M I D A D O O S S D A N E L I T
L E U A D N N H G H I O C Y B I E ?
I Y O A N A A M Y I T N E O U E V !
```

NOTES

ABBREVIATIONS

ESF   Elizebeth Smith Friedman
WFF   William Frederick Friedman
ESF COLLECTION Elizebeth S. Friedman Collection, George C.
                   Marshall Research Foundation (Lexington,
                   Virginia)
WFF COLLECTION William F. Friedman Collection, Marshall
                   Foundation
NARA   U.S. National Archives and Records
                   Administration (Washington)
NYPL   Bacon Cipher Collection, New York Public
                   Library, Manuscripts and Archives Division
                   (New York)
NSA   William F. Friedman Collection, U.S.
                   National Security Agency, 2015 release
                   (nsa.gov)
TNA   The National Archives of the UK (Kew,
                   United Kingdom)

AUTHOR'S NOTE

xii *"the world's greatest"* David Kahn, *The Codebreakers: The Comprehensive History of Secret Communication from Ancient Times to the Internet,* rev. ed. (New York: Scribner, 1997), 21.
"Singlehandedly, he made" Ibid., 392.
"CRYPTOLOGIC PIONEER" Program for "Dedication Ceremony, William F. Friedman Memorial Auditorium," May 21, 1975, box 14, file 12, ESF Collection.
"She and her husband" Memorandum from Chief of Communications to Chief (redacted), November 8, 1949, box 12, file 15, ESF Collection.
"Mrs. Friedman and her husband" U.S. Department of Justice, Federal Bureau of Investigation, memorandum, *Subject: Velvalee Dickinson*, R. A. Newby to D. M. Ladd, March 14, 1944. Obtained under the Freedom of Information Act from FBI; received December 2015.

xiii *"We try to tell people"* Jeffrey Kozak (director of library and archives at the Marshall Foundation) in discussion with the author, January 2015.

xiv *an elite codebreaking unit* "History of USCG Unit #387," Record Group 38, Crane Material, Inactive Stations, box 57, 5750/2, NARA. This is a 329-page technical history of ESF's coast guard unit between 1940 and 1945, a thick bound volume written in 1945 or 1946. The unit had multiple names over its lifetime—the Coast Guard Cryptanalytic Unit, Coast Guard Unit #387, then OP-20-GU, and later OP-G-70, after the unit was absorbed by the navy in 1941—but it's all the same organization, founded by ESF in 1931 and evolving as it faced different challenges through the end of the war. Every page of the technical history is stamped TOP SECRET ULTRA, including the cover. No author is listed; it was probably written by ESF's coast guard commander, Lieutenant Leonard T. Jones, in collaboration with her and other codebreakers on the team. It wasn't declassified until 2000.
tracked and exposed them ESF and her coast guard team preserved the decrypts that they generated during the war—the typed sheets of solved messages. These are located in two places at NARA. An incomplete set of decrypts is in RG 38, Records of the Office of the Chief of Naval Operations, CNSG Library, boxes 77–81. The bulk of the decrypts are in RG 457, Messages of German Intelligence/Clandestine Agents, 1942–1945, subseries SRIC, boxes 1–5. More than any others, these are the records that made it possible to figure out what ESF really did in the war and why it mattered. It is not good etiquette to cry

out in joy when you are researching in the National Archives, but I may have done that when I read the decrypts for the first time. I'm indebted to the Canadian historian John Bryden for flagging the importance of these documents in his excellent book *Best-Kept Secret: Canadian Secret Intelligence in the Second World War* (Toronto: Lester, 1993).

CHAPTER 1: FABYAN

3 *a female representative*   Transcript of ESF interview with Virginia T. Valaki, November 11, 1976, transcribed January 10, 2012, NSA Center for Cryptologic History. Obtained under the Freedom of Information Act from NSA; received October 2015; originally requested by G. Stuart Smith. Valaki was a cryptolinguist for the NSA and retired in 1994 after a forty-year career; she died in 2015. See "Virginia T. Valaki," obituary, *New Haven Register*, June 7, 2015, http://www.legacy .com/obituaries/nhregister/obituary.aspx?pid=175022791.
"Do you want a cigarette"   Ibid., 1.
4 *eighty-four years old*   ESF was born on August 26, 1892, in Huntington, Indiana. See Official Personnel Folder, box 7, folder 3.
"Nobody would believe it"   ESF interview with Valaki, November 11, 1976, transcribed February 16, 2012, 10.
"I'd be grateful"   ESF interview with Valaki, November 11, 1976, transcribed January 10, 2012, 1.
six slightly different answers   ESF interview with Valaki, November 11, 1976, transcribed February 16, 2012, 6–13.
5 *thought to tell the story*   Ibid., 6–7.
June 1916   Transcript of ESF interview with Forrest C. Pogue, May 16–17, 1973, box 16, folder 19, ESF Collection, 3.
a chauffeured limousine   Ibid., 2.
five foot three   ESF's ration book from the Second World War lists her as five foot three and 120 pounds, and she writes elsewhere that she was a bit smaller as a young woman.
dark-brown curls and hazel eyes   Though reporters sometimes called her eyes blue, and a 1930 oil painting of ESF shows her eyes to be a deep forest green, her children later insisted to a potential biographer of their mother that her eyes were really hazel. See Katie Letcher Lyle, "Divine Fire: Elizebeth Smith Friedman, Cryptanalyst," unpublished manuscript, July 4, 1991, ESF Collection, two PDF files, 175.
crisp gray dress   ESF interview with Pogue, 3.
6 *more than a foot*   Richard Munson, *George Fabyan: The Tycoon Who Broke Ciphers, Ended Wars, Manipulated Sound, Built a*

Levitation Machine, and Organized the Modern Research Center
(North Charleston, SC: Porter Books, 2013), 3. Fabyan was six
foot four, Elizabeth five three at most.

6 *impression of a windmill*   ESF interview with Valaki,
November 11, 1976, transcribed February 16, 2012, 8.
"Will you come to Riverbank"   ESF interview with Pogue, 2.
"Oh, sir"   Ibid.
"That's all right"   ESF interview with Valaki, November 11,
1976, transcribed February 16, 2012, 8.
lifting her by the arm   Ibid.
meek because she was small   ESF autobiography (unpublished
manuscript), ESF Collection, PDF file, 2.
"odious name of Smith"   ESF diary, April 22, 1913, box 21, folder
1, ESF Collection. She also wrote in this entry that she hated the
name Smith because it seemed terribly unfair for a lover of words
to be saddled with a name so lexically vanilla: "Call it vanity if
you will—but how should you like to have a name for which you
couldn't have even the fun of looking up the etymology?"
"I feel like snipping"   Ibid.

7 *John Marion Smith*   "Geneaology from notes of ESF," July 23,
1981, box 11, folder 21, ESF Collection.
served in local government   "Addenda and Corrections to
biographical data re Elizabeth Smith Friedman," box 11, folder 21,
ESF Collection.
"My Indiana family"   Lyle, "Divine Fire," 166.
Sopha Strock   ESF Personal History Statement, box 11, folder 16,
ESF Collection, 3.
grown up and scattered   Ibid., 13.
"We call a lot of things luck"   ESF diary, July 1, 1913.
"from her father"   Mary Goldman to Vanessa Friedman,
February 15, 1981, box 12, folder 14, ESF Collection.

8 *seamstress for hire*   ESF diary, February 27, 1913.
underlining the pages   ESF's volume of Tennyson, box 22, ESF
Collection.
Erasmus who "believed in one aristocracy"   ESF, "The Need for
Erasmianism," box 12, folder 8, ESF Collection.
"I sit stunned"   ESF, "After Senior Philosophy Course," 1915,
box 12, folder 9, ESF Collection.

9 *"passed away"*   ESF diary, March 20, 1913.
"I have marvelous abilities"   ESF diary, June 22, 1913.
"Very suggestive"   ESF, "The Need for Erasmianism."
"my musical heart was carried"   ESF diary, July 14, 1913.

10 *"it reveals the naked man-soul"*   Carleton Miller to ESF, July 22
[1915?], box 1, folder 44, ESF Collection.

10 *"mental question mark"*   ESF diary, January 29, 1916.
    substitute principal   ESF interview with Valaki, November 11,
    1976, transcribed February 16, 2012, 7.
    a county high school   It was the public high school in Wabash,
    Indiana. See "Education and Experience," ESF Personnel Folder.
    Almost 90 percent   Hans Joerg-Tiede, *University Reform: The
    Founding of the American Association of University Professors*
    (Baltimore: Johns Hopkins University Press, 2015), 14.
    939 women   National Center for Education Statistics, *120
    Years of American Education: A Statistical Portrait,* ed. Thomas
    D. Snyder (Washington, D.C.: U.S. Department of Education,
    Office of Educational Research and Improvement, 1993), 83.
    62 women   Ibid.
    "little, elusive, buried splinter"   ESF diary, October 10, 1914.
    "I am never quite so gleeful"   Ibid., July 2, 1913.
11 *More than a thousand*   *Official Report of the Proceedings of the
    Sixteenth Republican National Convention* (New York: Tenny
    Press, 1916), 11–13.
    rained most every day   Associated Press, "Republican Conclave
    Depressed by Weather; Shows Little Enthusiasm," *Chicago Daily
    Tribune,* June 9, 1916.
    the political delegates   Ibid.
    baseball parks   I. E. Sanborn, "Rain Stops Cubs; Double Bill
    Today with Herzog's Reds," *Chicago Daily Tribune,* June 21,
    1916. See also James Crusinberry, "Sox Lose Chance to Rise by
    Rain in Mack Series," *Chicago Daily Tribune,* June 9, 1916.
12 *died on a steamship*   Paul Finkelman, "Class and Culture in Late
    Nineteenth-Century Chicago: The Founding of the Newberry
    Library," *American Studies* 16 (Spring 1975): 5–22.
    had to be free to use   Ibid.
    wealthy Chicago businessmen   Ibid.
13 *dreamlike White City*   "World's Columbian Exposition of 1893,"
    Chicago Architecture Foundation, http://www.architecture.org
    /architecture-chicago/visual-dictionary/entry/worlds-columbian
    -exposition-of-1893/.
    a day of demonstrations   "Under 10,000 Wheels," *Chicago
    Tribune,* August 27, 1893.
    twice as large   The main building of the palace covered nine
    and a half acres and the U.S. Capitol building spreads across four
    acres. See *Encyclopaedia Brittanica,* New American Supplement
    to the New Werner Edition, s.v. "World's Fairs"; and Architect of
    the Capitol, "About the U.S. Capitol Building," https://www.aoc
    .gov/capitol-buildings/about-us-capitol-building.
    one hundred thousand people   "Under 10,000 Wheels."

13 *builders completed construction*   "History of the Newberry
    Library," https://www.newberry.org/newberry-library-history
    -newberry-library.
    "a select affair"   *Chicago Times,* July 17, 1887, cited in
    Finkelman, "Class and Culture in Late Nineteenth-Century
    Chicago."
    five-story building   Finkelman, "Class and Culture in Late
    Nineteenth-Century Chicago."
    fill out a slip   Ibid.
14 *hundreds of incunabula*   Ibid.
    Arabic script   "Frequently Asked Questions about Audrey
    Niffenegger's The Time Traveler's Wife," Newberry, https://
    www.newberry.org/time-traveler-s-wife. See also Lawrence
    S. Thompson, "Tanned Human Skin," *Bulletin of the Medical
    Library Association* 34, no. 2 (1946): 93–102.
    six thousand books   Finkelman, "Class and Culture in Late
    Nineteenth-Century Chicago."
    a haul that included   "Chicago Gets a Prize: Librarian Poole's
    Report on the Probasco Collection," *Chicago Daily Tribune,*
    November 22, 1890.
    Romanesque lobby   Finkelman.
    mounting exhibitions   Jo Ellen Dickie (reference librarian,
    Newberry Library), e-mail message to the author, January 4, 2017.
    13 inches tall and 8 inches wide   The name of this particular
    Folio is "Winsor 17" and it now resides in the special collections
    department of the Bryn Mawr College library in Pennsylvania.
    Anthony James West, *The Shakespeare First Folio: The History of
    the Book,* vol. 2 (New York: Oxford University Press, 2003), 233.
    an engraving of a man   The Bodleian First Folio: digital
    facsimile of the First Folio of Shakespeare's plays, Bodleian Arch.
    G c.7, http://firstfolio.bodleian.ox.ac.uk/.
    The text said   Ibid.
15 *"that an archaeologist has"*   ESF autobiography, 1.
    One of the librarians   ESF interview with Valaki, November 11,
    1976, transcribed February 16, 2012, 7.
    Richmond, Indiana   Ibid.
    "something unusual"   Ibid.
    reminded her of Mr. Fabyan   Ibid.
    "young, personable"   ESF autobiography, 1.
    too startled   Ibid.
    "Shall I call him up?"   ESF interview with Pogue, 2.
    "Well, yes"   Ibid.
    be right over   ESF interview with Valaki, November 11, 1976,
    transcribed February 16, 2012, 8.

16 *any minute*   Ibid., 8. Elizebeth recalled that Fabyan arrived "before you could have hit a button."
"This is Bert"   Ibid. Bert is spelled "Burt" in the NSA transcript but his name was Bert Williams, according to John W. Kopec, *The Sabines at Riverbank: Their Role in the Science of Architectural Acoustics* (Woodbury, NY: Acoustical Society of America, 1997), 29.
Chicago & North Western   ESF interview with Valaki, November 11, 1976, transcribed February 16, 2012, 8.
"Where am I"   Ibid.
she remained still   Ibid.
She smiled at him   Ibid., 6.

17 *within inches*   Ibid.
"WHAT IN HELL DO YOU KNOW"   Ibid.
something stubborn   Ibid., 9.
turned her head away   Ibid., 6.
"That remains, sir"   ESF interview with Pogue, 3.
most immoral remark   ESF interview with Valaki, November 11, 1976, transcribed February 16, 2012, 9.
a great roaring laugh   Ibid.
he began to talk of Shakespeare   ESF autobiography, 2.

18 *he believed what he was saying*   ESF eventually came to believe that Fabyan was deceptive in how he promoted his ideas but he did seem to earnestly believe them.
He said that a brilliant female scholar   ESF autobiography, 2.
350-acre estate   Munson, *George Fabyan*, 3.
Teddy Roosevelt, his personal friend   Ibid., 13.
P. T. Barnum   Ibid.
Famous actresses   Ibid. The actresses included Mary Pickford, Billie Burke, and Lillie Langtry.

19 *a second limousine*   ESF interview with Valaki, November 11, 1976, transcribed January 12, 2012, 6.
came to a stop   ESF autobiography, 3; ESF interview with Pogue, 3.
a two-story farmhouse   Ibid., 4; author's visit to the Fabyan Villa Museum, Geneva, Illinois, March 19, 2015.

CHAPTER 2: UNBELIEVABLE, YET IT WAS THERE

21 *A naked woman*   John W. Kopec, *The Sabines at Riverbank: Their Role in the Science of Architectural Acoustics* (Woodbury, NY: Acoustical Society of America, 1997), 36–37.
sign that read Fabyan   Ibid.

22 *satisfy his lust*   Ibid.

22 *two white flashes*   Norman Klein, "Building Supermen at Fabyan's Colony," *Chicago Daily News,* April 22, 1921.
The electric trolley   Richard Munson, *George Fabyan: The Tycoon Who Broke Ciphers, Ended Wars, Manipulated Sound, Built a Levitation Machine, and Organized the Modern Research Center* (North Charleston, SC: Porter Books, 2013), 48.
bombs exploding   Kopec, *The Sabines at Riverbank,* 42.
warplanes buzzing   Ibid.
"A Garden of Eden"   Mme. X, "A Visit to a Garden of Eden on Fox River," *Chicago Daily Tribune,* October 2, 1921.
"Fabyan's colony"   Klein, "Building Supermen at Fabyan's Colony."
"a wonder-working laboratory"   "A Wonder Working Laboratory Near Chicago," *Garard Review,* November 1928, 1.
"one of the strangest"   Klein, "Building Supermen at Fabyan's Colony."

23 *"one of the greatest"*   "Varying the List of Clubs . . ." *Cincinnati Star,* December 21, 1923, Box 14, "The Ideal Scrap Book," NYPL.
"one who has achieved"   "Scientist Spends Millions in Experiments to Develop Flapper into Perfect Woman," *Evening Public Ledger* (Philadelphia), July 18, 1922.
"the man of a thousand interests"   "War on Debutante Slouch Is Started by Col. Fabyan," July 5, 1922, Box 14, "The Ideal Scrap Book," NYPL.
"the lord and master"   Klein, "Building Supermen at Fabyan's Colony."
"Chicago inventor"   "Flywheel Discs Cut Resistance," *Kansas City Journal,* March 13, 1923.
multi-millionaire country gentleman   "Fabyan Tries to Rear Perfect Flapper on Farm," *Chicago Herald Examiner,* July 6, 1922.
"the seer"   Leroy Hennessey, "Twas Bill! Nay, Bacon! But Now E'en Fabyan Knows Not Who Did Shakespeare," *Chicago Evening American,* January 1922, Box 14, "The Ideal Scrap Book," NYPL.
"the caliph"   "Col. George Fabyan Declares War on Profiteers," Box 14, "The Ideal Scrap Book," NYPL.
"Credible persons"   Cinderella, "Chicagoan Wins Name at Sculpture," *Chicago Daily Tribune,* June 1, 1915. This seems to be a legend; staff at the Fabyan Villa Museum told me that Fabyan only rode in a zebra-drawn chariot once, not twice a day every day.
donations . . . board meetings   Munson, *George Fabyan,* 4.
The black sheep   Ibid., 20.
$3 million fortune   Ibid., 10.

23 *striped seersucker cloth*   Ibid., 22.
24 *"Ripplette"*   Ripplette ad, *Farmer's Wife* (St. Paul, Minnesota), January 1, 1927.
   "I ain't no angel"   George Fabyan to WFF, June 10, 1926, WFF and George Fabyan Correspondence, Item 734, WFF Collection.
   The steel magnates of Pittsburgh   Andrew Carnegie and Henry Clay Frick, who hated and trolled each other. Frick built his mansion one mile from Carnegie's and vowed to make his rival's home look "like a miner's shack" in comparison. Christopher Gray, "Carnegie vs. Frick, Dueling Egoes on Fifth Avenue," *New York Times*, April 2, 2000.
   a 165-room castle   "Other Features Around Hearst Castle," California State Parks, http://hearstcastle.org/history-behind -hearst-castle/the-castle/.
   "Some rich men"   Klein, "Building Supermen at Fabyan's Colony."
   Aspirin, vitamins   Aspirin was discovered in 1897, vitamins in 1912, blood types in 1900; medical X-rays began in 1895.
   Einstein's theory   He published his theory of general relativity in 1915. American Institute of Physics, "2015: The Centennial of Einstein's General Theory of Relativity," https://www.aip.org /history-programs/einstein-centennial-2015.
   swarm of bees   "Col. Geo. Fabyan Soon to be a Miller De Luxe," *Chicago Herald,* July 12, 1915, reprinted in Kopec, *The Sabines at Riverbank*, 30–32.
25 *"Do you ever think"*   Klein, "Building Supermen at Fabyan's Colony."
   "community of thinkers"   Ibid.
   an ultraquiet test chamber   Kopec, *The Sabines at Riverbank,* 59–73.
   the buzz of a stray mosquito   "A Wonder Working Laboratory."
   a pencil writing on paper   Ibid.
   "racket ogre"   "Fabyan May End Noises of City," *Aurora Beacon* (Aurora, Illinois), April 24, 1921.
   "Look through this telescope thing"   " 'Lord of Riverbank' Works in $100,000 Laboratory; Would Find Deafness Cure," box 14, manila folder of newspaper clippings, NYPL.
26 *one hundred or more*   Klein, "Building Supermen at Fabyan's Colony."
   "Over there in that hothouse"   Ibid.
   "bobbed blonde hair"   "Scientist Spends Millions."
   low-security juvenile prison   L. Mara Dodge, " 'Her Life Has Been an Improper One': Women, Crime, and Prisons in Illinois, 1835 to 1933" (Ph.D. diss., Univeristy of Illinois at Chicago, 1998), 535–41, 718–19.

26 *cottage built with a donation*　Kopec, *The Sabines at Riverbank*, 37.

required to undress　Ibid.

"The results of our experiments"　"Scientist Spends Millions."

27 *"in his effort to impress"*　Ibid.

told an Illinois historian　Munson, *George Fabyan*, 50.

"the beams would creak"　Ibid.

"recalled looking out the windows"　Ibid.

"The staff in charge"　Austin C. Lescarboura, "A Small Private Laboratory," *Scientific American*, September 1923, 154.

28 *an X-ray screen*　Kopec, *The Sabines at Riverbank*, 36.

$750,000 worth of radium　Munson, *George Fabyan*, 50.

discovered in 1895　"This Month in Physics History: November 8, 1895: Roentgen's Discovery of X-Rays," *American Physical Society News* 10, no. 10 (November 2001), https://www.aps.org/publications/apsnews/200111/history.cfm.

"Every so often the world"　Lescarboura, "A Small Private Laboratory."

then disappeared　ESF interview with Valaki, November 11, 1976, transcribed February 16, 2012, 9.

aristocratic appearance　ESF autobiography, 3; ESF interview with Valaki, November 11, 1976, transcribed January 12, 2012, 6.

29 *lived and worked here*　ESF interview with Valaki, November 11, 1976, transcribed February 16, 2012, 9.

freshen up　ESF interview with Valaki, November 11, 1976, transcribed January 12, 2012, 1.

striking new clothes　Ibid.

sat on the bannister　ESF interview with Pogue, 6.

a slim man　Ibid.

a neat bow tie　Ibid.

reminded Elizebeth of Beau Brummell　ESF interview with Pogue, 6.

polished his boots with champagne　"Fashions of Hunting," *Baily's Magazine of Sport and Pastimes* 65, nos. 431–36 (1896): 163.

Swedish and Danish servants　Transcript of ESF interview with Ronald Clark, handwritten note on page 7, March 25, 1975, box 16, file 22, ESF Collection.

30 *chickens, ducks, sheep, and turkeys*　Munson, *George Fabyan*, 63.

prize-winning livestock　Kopec, *The Sabines at Riverbank*, 30.

head of the table　ESF autobiography, 3.

J. A. Powell　ESF interview with Clark, 5.

"cause the University of Chicago"　"Here Are a Few Expert Suggestions for First Press Agent of U. of C.," *Chicago Tribune*, September 5, 1909.

30 *Bert Eisenhour*   ESF interview with Clark, 5.
a country bumpkin   ESF interview with Valaki, November 11, 1976, transcribed 16 February 2012, 5.
The dominant personality that night   ESF autobiography, 3.

31 *"Mrs. Gallup had dwelt"*   Ibid.
men's pajamas   ESF interview with Valaki, November 11, 1976, transcribed January 12, 2012, 7.
a pitcher of ice water . . . an enormous bowl of fresh fruit   Ibid.; ESF interview with Pogue, 5.
assigned an employee   I'm inferring this from the fact that Elizebeth doesn't say in her autobiography or later recollections that Fabyan was her tour guide. I think if he had done it himself, she would have said that.
a new laboratory   Kopec, *The Sabines at Riverbank,* 3–4.
Professor Wallace Sabine   Ibid.
ordnance building   Ibid., 42; ESF interview with Pogue, 3.
known as the Villa   Munson, *George Fabyan,* 25.

32 *suspended from the ceiling on chains*   ESF autobiography, 5.
Taxidermized animals   Personal visit to the Fabyan Villa Museum, Geneva, Illinois, March 19, 2015.
a life-size marble statue   F. Edwin Elwell, *Diana and the Lion* (sculpture, 1893), displayed in the Palace of Fine Arts in the White City, acquired by George Fabyan after 1917, according to a placard in the Fabyan Villa Museum.
A curving path   Munson, *George Fabyan,* 59–60; Kopec, *The Sabines at Riverbank,* 27–28.

33 *Tom and Jerry*   Ibid., 2.
flowing southward   Wikipedia, s.v. "Fox River (Illiois River tributary)," last modified May 1, 2017, https://en.wikipedia.org /wiki/Fox_River_(Illinois_River_tributary).
two bridges   "Fabyan Estate Viewed from the Southeast," map, in Kopec, *The Sabines at Riverbank;* Munson, *George Fabyan,* 5.
bought the windmill in Holland   As is often the case with Fabyan, the truth here is actually weirder than the legend. Fabyan didn't buy the windmill in Holland; he bought it from a German craftsman in Lombard, Illinois, paying the modern equivalent of $2 million to take it apart, lug it across the prairie, and reconstruct it on the opposite bank of the Fox River. See "Fabyan Windmill," Kane County Forest Preserve District, http://www .kaneforest.com/historicsites/fabyanwindmill.aspx.
Elizebeth sat down with Mrs. Gallup   ESF autobiography, 2.
two or three hours   Ibid.
oversize sheets of paper   Several of these large sheets are preserved in box 14, NYPL.

33 *rolled them out*   ESF interview with Valaki, November 11, 1976,
    transcribed January 12, 2012, 7.
    placed weights on the ends   This is my own inference, from
    having handled these scrolls myself at NYPL. They really are like
    window blinds; if you don't put the weights on them, they snap
    back into a scroll.
34 *would be twofold*   ESF autobiography, 5.
    popped in briefly   ESF interview with Valaki, November 11,
    1976, transcribed January 12, 2012, 7.
    another bowl of fresh fruit   Ibid.
    "a mixture of astonishment"   WFF and ESF, *The Shakespearean
    Ciphers Examined* (London: Cambridge University Press, 1958), 210.
    five thousand women   George Morris, "Clothing Wet, Ardor
    Undampened, 5,000 Women March," *Chicago Daily Tribune,*
    June 8, 1916.
35 *Water poured*   Ibid.
    "right of each state"   Republican Party Platform, June 7, 1916,
    http://www.presidency.ucsb.edu/ws/?pid=29634.
    idolized the suffrage pioneers   ESF League of Women Voters
    report on International "Equal Rights," April 6, 1933, box 7,
    folder 5, ESF Collection. See also ESF to Miss Belle Sherwin,
    President, National League of Women Voters, April 14, 1933, box
    7, folder 6, ESF Collection.
    "No woman's rights"   ESF diary, January 29, 1916.
    she reviewed her options   ESF interview with Valaki,
    November 11, 1976, transcribed January 12, 2012, 7.

CHAPTER 3: BACON'S GHOST

37 *a worksheet of white paper*   "Actors' Names—Shakespeare Folio
    1623," box 15, folder "Elizebeth Smith," NYPL.
38 *"The Names of the Principall Actors"*   The Bodleian First Folio:
    digital facsimile of the First Folio of Shakespeare's plays,
    Bodleian Arch. G c.7, http://firstfolio.bodleian.ox.ac.uk/.
    devout Christian   WFF and ESF, *The Shakespearean Ciphers
    Examined* (London: Cambridge University Press, 1958), 189.
    "Surprise followed surprise"   Elizabeth Wells Gallup, "Concerning
    the Bi-literal Cypher of Francis Bacon: Pros and Cons of the
    Controversy" (1902; Internet Archive, 2008), 60, https://archive
    .org/details/concerningbilite00gall.
    "The sole question is"   Ibid., 65.
39 The New Atlantis   Francis Bacon, *The New Atlantis* (1627;
    Project Gutenberg, 2008), https://www.gutenberg.org/files/2434
    /2434-h/2434-h.htm.

39 *Mark Twain believed it* Mark Twain, *Is Shakespeare Dead?* (1909; Project Gutenberg, 2008), https://www.gutenberg.org /files/2431/2431-h/2431-h.htm.
So did Nathaniel Hawthorne Nina Baym, "Delia Bacon: Hawthorne's Last Heroine," *Nathaniel Hawthorne Review* 20, no. 2 (Fall 1994): 1–10, http://www.english.illinois.edu/-people -/emeritus/baym/essays/last_heroine.htm.
can be anagrammed WFF and ESF, *Shakespearean Ciphers*, 110.
2+1+3+14+13 Ibid., 179, 181.
40 *Orville Ward Owen* Ibid., 63.
most scientific and plausible yet WFF and ESF, *Shakespearean Ciphers*, 188.
The method had been demonstrated Francis Bacon, *De Augmentis Scientarium,* translated by Gilbert Wats (Oxford, 1640); pages relevant to ciphers in Wells Gallup, "Concerning the Bi-literal Cypher," 23–27.
anything by means of anything Ibid.
the new alphabet Ibid.
41 *don't have to be* a *and* b Ibid.
42 *scoured photo enlargements* "A CATALOGVE," box 13, folder 11, NYPL.
Then she drew charts "Alphabets for the Catalogue of the Plays," box 14, NYPL.
"Queene Elizabeth is my true mother" Elizabeth Wells Gallup, *The Biliteral Cypher of Sir Francis Bacon Discovered in His Works and Deciphered by Mrs. Elizabeth Wells Gallup,* 3rd ed. (1901; Internet Archive, 2008), 166, http://www.archive.org /details/biliteralcyphero00gallrich/.
"Francis of Verulam is author" Ibid.
"Francis St. Alban, descended" Ibid.
"You will either finde" Ibid., 165.
43 *her 1899 book* Wells Gallup, *The Biliteral Cypher,* 1st ed.
a secret king WFF and ESF, *Shakespearean Ciphers*, 192–94.
clandestine society of engineers Richard Munson, *George Fabyan: The Tycoon Who Broke Ciphers, Ended Wars, Manipulated Sound, Built a Levitation Machine, and Organized the Modern Research Center* (North Charleston, SC: Porter Books, 2013), 103.
"Here are 360 pages" "Bacon-Shakespeare: Mrs. Elizabeth Wells Gallup Throws New Light Upon the Mystifying Question—The Bi-Literal Cipher," newspaper article, Box 14, clipping file in wooden box marked "California glace fruits," NYPL.
Skeptics questioned the veracity WFF and ESF, *Shakespearean Ciphers*, 196–99.

43 *"impossible to those"*   Ibid., 198.
   traveled to Oxford, England   Ibid., 202.
44 *Mrs. Gertrude Horsford Fiske*   Ibid., 196.
   Mr. Henry Seymour   Ibid.
   Mr. James Phinney Baxter   Ibid., 224.
   "acoustical levitation device"   John W. Kopec, *The Sabines at Riverbank: Their Role in the Science of Architectural Acoustics* (Woodbury, NY: Acoustical Society of America, 1997), 4–6.
   Eisenhour couldn't get it to work   Ibid.
   "The inheritance"   Four-page typewritten draft beginning "The use and the commixture," box 13, blank folder between folders 14 and 15, NYPL. The text in this draft appears similar to Fabyan's published introduction in *The First of the Twelve Lessons in the Fundamental Principles of the Baconian Ciphers, and Application to the Books of the Sixteenth and Seventeenth Centuries* (Geneva, IL: Riverbank Laboratories, 1916).
   a photo enlargement   ESF describes the process she was taught by Gallup in *Shakespearean Ciphers*, 209, and ESF's worksheets from her earliest deciphering tests at Riverbank—eight sheets total—are in box 15, folder "Elizebeth Smith," NYPL.
45 *particularly fond of puzzles*   ESF interview with Ed Meryl, March 1939, box 17, folder 14, ESF Collection.
   She got stuck   WFF and ESF, *Shakespearean Ciphers*, 210–11.
   eight hours   "Actors' Names," ESF wrote in pencil at the top of the worksheet, "8 hours' work," and marked the time and date of the solution: 10:30 A.M., June 5, 1916.
   twenty-four-word plaintext translation   Ibid.
46 *Elizebeth always asked Mrs. Gallup*   WFF and ESF, *Shakespearean Ciphers*, 210–11.
   Ragtime music   "Col. Geo. Fabyan Soon to be a Miller De Luxe," *Chicago Herald*, July 12, 1915, reprinted in Kopec, *The Sabines at Riverbank*, 30–32.
   a series of loudspeakers   Ibid.
47 *sisters from Chicago*   "An Investigation of the Newest Bacon-Shakespeare Cipher Theory," *St. Louis Post-Dispatch Sunday Magazine*, July 9, 1916, in box 14, "The Ideal Scrap Book," NYPL.
   "Our experience at Riverbank"   George Fabyan to chief of the MID, War Department, March 22, 1918, RG 165, Records of the Military Intelligence Division, Entry 65, box 2243.
   looking glass   WFF and ESF, *Shakespearean Ciphers*, 190.
   attempting to complete   Ibid., 208; Kate Wells to George Fabyan, n.d., box 14, NYPL.
   resembled a piece of art   Elizabeth Wells Gallup black notebook with red spine, box 14, NYPL.

47 *small wooden boxes*   Various boxes of news clippings, box 13 and 14, NYPL.
48 *"We lived hard and fast"*   ESF interview with Valaki, November 11, 1976, transcribed February 16, 2012, 10.
tiny salaries   ESF interview with Pogue, 5.
minor idle rich   Ibid.
invited Elizebeth to climb in   ESF interview with Valaki, November 11, 1976, transcribed February 16, 2012, 5.
"no billiard ball"   Mme. X, "A Visit to a Garden of Eden on Fox River," *Chicago Daily Tribune,* October 2, 1921.
49 *head would blow off*   ESF interview with Valaki, November 11, 1976, transcribed February 16, 2012, 5.
afraid he'd catch cold   Katie Letcher Lyle, "Divine Fire: Elizabeth Smith Friedman, Cryptanalyst," unpublished manuscript, July 4, 1991, ESF Collection, two PDF files, 44.
leisurely rides   ESF autobiography, 8–9; ESF interview with Pogue, 6.
sandhill cranes and red hawks   Gerald M. Haslam, "The Fox River Settlement Revisited: The Illinois Milieu of the First Norwegian Converts to Mormonism in the Early 1840s," *BYU Family Historian* 6 (2007): 59–82.
At twenty-five   WFF was born September 24, 1891, so he would have been just shy of twenty-five; they were a little more than one year apart in age.
old, creaky structure   Transcript of ESF interview with Marshall Research Library staff members Tony Crawford and Lynn Biribauer, Tape #5, June 6, 1974, 5.
teensy-weensy flies   Ibid.
quickly, then die   Ibid.
50 *one bottle of flies into another*   I am relying on my memory of performing this exact type of *Drosophila* experiment in high school. Thanks to my AP Biology teacher, Mr. Anderson.
150 workers   Norman Klein, "Building Supermen at Fabyan's Colony," *Chicago Daily News,* April 22, 1921.
Susumu Kobayashi   Kopec, *The Sabines at Riverbank,* 27.
Jack "the Sailor"   Ibid.
Belle Cumming   Ibid., 50.
Silvio Silvestri   Ibid., 4, 26.
"Achieve success!"   ESF interview with Valaki, November 11, 1976, transcribed January 12, 2012, 10.
handed out shiny dimes   Munson, *George Fabyan,* 6.
demonstrate how a snake   Ibid.
wearing red diapers   Kopec, *The Sabines at Riverbank,* 29.
51 *crops, genetics, and Francis Bacon*   Ibid., 13.

51 *so did Lillie Langtry*   ESF interview with Clark.
Billie Dove   Kopec, *The Sabines at Riverbank*, 23.
Richard Byrd   Munson, *George Fabyan*, 13.
the elegant Billie Burke   Ibid.
met and talked with Lillie Langtry   Ibid.
"star-complex and hero-worship"   ESF narrative of Gordon Lim case, 1937–38, box 6, manila folder of Lim case material, ESF Collection.
buy a new wardrobe   ESF autobiography, 6.
"so typically Fabyan"   ESF interview with Valaki, November 11, 1976, transcribed January 12, 2012, 14.
52 *bugler who played reveille*   Munson, *George Fabyan*, 58.
an honorary one   Kopec, *The Sabines at Riverbank*, 22–23.
the Fabyan Scouts   Ibid.
the Fox Valley Guards   Ibid.
screaming at the offender   ESF autobiography, 5.
stoking the coals   Ibid.
steel I-beams   Kopec, *The Sabines at Riverbank*, 52.
seventy-five plows   Klein, "Building Supermen at Fabyan's Colony."
Temple de Junk   Ibid.
He published a book   George Fabyan, *What I Know About the Future of Cotton and Domestic Goods*, 2nd ed. (Chicago, 1900).
one hundred blank pages   Munson, *George Fabyan*, 5.
53 *One day he walked past*   Ibid., 7.
"a very bright man"   ESF interview with Valaki, November 11, 1976, transcribed January 12, 2012, 1.
longer than a newspaper headline   ESF autobiography, 8.
repeat back verbatim   Ibid.
showed Elizebeth a prototype   ESF interview with Marshall staff, Tape #5, June 6, 1974, 5.
crawled on their stomachs   Klein, "Building Supermen at Fabyan's Colony."
54 *Professor So-and-So*   ESF interview with Valaki, November 11, 1976, transcribed January 12, 2012, 8.
"We'll get along fine"   Ibid.
all expenses paid   WFF and ESF, *Shakespearean Ciphers*, 205–6.
"useless Bacon-Shakespeare controversy"   Letter on Bliss Fabyan & Company letterhead, September 1916, box 13, blank folder between folders 14 and 15, NYPL.
getting to the bottom   "An Investigation of the Newest Bacon-Shakespeare Cipher Theory."
"hard, cold facts"   Letter on Bliss Fabyan letterhead, 1916.
disorient the guest   WFF and ESF, *Shakespearean Ciphers*, 206.

55 *convinced that the work was solid* "An Investigation of the Newest Bacon-Shakespeare Cipher Theory." In this article the *Post-Dispatch* reporter briefly describes meeting twenty-two-year-old Elizebeth: "Miss Smith told me that when she went to Riverside [*sic*] she did not believe there was anything to the bi-literal cipher theory. Now, she says, she hasn't the slightest doubt."
beginning to doubt WFF and ESF, *Shakespearean Ciphers*, 211.
John Matthews Manly Eric Powell, "A Brief History of the English Department at the University of Chicago," September 2014, https://english.uchicago.edu/about/history.
"wrassle" ESF interview with Valaki, November 11, 1976, transcribed January 12, 2012, 12.
Manly pushed her on the shoulder Ibid., 13.
"Oh, my!" Ibid.
strained credulity WFF and ESF, "Elizabethan Printing and Its Bearing on the Biliteral Cipher," in *Shakespearean Ciphers*, 216–29.
"See how she leans" William Shakespeare, *Romeo and Juliet*, ed. Brian Gibbons (New York: Bloomsbury, 1980), 2.2.23–25.

56 *never once suspected* WFF and ESF, *Shakespearean Ciphers*, 264.
"She could go through the texts" Ibid.
comparing herself to Galileo Elizabeth Wells Gallup, "Bacon's Lost Manuscripts, A Review of Reviews," box 14, clipping file in wooden box marked "California glace fruits," NYPL.

57 *business cards* "Francis Bacon," February 10, 1917, box 14, magenta-colored scrapbook, NYPL.
"Riverbank Laboratories are a group" Letter on Bliss Fabyan letterhead, 1916.

58 *a town in Russia called Kishinev* Ronald Clark, "Preparation," in *The Man Who Broke Purple: Life of Colonel William F. Friedman, Who Deciphered the Japanese Code in World War II* (Boston: Little, Brown, 1977), 7–26.
fluent in eight languages Ibid.
"research and ingenuity" Ibid.
an expert in heredity ESF interview with Pogue, 5.
"an agricultural expert" George Fabyan to WFF, June 14, 1915, Item 734, WFF Collection.
"I realize the value" WFF to George Fabyan (undated), Item 734, WFF Collection.

59 *vague, long-winded riff* George Fabyan to WFF, August 12, 1915, Item 734, WFF Collection.
"I want the father of wheat" Ibid.
"Jewish Invasion" Burton J. Hendrick, "The Jewish Invasion of America," *McClure's Magazine*, March 1913, 125.

59 *"nervous, restless ambition"* Ibid., 127.
60 *made him miserable* Clark, "Preparation," 16–17.
"My idea of real love-making" Lyle, "Divine Fire," 59.
youthful fascination WFF, "Edgar Allan Poe, Cryptographer,"
in *On Poe*, ed. Louis J. Budd and Edward Harrison Cady
(Durham, NC: Duke University Press, 1993), 40–54.
The plot of the story Edgar Allan Poe, "The Gold-Bug," *Dollar
Newspaper* (Philadelphia), June 23, 1843, 1 and 4, https://www
.eapoe.org/works/tales/goldbga2.htm.
61 *Americans associated codebreakers* WFF, "Edgar Allan Poe,
Cryptographer."
a sketch of a long-stemmed plant WFF, "Cipher Baconis
Gallup," box 13, folder 5, NYPL.
a gray duration of pitiless wind Kopec, *The Sabines at
Riverbank*, 3.
sit on his lap Lyle, "Divine Fire," 53, 60.
wondering the same ESF autobiography, 9: "As we were thrown
together so much in our examination of the cipher proofs, we
had many . . . talks ourselves. Even that first summer we began to
wonder about the authenticity of Mrs. Gallup's 'solution.'"
There are no hidden messages The Friedmans, as careful
scientists, never really said it this starkly, but I think this is
what they believed. On the last page of *Shakespearean Ciphers*
(288) they suggest that if the people looking for hidden
messages in Shakespeare taught themselves the true science of
cryptology and applied it to their efforts, the whole dispute
"might cease altogether"—in other words, if the seekers really
understood cryptology, they'd also understand that their quest
is doomed.

CHAPTER 4: HE WHO FEARS IS HALF DEAD

63 *"and then begins step step leap"* Anne Carson, *Float* (New
York: Knopf, 2016), 138.
burned its way from hand to hand Barbara W. Tuchman, *The
Zimmermann Telegram* (New York: Ballantine, 1958), 160–72.
At 11 A.M. on February 27 Ibid., 172.
"Good Lord!" Ibid.
Germany to Mexico on January 16 Ibid., 145.
a series of number blocks Ibid., 201.
toiled for a month in a secret office Ibid., "A Telegram Waylaid,"
3–24.
64 *"We intend to begin"* Ibid., 146.
outrage against Germany Ibid., 184–86.

64 *Her father was there*   ESF to WFF, January 31, 1917, ESF Collection.
pinkish fluid   Ibid.
"My book-bag lies here unopened"   Ibid.
65 *"yours, Elsbeth"*   ESF to WFF, February 7, 1917, ESF Collection.
"one of the truest friends"   Ibid.
"rocking"   Katie Letcher Lyle, "Divine Fire: Elizabeth Smith Friedman, Cryptanalyst," unpublished manuscript, July 4, 1991, ESF Collection, two PDF files, 53, 60. Lyle notes that at some point William also provided this "rocking" to Elizebeth's elder sister, Edna.
"I love you / Elsbeth"   ESF to WFF, January 31, 1917.
seized by a new impatience   ESF to WFF, February 7, 1917.
seemed a little cruel   ESF interview with Marshall staff, Tape #2, June 4, 1974, 14.
they buttonholed him   ESF autobiography, 10.
He shouted them down   Ibid.
66 *length of three miles*   Richard Munson, *George Fabyan: The Tycoon Who Broke Ciphers, Ended Wars, Manipulated Sound, Built a Levitation Machine, and Organized the Modern Research Center* (North Charleston, SC: Porter Books, 2013), 8.
"Gentlemen," he wrote   George Fabyan to War Department Intelligence Office, 15 March 1917, Item 734, WFF Collection.
67 *"There were possibly three"*   ESF autobiography, 11.
nine-year-old   "A Brief History: The Nation Calls, 1908–1923," FBI, https://www.fbi.gov/history/brief-history.
only three hundred agents . . . half a million dollars   Regin Schmidt, *Red Scare: FBI and the Origins of Anticommunism in the United States, 1919–1943* (Copenhagen: Museum Tusculanum, 2000), 83.
April 6, 1917   "April 6, 2017: The 100th Anniversary of the American Entry into World War I," American Battle Monuments Commission, https://www.abmc.gov/news-events/news/april -6-2017-100th-anniversary-american-entry-world-war-i.
seventeen officers   Joseph W. Bendersky, *The Jewish Threat: Anti-Semitic Politics of the U.S. Army* (New York: Basic Books, 2000), 49.
codes and ciphers an "emergency"   Ralph Van Deman to Acting Commandant, Army Service Schools, Fort Leavenworth, Kansas, April 17, 1917, Item 734, WFF Collection.
a radio signal from a plane   Paul W. Clark and Laurence A. Lyons, *George Owen Squier: U.S. Army Major General, Inventor, Pioneer, Founder of Muzak* (Jefferson, NC: McFarland, 2014), 187.

68 *stand eye to eye*   ESF interview with Valaki, transcribed February 16, 2012, 2.

"the two greatest people"   Joseph Mauborgne to WFF and ESF, January 8, 1956, box 1, folder 21, ESF Collection.

"to take immediate advantage"   Joseph Mauborgne to chief of the War College Division, April 11, 1917, Item 734, WFF Collection.

"your exceedingly kind"   Ralph Van Deman to George Fabyan, April 18, 1917, Item 734, WFF Collection.

69 *intercepted by covert means*   WFF, Lecture V, 107, in *The Friedman Legacy, Sources on Cryptologic History,* no. 3 (Center for Cryptologic History: 2006).

BGVKX   ESF and WFF, "Riverbank Problems in Cryptanalysis," no. 1, Item 290, WFF Collection.

70 *403,291,461,126,605,635,584,000,000*   This is the number of permutations of 26 letters, a quantity written as 26! and calculated by multiplying 26 x 25 x 24 x 23 x 22 x 21 x 20 x 19 x 18 x 17 x 16 x 15 x 14 x 13 x 12 x 11 x 10 x 9 x 8 x 7 x 6 x 5 x 4 x 3 x 2 x 1. See Wolfram Alpha, https://www.wolframalpha.com/input/?i=26!.

A thousand computers   Lambros D. Callimahos, "Summer Institute for Mathematics and Linguistics," lecture, NSA, Fort Meade, Maryland, 1966, NSA Reading Room, https://www.nsa.gov/resources/everyone/foia/reading-room.

Monks, librarians, linguists   David Kahn, *The Codebreakers: The Comprehensive History of Secret Communication from Ancient Times to the Internet,* rev. ed. (New York: Scribner, 1997), is the definitive history of the field, with hundreds of pages of this stuff.

71 *a Belgian countess named Alexandrine*   Nadine Akkerman, "The Postmistress, the Diplomat, and a Black Chamber? Alexandrine of Taxis, Sir Balthazar Gerbier and the Power of Postal Control," in Robyn Adams and Rosanna Cox, *Diplomacy and Early Modern Culture* (London: Palgrave Macmillan, 2011), 172–88.

an early example   Ibid. Akkerman argues that Alexandrine's Chamber of the Thurn and Taxis might have been the very first black chamber in Europe.

"What if this countess"   Ibid.

Parker Hitt and Genevieve Hitt   Betsy Rohaly Smoot, "Pioneers of U.S. Military Cryptology: Colonel Parker Hitt and His Wife, Genevieve Young Hitt," *Federal History* no. 4 (2012): 87–100.

"This is a man's size job"   Ibid.

"Good work, old girl"   Ibid.

72 *a serious book*   Parker Hitt, *Manual for the Solution of Military Ciphers* (Fort Leavenworth, KS: Press of the Army Service Schools, 1916).

72 *Aimed at Army units*   Ibid., 1–3.
letter is E   WFF and ESF, *An Introduction to Methods for the Solution of Ciphers,* Riverbank, no. 17 (Geneva, IL: Riverbank Laboratories, 1918), 2.
a unique signature   Hitt, *Manual for the Solution of Military Ciphers,* 4–14.
It looks like this   The frequency table here is one I made myself while playing around with the techniques described in ESF's codebreaking book for young adults.
74 *"certain internal relations"*   WFF and ESF, *An Introduction to Methods for the Solution of Ciphers,* 6.
"There lives more faith"   Alfred, Lord Tennyson, "In Memoriam A.H.H.," https://www.poetryfoundation.org/poems -and-poets/poems/detail/45349.
75 *TZYTV*   ESF and WFF, "Riverbank Problems in Cryptanalysis," no. 5b, Item 290, WFF Collection.
"The thrill of your life"   ESF codebreaking book (unpublished manuscript), box 9, file 12, ESF Collection.
Able for the letter A   WFF and ESF, *An Introduction to Methods.*
A single miscopied letter   Ibid., 3–4.
76 *The less you had to think about*   Ibid.
never deviated   Ibid.
black with white erasers   One of the old pencils is on display under glass inside the Fabyan Villa, now maintained as a museum by the Forest Preserve District of Kane County, Illinois.
graph paper   WFF and ESF, *An Introduction to Methods,* 3.
never threw anything out   Ibid.
"a group of two operators"   Ibid., 4–5.
Elizebeth filled the margins   Hitt, *Manual for the Solution of Military Ciphers,* ESF's copy with annotations, Item 150, copy no. 3, WFF Collection.
77 *eight pamphlets*   The first seven were written before William deployed to France in 1918 and the eighth, *The Index of Coincidence,* was published in 1920.
"rise up like a landmark"   Kahn, *The Codebreakers,* 374.
Methods for the Reconstruction of Primary Alphabets   WFF and ESF, *Methods for the Reconstruction of Primary Alphabets,* Riverbank No. 21 (Geneva, IL: Riverbank Laboratories,1918).
Methods for the Solution of Running Key Ciphers   This is the Running Key paper, Riverbank No. 16. In ESF's interview with Valaki, Valaki asks her, "You participated in writing one of the manuals though, didn't you . . . one of the, ah, Riverbank books?" Elizebeth replies, "Yeah, the running-key cipher. Ah, admitted . . . Even in those days I was admitted to have been one

of the authors." ESF interview with Valaki, November 11, 1976, transcribed January 12, 2012, 8.

78 *the drafts marked up* "Chapter II: On the Flexibility of Mind Necessary for Cryptographic Analysis," box 14, folder 2, NYPL. This is a partial typed draft of the eventual Riverbank No. 17 with editing marks in both WFF's and ESF's handwriting.

the historical sections "Appendix I, Historical and General," box 14, folder 2, NYPL. This is a typed draft of pages 7 and 8 from Riverbank No. 17. Whoever typed the draft didn't include the name of the author, but it's fascinating to note that at the top of the draft's first page, the words "By Elizabeth Smith Friedman" have been added in pencil—in WFF's handwriting.

"our pamphlets" WFF to ESF, 15 July 1918, box 2, folder 14, ESF Collection.

"a piece of work" ESF interview with Valaki, November 11, 1976, transcribed January 10, 2012, 4.

"Mrs. Friedman had a tendency" Marshall Foundation to Vanessa Friedman, October 6, 1981, box 12, folder 15, ESF Collection.

"It may be egotism on my part" George Fabyan to WFF, January 12, 1922, Item 734, WFF Collection.

79 *Seven of the eight* Kahn, *The Codebreakers*, 374.

"That World War I leapt on" ESF interview with Valaki, transcribed February 16, 2012, 2.

"Nothing was ever" Ibid., 10.

"I don't think I remember" ESF interview with Valaki, transcribed January 12, 2012, 12.

"I feel no confidence" Ibid., 8.

80 *all the way from Scotland Yard* ESF autobiography, 18–24.

"quite baffling" WFF to Travis Hoke (reporter at *Popular Science Weekly*), January 21, 1920, box 6, folder 13, ESF Collection. This is a detailed narrative of the process used by WFF and ESF to solve the conspirators' cryptograms.

81 *Of 100,000 total words* Ibid.

82 *"I challenge anybody"* Thomas J. Tunney, *Throttled: The Detection of the German and Anarchist Bomb Plotters in the United States* (Boston: Small, Maynard & Co., 1919), 89. See also Rose Mary Sheldon, "The Friedman Collection: An Analytical Guide," rev. October 2013, Marshall Foundation, PDF file, 167, where the Friedmans point out that their solutions were included in this book without credit.

"someone had to stay behind" ESF autobiography, 24.

83 *erupted in spectacle* Tunney, *Throttled*, 103–4; ESF autobiography, 23.

83 all of the codebreaking   ESF autobiography, 13.
   to be indecipherable   R. H. Van Deman, "Memorandum for
   Chief Signal Officer: Subject: Cipher with Running Key,"
   March 16, 1918, Item 734, WFF Collection.
84 *"What Colonel Fabyan"*   WFF and ESF, *The Shakespearean
   Ciphers Examined* (London: Cambridge University Press, 1958),
   287.
   invented a new word   WFF, "On the Flexibility of Mind," box
   14, folder 2, NYPL. This is a typewritten draft of a Riverbank
   Publication passage where you can actually see WFF cross out the
   word "decipherer" and write "cryptanalyst" above it.
85 *Callimahos took up snuffing*   Lambros D. Callimahos, "The
   Legendary William F. Friedman," *Cryptologic Spectrum* 4, no. 1
   (Winter 1974): 9–17.
   "cursed by luck"   Callimahos, "Summer Institute."
   "Even if he computed odds"   Callimahos, "The Legendary
   William F. Friedman."
   "Everything he touched"   Ibid.
   "God-given"   Fred Friendly, remarks at ESF's funeral,
   November 5, 1980, in "Elizebeth Smith Friedman," *Cryptologic
   Spectrum* 10, no. 1 (Winter 1980): box 16, file 24, ESF Collection.
   "I was never able to decide"   J. Rives Childs to Vanessa
   Friedman, September 28, 1981, box 12, folder 14, ESF Collection.
86 *There's a now-famous story*   ESF autobiography, 27–30.
87 *might have used key words*   Ibid.
   "I was sitting across the room"   Ibid.
   "springlike elasticity"   Ibid.
   "it did not occur"   Ibid.
88 *"The female mind"*   WFF, "Second Period, Communications
   Security" (lecture), 50, NSA.
   "I came to the end of my rope"   Ibid., 49.
   "a wonderfully warm man"   ESF to Barbara Tuchman
   (undated), box 14, folder 11, ESF Collection.
   "the smartest man who ever lived"   ESF codebreaking book, 65.
   an article about ciphers   John Holt Schooling, "Secrets in Cipher
   IV: From the Time of George II to the Present Day," *Pall Mall
   Magazine* 8 (January–April 1896): 609–18.
   This "Nihilist" cryptogram   Ibid., 618.
89 *"The meaning of the cipher"*   Ibid.
   "met up with that message"   ESF codebreaking book, 65–66.
   "courage"   Craig P. Bauer, *Unsolved! The History and Mystery
   of the World's Greatest Ciphers from Ancient Egypt to Online
   Secret Societies* (Princeton, NJ: Princeton University Press, 2017),
   145–46.

89 *"Of course, when I learned"* ESF codebreaking book, 65–66.
wanted her all the time WFF to ESF, December 21, 1938,
large blue binder of letters donated by John Ramsay Friedman,
Marshall Research Library, ESF Collection. See also WFF to
ESF, September 9, 1918, box 2, folder 16, ESF Collection.
removed the pin WFF to ESF, September 9, 1918.
imagined a life with her Ibid.
"The glacial undercurrents" "Intermarriage of Jews Presents New
Angle of Problem," *Jewish Criterion* (Pittsburgh), March 9, 1917.
90 *"A part cannot become merged"* " 'Harper's Weekly' Weakness,"
Jewish Criterion (Pittsburgh), February 24, 1905.
"WILL THE JEWS COMMIT SUICIDE" Charles Fleisher,
"Will the Jews Commit Suicide Through Mixed Marriages?"
Jewish Criterion (Pittsburgh), October 25, 1907.
"your soul and spirit and heart" Lyle, "Divine Fire," 85–86.
"skinned to a frazzle . . . You're lots smarter" WFF to ESF,
August 7, 1917, box 2, folder 15, ESF Collection.
"You can soar away" Lyle, "Divine Fire," 85–86.
"Oh Divine Fire Mine" WFF to ESF, undated letter, box 2,
folder 13, ESF Collection.
whose fire is in Zion Isaiah 31:9.
91 *realized that electronic circuits* C. E. Shannon, "A Symbolic
Analysis of Relay and Switching Circuits," *Transactions of the
AIEE* 57, no. 12 (1938): 713–23.
secret NSA projects Transcript of Solomon Kullback oral
history interview with NSA, August 26, 1982. Kullback discusses
the NSA's interest in Shannon's research and says, "We had very
close contacts with the Bell Laboratories. They were very, let's
say, willing to work along with us."
communicating through a noisy system C. E. Shannon, "A
Mathematical Theory of Communication," *Bell System Technical
Journal* 27, no. 3 (July 1948), http://ieeexplore.ieee.org
/document/6773024/.

CHAPTER 5: THE ESCAPE PLOT

93 "To be your North Star" ESF to WFF, February 7, 1917, box 2,
folder 1, ESF Collection.
94 *"I miss you infinitely"* Ibid.
"I shall work for you" Ibid.
I have dreamed about you Ibid.
"Anyway, Billy Boy" Ibid.
"Work hard on the letter tests" ESF to WFF, January 31, 1917,
box 2, folder 1, ESF Collection.

94 *"There was a time"* ESF diary, 46.
a careful, unemotional tone WFF to ESF, September 9, 1918,
box 2, folder 16, ESF Collection.
He confessed later Ibid.
95 *getting married would be silly* Ibid.
fetch a newspaper ESF autobiography, 14–15.
forced William to change Ibid.
"It just didn't go down" ESF interview with Clark, 11.
96 *"make a mark in something"* WFF to ESF, July 24, 1918, box 2,
folder 14, ESF Collection.
they went missing "William Friedman and Miss Elizabeth [*sic*]
Smith Were Married Monday," *Geneva Republican* (Geneva,
Illinois), May 23, 1917.
light-colored striped pants "Bride and Groom William F.
Friedman and Elizebeth S. Friedman," photograph, 1917, ESF
Collection.
A rabbi named Hersh Ibid.
The wedding announcement Ibid.
a story about a Selective Service bill "Sheriff Richardson Gets
Official Notice: The Sheriff Has Received Plans for Draft of
Eligibles," ibid.
"Mr. Friedman came to Riverbank" "William Friedman and
Miss Elizabeth [*sic*] Smith."
admitted this later ESF diary, 62.
"Splash!" Arthur Stringer, *The Prairie Wife* (Indianapolis:
Bobbs-Merrill, 1915), 3.
97 *"I am learning"* ESF diary, June 20, 1917.
a wire to Riverbank ESF to WFF, May 8, 1917, box 2, folder 1,
ESF Collection.
"I am cast into a whirl" Ibid.
"You would have thought" Ronald Clark, *The Man Who Broke
Purple: Life of Colonel William F. Friedman, Who Deciphered the
Japanese Code in World War II* (Boston: Little, Brown, 1977), 39.
She moved from Engledew Cottage into the windmill WFF
to ESF, June 1918, "Installment #3," box 2, folder 13, ESF
Collection.
98 *invited the Army to Riverbank* Jack Lait, "Recruit Rally Thrills
Throng," *Chicago Herald*, July 9, 1917.
A U.S. Army captain Ibid.
"Better to go and die" Ibid.
three hundred and fifty dollars Ibid.
He began to pester Fabyan ESF interview with Pogue, 65
Fabyan always waved him off Ibid.
99 *intercepting the Friedmans' mail* Ibid; ESF autobiography, 16.

99 *secret listening devices*   ESF interview with Valaki, transcribed January 12, 2012, 4.

Mr. Powell   ESF diary, August 13, 1917.

"My dearest"   ESF diary. The slip of paper from WFF is inserted between pages 42 and 43.

"My heart sang"   ESF diary, August 13, 1917.

throwing her arms   WFF to ESF, December 21, 1938, ESF Collection.

"My Lover-Husband"   ESF diary, August 13, 1917.

100 *started to dry up*   ESF autobiography, 16, 26.

named Herbert O. Yardley   David Kahn, *The Reader of Gentlemen's Mail: Herbert O. Yardley and the Birth of American Codebreaking* (New Haven, CT: Yale University Press, 2004).

"Why did America"   Herbert O. Yardley, *The American Black Chamber* (Indianapolis: Bobbs-Merrill, 1931), 20.

a shark at poker   Kahn, *The Reader*, 3.

Known officially as MI-8   Ibid., "Staffers, Shorthand, and Secret Ink," 28–35.

101 *"just off the street"*   ESF interview with Valaki, transcribed January 12, 2012, 4.

"What was taught was taught"   ESF interview with Valaki, transcribed February 16, 2012, 11.

booked the largest hotel   ESF interview with Marshall staff, Tape #2, June 4, 1974, 11.

William and Elizabeth taught class   ESF interview with Pogue, 64–66; ESF interview with Clark, 6.

stationed in paradise   John W. Kopec, *The Sabines at Riverbank: Their Role in the Science of Architectural Acoustics* (Woodbury, NY: Acoustical Society of America, 1997), 41.

lavish military ball   Ibid.

four of the officers' wives   George Fabyan to chief of the MID, March 22, 1918.

102 *gathered outside the hotel*   Clark, *The Man Who Broke Purple*, 47; Kopec, *The Sabines at Riverbank*, 47–48.

Each person stood for a letter   Ibid.

the glass surface of his work desk   WFF's desk is preserved at the Marshall Foundation, glass removed.

In May 1918   ESF diary, 43.

boarded a train to Chicago   WFF to ESF, June 8, 1918, box 2, folder 13.

103 *"I, a mere woman"*   ESF, "Pure Accident," *The ARROW*, box 12, folder 9, 401, ESF Collection.

"heartache of separation"   ESF diary, July 1918 entry, 44.

"a calm Whole"   Ibid.

103 *"The work is so hard"* WFF to ESF, July 24, 1918, box 2, folder 14, ESF Collection.
"out of the clear blue" WFF to ESF, August 26, 1918, box 2, folder 15, ESF Collection.
"On Saturday Col. M" WFF to ESF, November 10, 1918, box 2, folder 18, ESF Collection.
Frank Moorman WFF, "Six Lectures on Cryptology by William F. Friedman," Lecture V, 117, in *The Friedman Legacy, Sources on Cryptologic History,* no. 3 (Center for Cryptologic History: 2006).
"Love-girl" WFF to ESF, July 21, 1918, box 2, folder 14, ESF Collection.

104 *the .45 pistol* WFF to ESF, July 23, 1918, box 2, folder 14, ESF Collection.
French woman he called Madame WFF to ESF, October 6, 1918, box 2, folder 17, ESF Collection.
cigarettes as torches WFF to ESF, September 9, 1918, box 2, folder 16, ESF Collection.
based on six letters WFF, Lecture V, 109, in *The Friedman Legacy.*
"how much 'group work'" WFF to ESF, August 4, 1918, box 2, folder 15, ESF Collection.
didn't care for the taste WFF to ESF, July 6, 1918, box 2, folder 14, ESF Collection.
Lemonade Ibid.
nursing highballs WFF to ESF, October 15, 1918, box 2, folder 17, ESF Collection.
always regretted WFF to ESF, July 6, 1918, box 2, folder 14, ESF Collection. "Before I left home the Colonel's advice was that I forget poker completely."
spent time in France Kahn, *The Reader,* 45–49.
"I must confess" WFF to ESF, December 16, 1918, box 2, folder 19, ESF Collection.
each time he struck a match WFF describes this ritual in WFF to ESF, December 19, 1918, box 2, folder 19, ESF Collection.

105 *"Do you miss your Biwy Boy"* WFF to ESF, October 6, 1918, box 2, folder 17, ESF Collection.
"the many imperfections" Katie Letcher Lyle, "Divine Fire: Elizebeth Smith Friedman, Cryptanalyst," unpublished manuscript, July 4, 1991, ESF Collection, two PDF files, 86.
"good lover" Ibid.
"towered above me" Ibid., 87.
with the windows open WFF to ESF, October 6, 1918.
a recurring dream WFF to ESF, July 21, 1918, box 2, folder 14,

ESF Collection. "Most of my dreams of you have pictured me as losing you, and I awoke trembling and with a deep fear in my heart."

105 *"You didn't yike me"*   WFF to ESF, undated letter (December 1918), box 2, folder 19, ESF Collection.

"no money and a lot of debts"   WFF to ESF, August 4, 1918, box 2, folder 15, ESF Collection.

fixing rare grammatical mistakes   WFF to ESF, November 3, 1918, box 2, folder 18, ESF Collection. WFF circled the word "vastly" and scrawled next to it, "Why Billy! Don't you know better than to split an infinitive or something!"

106 *"This cable will read"*   WFF to ESF, September 20, 1918, box 2, folder 16, ESF Collection.

a lock of her hair   WFF to ESF, December 26, 1918, box 2, folder 19, ESF Collection.

It's likely she destroyed them   This is Lyle's conclusion in "Divine Fire," 84, and I tend to agree; Lyle was able to interview Elizebeth when she was alive, and I think "Divine Fire" is excellent on the personal issues Elizebeth faced in her twenties.

Riverbank as R.   WFF to ESF, January 28, 1919, box 2, folder 20, ESF Collection.

G.F.   Ibid.

B.C.   Ibid.

Fabyan's "excesses"   Ibid.

107 *"You are perfectly right"*   WFF to ESF, November 10, 1918.

she revealed something   WFF to ESF, October 7, 1918, box 2, folder 17, ESF Collection.

"Honey, I could have committed"   Ibid.

later confided to friends   Lyle, "Divine Fire," 96.

"Honey, don't be afraid"   WFF to ESF, August 30, 1918, box 2, folder 15, ESF Collection.

German prisoners of war   Heber Blankenhorn, *Adventures in Propaganda: Letters From an Intelligence Officer in France* (Boston: Houghton Mifflin, 1919), 82. Blankenhorn was a MID captain working at GHQ in Chaumont, same as William, and he wrote beautifully about life there.

a group of American soldiers   Ibid., 135.

blew up bombs   Ibid., 136.

hung lanterns   Ibid.

he stayed indoors   WFF to ESF, November 10, 1918, box 2, folder 18, ESF Collection.

"Home does not entail"   Ibid.

108 *"Elsbeth, my Dearest"*   Ibid.

108 *"The signing of the Armistice"* WFF to ESF, December 16, 1918, box 2, folder 19, ESF Collection.
"What shall I say" Ibid.
He had to stay in Chaumont WFF to ESF, November 26, 1918, box 2, folder 18

109 *its Code and Signal Section* Office of Naval Intelligence to ESF, n.d. [fall 1918?], Item 734, WFF Collection.
"of the greatest value" John M. Manly to ESF, September 12, 1918, Item 734, WFF Collection.
how small an electron is WFF to ESF, January 2, 1919, box 2, folder 20, ESF Collection.
"Can't two perfectly" WFF to ESF, December 16, 1918, box 2, folder 19, ESF Collection.
"a long enough vacation" George Fabyan to WFF, November 13, 1918, Item 734, WFF Collection.
"I refuse to have anything" WFF to ESF, January 28, 1919.
"I don't want to flatter ourselves" Ibid.
a love note in cipher WFF to ESF, undated letter beginning "Good Morning, Flower Face Mine," box 2, folder 20, ESF Collection.

110 *"I am wondering how you are"* George Fabyan to ESF, November 2, 1918, box 1, folder 42, ESF Collection.
blue colored pencil George Fabyan to ESF, September 26, 1918, Item 734, WFF Collection.

111 *divide-and-conquer* George Fabyan to ESF, January 6, 1919, box 1, folder 42, ESF Collection.
"Does he suppose" WFF to ESF, January 28, 1919.
"see them in hell" Fabyan to ESF, January 6, 1919.
"old man going down hill" Ibid.
"I am inclined to agree" ESF to George Fabyan, January 9, 1919, box 1, folder 43, ESF Collection.
"Won't our reunion" WFF to ESF, February 5, 1919, box 2, folder 20, ESF Collection.
They stayed in the East ESF autobiography, 33.
couldn't go back to Riverbank Ibid.

112 *"The War will not make"* WFF to ESF, October 6, 1918, box 2, folder 17, ESF Collection.
return to his first love, genetics Ibid.
"extraordinary gift" ESF autobiography, 34.
"Everybody said" ESF interview with Clark.
"Come back to Riverbank" ESF interview with Marshall staff, Tape #2, June 4, 1974, 12.
"He had us followed" ESF interview with Pogue, 70.
in their own house Ibid., 72.
raises never materialized Ibid.

112 *shoved the report into a drawer* WFF and ESF, *The Shakespearean Ciphers Examined* (London: Cambridge University Press, 1958), 217–21.

113 *a crowning scientific achievement* David Kahn, *The Codebreakers: The Comprehensive History of Secret Communication from Ancient Times to the Internet,* rev. ed. (New York: Scribner, 1997), 376–85.
"The Index of Coincidence" WFF, *The Index of Coincidence and Its Applications in Cryptography* (Washington, D.C.: US Government Printing Office, 1925).
6.67 percent James R. Chiles, "Breaking Codes Was This Couple's Lifetime Career," *Smithsonian* (June 1987): 128–44.
statistics with cryptology Kahn, *The Codebreakers,* 376–85.
first in France WFF to Nelle Fabyan, October 30, 1937, Item 734, WFF Collection; ESF autobiography, 37–38.
"Fabyan's skullduggery" Herbert O. Yardley to WFF, August 14, 1919, Item 734, WFF Collection.
leapt at the chance Joseph Mauborgne to WFF, November 27, 1920, Item 734, WFF Collection.
"a great misfortune" Ibid.
"as powerful as he is ruthless" WFF to Joseph Mauborgne, November 29, 1920, Item 734, WFF Collection.
"I expect a lively row" Joseph Mauborgne to WFF, December 16, 1920, Item 734, WFF Collection.

114 *overly cruel* ESF interview with Clark, 11.
just as tricky ESF interview with Pogue, 72.
"our secret plot" Ibid., 73.
One morning Ibid.
the three o'clock train Ibid., 72.
an eerie calmness WFF to Joseph Mauborgne, December 16, 1920, Item 734, WFF Collection.
William assumed Ibid.
"after a very limited" WFF to ESF, January 28, 1919.
"Oh, you are some partner" WFF to ESF, January 28, 1919, "P.S." on a separate page, ESF Collection.
"By the end of the war" ESF, "Pure Accident."

PART II: TARGET PRACTICE

119 *"To work in this field"* Niels Ferguson, Bruce Schneier, and Tadayoshi Kohno, *Cryptography Engineering: Design Principles & Practical Applications* (Indianapolis: Wiley, 2010), 8.
"completely inadequate" WFF, "Second Period, Communications Security" (lecture), 45, NSA.

119 *"Military, naval, air"*   Ibid., 45–46.
    two sides of the same coin   WFF, "Communications Intelligence
    and Security Presentation Given to Staff and Students" (lecture,
    Breckinridge Hall, Marine Corps School, April 26, 1960), 5,
    NSA.
120 *"All the countries of the world"*   ESF interview with Clark, 16.
    1,400 newly hired Prohibition agents   Thomas V. DiBacco,
    "Prohibition's First 'Dry' New Year's Eve," *Washington Times,*
    December 30, 2015, http://www.washingtontimes.com/news/2015
    /dec/30/thomas-dibacco-prohibitions-first-dry-new-years-ev/.
    "to make the celebration"   Ibid.
    the Munitions Building   "Main Navy and Munitions Buildings,"
    Histories of the National Mall, http://mallhistory.org/items
    /show/57.
    fourteen thousand army and navy workers   Ibid.
    His starting salary   "1921: William Friedman Joined War
    Department," National Cryptologic Museum Foundation,
    https://cryptologicfoundation.org/m/cch_calendar_mobile.html
    /event/2016/07/01/1467349200/1921-william-friedman-joined
    -war-department/74534.
    Elizebeth's was $2,200   ESF Personnel Folder, "Personal
    History Statement," July 1, 1930.
121 *a piano studio above a bakery*   ESF interview with Pogue, 77.
    Mauborgne visited with his cello   ESF interview with Clark.
    pedestrians stopping   Ibid.
    with two dozen people   "Morning in New York," in David
    Kahn, *The Reader of Gentlemen's Mail: Herbert O. Yardley
    and the Birth of American Codebreaking* (New Haven, CT: Yale
    University Press, 2004), 50–62.
    the messages of Japanese diplomats   Ibid., 63–71.
    wasn't skilled enough to go further   John Bryden, *Best-Kept
    Secret: Canadian Secret Intelligence in the Second World War*
    (Toronto: Lester, 1993), 88–89.
    apartment for his mistress   Kahn, *The Reader,* 48.
    "an edge on her"   ESF to WFF, "Friday, 2:30 P.M.," summer
    1921, box 2, folder 2, ESF Collection.
122 *first scientifically constructed*   ESF autobiography, 39–40.
    "all the difference"   ESF interview with Clark, 16.
    designed to survive capture   David Kahn, *Seizing the Enigma:
    The Race to Break the German U-boat Codes, 1939–1943*
    (Boston: Houghton Mifflin, 1991), 33.
123 *the application of electrical current*   WFF, Lecture V, 156–57,
    in *The Friedman Legacy, Sources on Cryptologic History,* no. 3
    (Center for Cryptologic History: 2006).

123 *William asked Hebern*   WFF, "Second Period, Communications Security" (lecture), 18, NSA.
"discouraged to the point of blackout"   ESF interview with Pogue, 43.
"As I was tying my black tie"   WFF, "Communications Intelligence and Security," 34.
"P.S."   WFF to George Fabyan, March 10, 1924, Item 734, WFF Collection.
"It's a striking paradox"   WFF to George Fabyan, June 23, 1926, Item 734, WFF Collection.

124 *two discs of alphabets*   WFF, "Six Lectures on Cryptology by William F. Friedman," Lecture VI, 149, in *The Friedman Legacy, Sources on Cryptologic History,* no. 3 (Center for Cryptologic History: 2006)
2.29 x 10^{82}   Lambros D. Callimahos, "The Legendary William F. Friedman," *Cryptologic Spectrum* 4, no. 1 (Winter 1974): 9–17.
the number of atoms   Thought to be 10^{80}. Wikipedia, s.v. "Observable universe," last modified April 22, 2017, https://en.wikipedia.org/wiki/Observable_universe.
"The number of permutations"   WFF, "Communications Intelligence and Security," 30.
demonstrated his mastery   Callimahos, "The Legendary William F. Friedman."

125 *"but it helps"*   Rose Mary Sheldon, "William F. Friedman: A Very Private Cryptographer and His Collection," *Cryptologic Quarterly* 34, no. 1 (2015): 20.
Rochefort recalled later   Captain Joseph J. Rochefort, interview by U.S. Naval Institute, Annapolis, Maryland, 1983, 45–47.
"Here is a bunch of messages"   Ibid., 47.
had to stop breaking codes   Ibid., 45.
"no one that could compare"   Ibid., 40.
a young German engineer   "The Man, the Machine, the Choice," in Kahn, *Seizing the Enigma,* 31–48.
"clever inventor"   WFF, "Six Lectures on Cryptology by William F. Friedman," Lecture VI, 153, in *The Friedman Legacy.*

126 *"stay home and write some books"*   ESF interview with Marshall staff, Tape #5, June 6, 1974, 8.
excited about her books   WFF to George Fabyan, August 10, 1926, Item 734, WFF Collection.
"I am all alone"   WFF to John M. Manly, February 4, 1922, box 13, folder 22, ESF Collection.
"I drove 36 miles"   ESF to WFF, letter marked "Monday,

9:30 P.M." by ESF and "Washington era between 1921 and 1923"
by archivists, box 2, folder 2, ESF Collection.

126 *midwestern stranger*   Ibid.
"My dear"   ESF to WFF, "Monday, 9:30 P.M."

127 *move out of the city*   Katie Letcher Lyle, "Divine Fire: Elizebeth
Smith Friedman, Cryptanalyst," unpublished manuscript, July 4,
1991, ESF Collection, two PDF files, 126.
kryptos   Dictionary.com, s.v. "crypt," accessed May 10, 2017,
http://www.dictionary.com/browse/crypt.

128 *Pinklepurr*   Lyle, "Divine Fire," 218.
A. A. Milne poem   A. A. Milne, "Pinkle Purr," in *Now We Are
Six* (New York: Puffin Books, 1992), 89.
in the morning   Lyle, "Divine Fire," 128.
a former boxer   WFF, "Second Period, Communications
Security" (lecture), 20, NSA.
a pug nose   Ibid.
"Now," Elizebeth wrote   ESF codebreaking book.
"little book"   Ibid.
a children's history of the alphabet   ESF children's history
of the alphabet (unpublished manuscript), box 9, file 14, ESF
Collection.

129 *knock on her door*   ESF interview with Marshall staff, Tape #5,
June 6, 1974, 8.
A retired astronomy professor   Craig P. Bauer, *Unsolved!
The History and Mystery of the World's Greatest Ciphers
from Ancient Egypt to Online Secret Societies* (Princeton, NJ:
Princeton University Press, 2017), 500–3.
A man sent a bomb to Huey Long   WFF backfile part IV, "(FBI
J Edgar Hoover) Bank Robbery. Cases with Dept. of Justice,
Ohio State Penitentiary, and Post Office Inspection Service,"
Item 849, WFF Collection.
a plot   Ibid.
a congressional committee   ESF autobiography, 41–42.
caught the eye   Ibid.

130 *notes scrawled with code*   Ibid.
refused to pay him   ESF autobiography, 43.
she wore it around her neck   ". . . the one-of-a-kind Evalyn
Walsh McLean," PBS Treasures of the World, http://www.pbs
.org/treasuresoftheworld/hope/hlevel_1/h3_ewm.html.
an anti-Semitic weekly newspaper   "The International Jew:
The World's Problem," *Dearborn Indepdendent* (Dearborn,
Michigan), May 22, 1920.
intelligence reports about Jewish activities   Joseph W. Bendersky,

The Jewish Threat: Anti-Semitic Politics of the U.S. Army (New York: Basic Books, 2000), xiii–xiv.

130 *"When they couldn't get him"*   ESF interview with Marshall staff, Tape #5, June 6, 1974, 8.
"second-hand"   ESF interview with Clark.
"Sad for me"   ESF interview with Marshall staff, Tape #5, June 6, 1974, 8.

131 *"I didn't want to work for the Navy"*   ESF interview with Pogue, 41–42.
She took Driscoll's place   Ibid.
left the navy after five months   Ibid.
suppressing her own desires   "Sopha was a shadowy figure, her life obviously deeply and exclusively involved in her anatomical destiny, that of bearing ten children over an eighteen year period." Lyle, "Divine Fire," 165.
"Often I feel"   WFF to ESF, July 31, 1918, box 2, file 14, ESF Collection.
"Sometimes I wish"   Ibid.
"a queer sensation"   Ibid.
"we could help her"   George Fabyan to WFF, February 24, 1924, Item 734, WFF Collection.

132 *she would be all right*   WFF to George Fabyan, February 25, 1924, Item 734, WFF Collection.
black woman named Cassie   ESF autobiography, 63.
constantly perplexed   ESF copy of George Fabyan's *What I Know About the Future of Cotton and Domestic Goods,* 2nd ed. (Chicago, 1900), containing diary entries about Barbara and John Ramsay, box 21, folder 1, ESF Collection.
shared her dolls with Krypto   Ibid.
his mirth excited   Ibid.
top volume on the Victrola   Ibid.
a doctrine of no doctrines   Ibid.
"let the rest take care of itself"   Ibid.
"She strings together consonants"   Fabyan, *What I Know About the Future,* Item 602, WFF Collection.

133 *3932 Military Road*   ESF Personnel Folder.
203 slow small patrol boats   David P. Mowry, "Listening to the Rum-Runners: Radio Intelligence During Prohibition," 2nd ed., Center for Cryptologic History, 2014, 16.
five thousand miles of coastline   Ellen NicKenzie Lawson, *Smugglers, Bootleggers, and Scofflaws: Prohibition and New York City* (Albany: State University of New York Press, 2013), 7.
in radio prowess   Commander J. F. Farley, "Radio in the Coast Guard," *Radio News* (January 1942): 43–48.

134 *a ninety-day contract*   ESF, "Personal History," Personnel
Folder, July 1, 1931.
work from home   ESF interview with Clark, 15.
Fifteen thousand people   U.S. Treasury Department, *Annual
Report of the Secretary of the Treasury on the State of the
Finances for the Fiscal Year Ended June 30, 1926* (Washington,
D.C.: Government Printing Office, 1927).
six separate law enforcement agencies   "SA Eliot Ness, a
Legacy ATF Agent," Bureau of Alcohol, Tobacco, Firearms and
Explosives, https://www.atf.gov/our-history/eliot-ness.
135 *a soft-spoken father of two*   Robert G. Folsom, *The Money
Trail: How Elmer Irey and His T-men Brought Down America's
Criminal Elite* (Washington, D.C.: Potomac Books, 2010), 313.
"the Treasury fist"   "T-men (1947) Quotes," IMDb.com, http://
www.imdb.com/title/tt0039881/quotes.
solution in the margin   "Elmer Irey Retires: Boss of Treasury
T-men Was One of World's Greatest Detectives," *Life*
(September 2, 1946).
independent sea captains   Frederick Van de Water, *The Real
McCoy* (Mystic, CT: Flat Hammock Press, 2007).
the envy of many small nations   ESF, "History of Work in
Cryptanalysis," April 27–June 1930," box 4, folder 17, ESF
Collection; ESF, "West Coast," narrative of West Coast
smuggling operations, box 4, folder 23, ESF Collection.
"The whole half of the world"   ESF interview with Clark.
136 *"Mrs. Friedman is the only person"*   Roy A. Haines, Acting
Prohibition Commissioner, to Civil Service Commission,
April 22, 1927, box 4, folder 16, ESF Collection.
"see what you can do"   Office of Chief Prohibition Investigator
to ESF, February 1, 1926, box 4, folder 10, ESF Collection.
sent from Halifax, Nova Scotia   "1,000, Following for Your
Information," January 29, 1926, box 4, folder 10, ESF Collection.
"most secret communications"   ESF, "History of Work in
Cryptanalysis," April 27–June 1930," box 4, file 17, ESF Collection.
used different code systems   ESF, "History of Chief Smuggling
Interests on the Pacific Coast," box 4, folder 23, ESF Collection.
137 *"If I may capture a goodly number"*   ESF, "A Cryptanalyst,"
Arrow (February 1928), box 12, folder 9, 531–34, ESF Collection.
maps of the radio traffic   ESF, "Chart Showing Operations of
Liquor Smuggling Vessels as Directed by Short Wave Radio
Through Secret Systems of Communication, Pacific Coast,"
August 1933, box 6, file 1, ESF Collection.
"I sort of floated around"   ESF interview with Clark.
138 *She traveled to the West Coast*   ESF autobiography, 52.

138 *sailing up the Hudson River* Transcript of ESF interview with Margaret Santry, NBC national radio broadcast, May 25, 1934, box 19, folder 6, ESF Collection.

brothers named Hobbs ESF, "History of Chief Smuggling Interests."

"to Joseph Kennedy, Ltd." Ibid. For more on Joseph Kennedy's connections to Vancouver rum interests, see Stephen Schneider, *Iced: The Story of Organized Crime in Canada* (Mississauga, Ontario: Wiley, 2009), 207.

Elizebeth wasn't afraid ESF interview with Ed Meryl. Asked if she was ever in physical danger due to her work, ESF responded dismissively—"Not that I know of"—despite the fact that she needed bodyguard protection during the *I'm Alone* proceedings.

She went to Houston, Texas ESF autobiography, 52, 92–95.

650 messages in 24 different code systems ESF, "History of Work in Cryptanalysis."

a one-legged cabdriver "Bond for Armatou Lowered to $250," *Galveston Daily News*, March 26, 1930.

139 *"he was in a very mean mood"* ESF interview with BBC for "Codebreakers" TV special, box 15, folder 5, ESF Collection.

"Only woman on plane" ESF Vancouver trip log, written for her children while flying cross-country, October 16–17, 1937, box 16, folder 2, ESF Collection.

out for afternoon tea ESF to WFF, 1932, box 2, folder 3, ESF Collection.

"messages unearthed from a safe" ESF to WFF, October 21, 1937, box 2, folder 4, ESF Collection.

"under the press of duties" ESF to Josephine Coates, January 23, 1930, box 1, folder 2, ESF Collection.

two thousand messages per month ESF, "History of Work in Cryptanalysis."

a single clerk-typist ESF, "Memorandum upon a Proposed Central Organization at Coast Guard Headquarters for Performing Cryptanalytic Work," box 5, file 6, ESF Collection.

twelve thousand rum messages ESF, "History of Chief Smuggling Interests."

a seven-page memo Ibid.

140 *thirty of these books* Ibid.

a refuge from problems ESF interview with Clark, 15–16.

"Position," "Landing boat" See intercepted rum messages and worksheets in box 4, folder 14, ESF Collection.

Elizebeth came home ESF interview with Clark, 15.

"pair of shoes, size 15" Ibid.; ESF, "History of Chief Smuggling Interests."

141 *a codebreaking team of her own* "History of USCG Unit #387,"
Foreword.
Cryptanalyst-in-Charge ESF, "Memorandum upon a Proposed
Central Organization."; U.S. Treasury Department, Office of the
Secretary to ESF, June 30, 1931, ESF Personnel Folder.
Scouring civil service lists ESF autobiography, 53.
worried that her new employees Ibid., 56.
the top scorer Ibid., 53.
"he did not comprehend" Ibid., 54.
Hyman Hurwitz Hyman Hurwitz, Official Personnel Folder,
National Personnel Records Center, National Archives at
St. Louis, requested September 2016.

142 *Vernon Cooley* Vernon E. Cooley, Official Personnel Folder,
National Personnel Records Center, National Archives at St.
Louis, requested September 2016.
Robert Gordon Robert E. Gordon, Official Personnel Folder,
National Personnel Records Center, National Archives at
St. Louis, requested September 2016.
"able, agreeable, and cooperative" ESF autobiography, 55.
a productive routine Ibid.
pound for pound the best This is my own conclusion. It was
shared by the future officers of British Security Coordination,
who arrived in the United States in 1940 and determined that
Elizebeth's coast guard unit was the most effective in the country.
See British Security Coordination, *The Secret History of British
Intelligence in the Americas, 1940–1945* (New York: Fromm
International, 1999), 469–70.
"The world is a mess" George Fabyan to WFF, January 23, 1932,
Item 734, WFF Collection.
Elizebeth took her daughter ESF untitled two-page narrative
about her work with the League of Women Voters, box 7, folder 6,
ESF Collection.
"The only thing we have to fear" Franklin Delano Roosevelt,
"Inaugural Address," March 4, 1933, http://www.presidency
.ucsb.edu/ws/?pid=14473.

143 *outside in the freezing wind* James A. Hagerty, "Roosevelt
Address Stirs Great Crowd," *New York Times,* March 5, 1933.
handed out League of Women Voters fliers ESF League of
Women Voters narrative.
"ardent worker" Ibid.
Eighteen days later, in Germany "Dachau Opens," United
States Holocaust Memorial Museum, https://newspapers.ushmm
.org/events/dachau-opens.
at a press conference "Himmler sets up Dachau," The Nazi

Concentration Camps, Birbeck University of London, http://
www.camps.bbk.ac.uk/documents/003-himmler-sets-up-dachau
.html.

143 *"I pack my bag"* ESF, "Pure Accident."
"Please state your name" *United States v. Albert M. Morrison et
al.* (E.D. La. 1933), No. 16,981, May 2, trial transcript vol. 1, 141.
twenty-three suspected agents "Wireless Station Operator
Called in Rum Ring Trial," *Times-Picayune* (New Orleans),
May 2, 1933. The *Times-Picayune* says there were twenty-
four defendants; ESF says twenty-three, on page 80 of her
autobiography. ESF never miscounts things, so I am going with
her number. A few of the indicted men, including Isadore "Kid
Cann" Blumenfeld, a gangster from Minneapolis, never appeared
in court, creating some confusion.
a fleet . . . a pirate radio station *United States v. Morrison,* No.
6,981, May 1, trial transcript vol. 1, 47–62. This is Woodcock's
opening statement, which gives an overview of the case.
at least thirty-two coded radio messages Ibid., Bill of
Exceptions, Exhibit X 29.
mailed them Ibid., May 2, trial transcript vol. 1, 170.
144 *their system* Ibid., Bill of Exceptions.
"the greatest rum-running conspiracy" "Greatest Liquor Plot
Case Trial Delayed 35 Days," *Times-Picayune* (New Orleans),
April 15, 1932.
a hat with a flower "Code Expert Testifies at Trial,"
Times-Picayune (New Orleans), May 3, 1933.
a stack of yellow papers *United States v. Morrison,* No. 16,981,
May 2, trial transcript vol. 1, 150–55.
quietly switching seats Ibid., May 1, trial transcript vol. 1, 64–76.
a rawboned man Ibid.
the aliases Ibid., May 1, trial transcript vol. 1, 47.
"Mr. Burk" Ibid., May 1, trial transcript vol. 1, 64–76. "Mr. Burk"
was what Morrison called himself when he dealt with the
shortwave radio operator in New Orleans, Charles Andres.
Nathan Goldberg, Al Hartman, and Harry Doe "Bond
Reductions Ordered for Six Held in Rum Plot," *Times-Picayune*
(New Orleans), April 15, 1931.
$500,000 and more than two years ESF autobiography, 80; the
first radio messages from the ring were intercepted in March 1931.
methodical former Army colonel S. J. Woolf, "Col. Woodcock:
Leader of the Dry Army," *New York Times,* November 2, 1930.
"a steady attack" Ibid.
"Have you a message" *United States v. Morrison,* No. 16,981,
May 2, trial transcript vol. 1, 143.

144 *"QUIDS, ABGAH"*   Ibid., Bill of Exceptions, Exhibit X 29.
"incompetent, irrelevant"   Ibid., May 2, trial transcript vol. 1, 145.
Eight other defense attorneys   Ibid., May 2, trial transcript vol. 1, 1.

145 *handled Capone's appeals*   "Capone Renews Fight for Freedom,"
Town Talk (Alexandria, Louisiana), April 30, 1934.
Walter J. Gex Sr.   "Leader of City and County Passes Away;
Last Rites Monday P.M.," *Sea Coast Echo* (Bay St. Louis,
Mississippi), February 1937.
"I believe I am asked"   Ibid., 145–47.
"That certainly is some information"   Ibid., 162.
"GD (HX) gm"   *United States v. Morrison,* No. 16,981, Bill of
Exceptions, Exhibit X 6.
"SUBSTITUTE FIFTY CANADIAN"   Ibid., May 2, trial
transcript vol. 1, 150.
"How shall I address you"   Ibid., 164–66.

146 *"I move that all of the testimony"*   Ibid., 168–69.
the jury convicted five   "Five Men Found Guilty of Liquor
Plotting Charge," *Times-Picayune* (New Orleans), May 7, 1933.
two years in prison   "Woodcock Move Clears Five Men in Rum
Plot Case," *Times-Picayune* (New Orleans), May 9, 1933.
"made an unusual impression"   ESF autobiography, 88.
"a pretty government"   Lyle, "Divine Fire," 174.
"a pretty middle aged woman"   ESF autobiography, 70.
"a pretty young woman with a filly pink dress"   Ibid.
"a pretty little woman who protects"   Lyle, "Divine Fire," 174.
young enough to resent   ESF autobiography, 70–71.
badly written   Ibid.
Facing off against Edwin Grace   ESF autobiography, 84–85.
"CLASS IN CRYPTOLOGY"   Ibid., 82.

147 *She hoped the attention*   Ibid., 83.
"He never put into words"   ESF interview with Valaki,
November 11, 1976, transcribed January 10, 2012, 16.
"a certain grim look"   Ibid.
A former artillery officer   "About Henry L. Stimson," Stimson
Center, https://www.stimson.org/content/about-henry-l
-stimson.
"Gentlemen do not read each other's mail"   Kahn, *The Reader,*
98.

148 *a new army codebreaking unit*   Frank Rowlett, *The Story of
Magic: Memoirs of an American Cryptologic Pioneer* (Laguna
Hills, CA: Aegean Park Press, 1998), 6–33.
adjacent desks   Ibid.
he popped his head in the vault   "We Discover the Black
Chamber," ibid., 34–39.

148 *"Welcome, gentlemen"*   Ibid.
149 *challenging them to solve different machines*   Rowlett, *The Story of Magic,* 59–76.
    Angooki Taipu A   Craig P. Bauer, *Secret History: The Story of Cryptology* (Boca Raton, FL: CRC Press, 2013), 296–300.
    Angooki Taipu B   Ibid., 301–4.
    Converter M-134   WFF, "Important Contributions to Communications Security, 1939–1945," 1, NSA.
150 *held secret for many years*   Fischer, Willis and Panzer, "Memorandum Concerning a Bill for the Relief of William F. Friedman," August 21, 1950, NSA.
    an enduring frustration   Ibid.
    evolved into the SIGABA   WFF, "Important Contributions," 3–5.
    Up to four rotors   Timothy J. Mucklow, "The SIGABA/ECM II Cipher Machine: 'A Beautiful Idea,'" Center for Cryptologic History, National Security Agency, 2015.
    10,060 SIGABA machines   Ibid.
    "the absolute security"   WFF, "Important Contributions," 5.
    "Never said a word to me"   ESF interview with Pogue, 24.
    "never opened his mouth"   Ibid., 53.
151 *"was hugging me warmly"*   Barbara Friedman to Ronald Clark, September 26, 1976, and "P.S.," October 6, 1976, box 14, folder 14, ESF Collection.
    make himself a sandwich   ESF interview with Pogue, 27.
    "nothing more or less than exhaustion"   Ibid., 46.
    the ice pick lobotomy   Michael M. Phillips, "The Lobotomy Files: One Doctor's Legacy," *Wall Street Journal,* http://projects.wsj.com/lobotomyfiles/?ch=two.
    eventually, William did   His psychiatrist starting in the 1940s was Dr. Zigmond Lebensohn of George Washington University, a junior colleague of Freeman. See Zigmond Lebensohn, "The History of Electroconvulsive Therapy in the United States and Its Place in American Psychiatry: A Personal Memoir," *Comprehensive Psychiatry* 40, no. 3 (1999): 173–81.
    "Any story of my experiences"   ESF, "A Cryptanalyst," *Arrow* (February 1928), box 12, folder 9, 531–34, ESF Collection.
152 *a letter years later to Barbara*   ESF to Barbara Friedman, February 12, 1945, box 3, folder 26, ESF Collection.
    Immortal Wife   Irving Stone, *Immortal Wife* (Garden City, NY: Doubleday, 1944).
    "to be a good wife"   Ibid., 33.
    "gray-haired"   Ibid., 134.
    "Everything that happened to John"   Ibid., 140–41.

153 *"In many ways it parallels"* WFF to John Ramsay Friedman, August 13, 1945, box 4, folder 8, ESF Collection.
a square of red paper 1928 Friedman holiday card, Item 568, WFF Collection.

154 *"Friedman's Wishing Tree"* "Friedman's Wishing Tree," box 13, folder 9, ESF Collection.
scavenger hunts "The Single Intelligence Skool, Pap Problem No. 1," and other scavenger-hunt materials (envelopes, cryptograms, etc.), box 13, folder 9, November 6, 1938.
"The first item was a series of dots" Virginia Corderman to Vanessa Friedman, October 2, 1981, box 12, folder 15, ESF Collection.
Elizebeth designed a menu Ibid.

155 *The Crypto-Set Headquarters Army Game* Item 2097, WFF Collection.
A second prototype, Kriptor Item 2098, WFF Collection.
sent it to Milton Bradley WFF to ESF, November 29, 1938.
beguiled and frustrated Modernist fiction baffled WFF, but he collected it, read it, tried to make sense of it, thought the sentences were often beautiful, and corresponded with Joyceans, including J. F. Byrne, Joyce's real-life schoolmate and the inspiration for "Cranly" in *A Portrait of the Artist as a Young Man*. Byrne thought he had invented an unbreakable cipher that he called the "Chaocipher." WFF told him his cipher was worthless. See Sheldon, "The Friedman Collection: An Analytical Guide," 466, and Item 1405.1 in the WFF Collection.
with a burgundy binding Giambattista della Porta, *De furtivis literarum notis, vulgo de Ziferis* (Naples, 1563; forgery, London, 1591), Item 119, WFF Collection.

156 *"He came up to me one morning"* WFF, "Communications Intelligence and Security," 9.
American University in 1929 ESF's graduate-school course sheets, box 4, folder 23, ESF Collection.
his copy of the Voynich Manuscript Bauer, *Unsolved!* 83–86, includes a great account of the Voynich study group that WFF started at NSA in 1944 and his attempts to make sense of the weird manuscript using NSA technology.
"worked on the cryptogram" WFF speech to NSA, 1958, box 8, folder 4, ESF Collection.

157 *"for my work in the FBI"* FBI, "The Hoover Legacy, 40 Years After: Part 2: His First Job and the FBI Files," June 28, 2012, https://www.fbi.gov/news/stories/copy_of_the-hoover-legacy -40-years-after.
a professor friend Sheldon, "The Friedman Collection: An Analytical Guide," 434.

157 *a crimson warrior swinging an axe*   This image is pasted inside almost every book in the WFF Collection.

158 *her death in 1934*   WFF and ESF, *The Shakespearean Ciphers Examined* (London: Cambridge University Press, 1958), 208.
"I have no facilities"   Leroy Hennessey, "Twas Bill! Nay, Bacon! But Now E'en Fabyan Knows Not Who Did Shakespeare," *Chicago Evening American* (January 1922): box 14, "The Ideal Scrap Book," NYPL.
called Yardley an "ass"   George Fabyan to WFF, September 16, 1931, Item 734, WFF Collection.
Fabyan seemed glum and tired   Ibid., May 10, 1934.

159 *complained of a hernia*   Ibid., May 31, 1935.
"Always the same old, GF"   Ibid.
age of sixty-nine   Richard Munson, *George Fabyan: The Tycoon Who Broke Ciphers, Ended Wars, Manipulated Sound, Built a Levitation Machine, and Organized the Modern Research Center* (North Charleston, SC: Porter Books, 2013), 141–42.
a letter from an old Riverbank colleague   Cora Jensen to WFF and ESF, May 29, 1936, box 7, folder 18, ESF Collection.
It went dark   Munson, *George Fabyan*, 141–42.
far less money   Jensen to WFF and ESF.
$175,000 to his widow   "Will of Col. Fabyan Filed Tuesday Leaves $175,000 to Widow," 1936 newspaper clipping from Jensen, box 7, folder 18, ESF Collection.
Nelle died   John W. Kopec, *The Sabines at Riverbank: Their Role in the Science of Architectural Acoustics* (Woodbury, NY: Acoustical Society of America, 1997), 56.
Cumming was killed   Ibid.
$70,500   "Buy $800,000 Fabyan Estate as Playground," 1936 newspaper clipping from Jensen, box 7, folder 18, ESF Collection.
"hands it might fall into"   George Fabyan to WFF, July 29, 1935, Item 734, WFF Collection.

160 *wrote to his widow, Nelle*   WFF note in red pencil beginning "Colonel Fabyan's last letter to me," Item 734, WFF Collection.
ordered many records to be burned   ESF interview with Valaki, transcribed January 12, 2012, 2–3.
prevent embarrassing revelations   Ibid.
"A lie!"   WFF, annotated copy of Herbert O. Yardley, *The American Black Chamber* (Indianapolis: Bobbs-Merrill, 1931), Item 604, 43, WFF Collection.
"This is a patchwork"   Ibid., 44.
"Lies, lies, lies"   Ibid.

161 *unable to find a new job*   Kahn, *The Reader*, 104.

161 *"a glimpse behind the heavy curtains"* Yardley, *The American Black Chamber*, Foreword.
there was no harm Ibid.
31,119 copies Kahn, *The Reader*, 131.
"A lovely girl dances" Yardley, *The American Black Chamber*, Foreword.
"the beautiful blonde woman" Ibid., 90–119.
"It did seem to me" Yardley, *The American Black Chamber*, 329.
searched her desk Ibid., 331.
162 *"damned lie"* WFF copy of ibid.
Yardley embellished Kahn, *The Reader*, 113.
compressed time, invented dialogue Ibid., 117.
startle adversaries WFF, "World War I Codes and Ciphers" (lecture, SCAMP, 1958), 22–23, NSA.
"A lie! Which can be so proved" WFF copy of Yardley, *The American Black Chamber*, 45.
163 *four of his colleagues in the army* Ibid., front matter.
Hollywood movies Kahn, *The Reader*, 173–86.
Congress passed a law T. M. Hannah, "The Many Lives of Herbert O. Yardley," *Cryptologic Spectrum* 11, no. 4 (1981): 5–29.
"Widespread interest in the romantic stories" "Extract from R.I.P. No. 98," April 5, 1943, 118–23, NSA.
"I'll confess, Mrs. Friedman" ESF interview with Santry.
164 *"I never thought of my job"* Ibid.
from the glassed-in control room ESF autobiography, 76.
"Does the habit?" ESF interview with Santry.
165 *preferred speaking with female reporters* ESF autobiography, 71.
166 *code books designed for Chinese merchants* Chinese Telegraph Code book used in ESF's Gordon Lim case, 342.3, WFF Collection.
"The whole deciphering science" ESF interview with Santry.
intercepted letters and telegrams U.S. Customs Service, memorandum about the Ezra case, Frederick S. Freed, supervising customs agent, June 8, 1933, box 11, folder 10, ESF Collection; see also Leah Stock Helmick, "Key Woman of the T-men," *Reader's Digest* (September 1937): 51–55.
Green Gang of criminal warlords ESF's Ezra case files list "Paul A. Yip" as the Shanghai contact of the Ezra twins. For background on Yip, see Kathryn Meyer and Terry Parssinen, *Webs of Smoke: Smugglers, Warlords, Spies, and the History of the International Drug Trade* (Plymouth, UK: Rowman & Littlefield, 1998), 159. The book describes Yip as "Green Gang member, opium trafficker, and double agent." The Ezra case files are in box 6, folder 25, ESF Collection.

166 *520 tins of smoking opium*   Freed memorandum, June 8, 1933.
EZRA GANG FALLS IN TRAP   "Ezra Gang Falls in Trap of Woman
Expert in Puzzles," *San Francisco Chronicle,* September 28 or 29,
1933, box 18, folder 5, ESF Collection.
"supplied the Federals"   Ibid.

167 *"We have to keep our ideas secret"*   Ibid.
begged reporters to credit   "Woman Jails Dope Runners,"
Universal Service, box 18, folder 5, ESF Collection.
"The mystery-lure"   ESF to F. E. Pollio, February 14, 1938, box
1, folder 9, ESF Collection.
"She is entrusted"   Helmick,"Key Woman of the T-men."
more than a million subscribers   Trusted Media Brands,
"Expansion (1930s–70s)," http://www.tmbi.com/history/.
"could lead to only one conclusion"   "Extract from R.I.P. No. 98."
stormed into William's office   Ronald Clark, *The Man Who
Broke Purple: Life of Colonel William F. Friedman, Who
Deciphered the Japanese Code in World War II* (Boston: Little,
Brown, 1977), 179.

168 *"I lost my tongue completely"*   Ibid., 180.
Elizebeth flew to Vancouver   ESF Vancouver trip log.
thirteen Luger pistols   A. H. Williamson, "Woman Translates
Code Jargon in Assizes at Trial of Five Chinese," newspaper
clipping, box 18, folder 6, ESF Collection.
that were translated into English   Ibid.
"select fully Wat list"   Ibid.
"hams," "presses," and "tails"   "Woman Helps Canada Break Big
Opium Ring," *New York Times,* February 8, 1938.

169 *CANADA SMASHES*   Staff Correspondent, "Canada Smashes Opium
Ring with U.S. Woman's Aid," *Christian Science Monitor,*
February 9, 1938.
WOMAN TRANSLATES CODE JARGON   A. H. Williamson, "Woman
Translates Code Jargon in Assizes at Trial of Five Chinese," box
18, folder 6, ESF Collection.
"careers unusual for their sex"   "These Women Make Their
Hobbies Pay," *Look,* February 15, 1938, 46.
printed fourteen pages   James W. Booth, "Lady Manhunter,"
Detective Fiction Weekly, September 28, 1940, 60–73.
a newspaper editor   A list of Booth's stories is available at the
Crime, Mystery & Gangster Fiction Magazine Index maintained
by Phil Stephenson-Payne, http://www.philsp.com/homeville
/cfi/s116.htm#A2001.
"PLEASE COOPERATE"   Theodore Adams to ESF (telegram), box 6,
folder 4, ESF Collection.
"Ad Absurdum!"   Ibid.

169 *she had taught her husband everything*   ESF autobiography, 73.
regale him with stories   WFF to ESF, December 29, 1938.
"When people introduce me"   Ibid.

170 *never wanted to see her name*   ESF to Mrs. T. N. Alford,
October 19 (no year given, 1938 or 1939), box 1, folder 9, ESF
Collection.
Hawaii in October 1938   WFF, "Important Contributions to
Communications Security, 1939–1945," 1, NSA.
"highly secret communications"   Ibid.
install and test them   Rowlett, *The Story of Magic,* 142.
six bulky cipher machines   WFF, "Important Contributions."
books of essays and poetry   WFF to ESF, January 6, 1939, ESF
Collection.
Nazi mobs in a thousand German towns   Martin Gilbert, "The
Night of Broken Glass," in *Kristallnacht: Prelude to Destruction*
(New York: Harper Perennial, 2006), 23–41.
hacked apart a grand piano   Ibid., 47.
her sister was simply overworked   Edna Dinieus to ESF,
January 1, 1939, small blue binder.

171 *sometimes able to sneak on*   WFF to ESF, January 6, 1939, ESF
Collection.
got the sense his wife was struggling   Ibid., December 21, 1938,
large blue binder.
"I can almost forsee"   Ibid.
a Christmas dance for passengers   Ibid.
"simple personal habits"   "Munich Cardinal Praises Hitler's
'Personal Habits,'" *Washington Post,* January 1, 1939.
"far-reaching consequences"   Voices of the Manhattan Project,
"Columbia University," Atomic Heritage Foundation and Los
Alamos Historical Society, http://manhattanprojectvoices.org
/location/columbia-university.

172 *the moon rose fat and white*   WFF to ESF, January 6, 1939, 26.
He played shuffleboard   Ibid., November 29, 1938.
he sent to Milton Bradley   Ibid.
ran to thirty pages   Ibid.
"I hope these sheets"   Ibid.
love poems by Tennyson   Ibid., 21. The Tennyson poem he quoted
was an erotic one, "The Miller's Daughter": "And I would be the
necklace / and all day long to fall and rise / Upon her balmy bosom
/ With her laughter or her sighs." https://www.poetryfoundation
.org/poems-and-poets/poems/detail/50267.
"shelter'd here upon a breast"   William Makepeace Thackeray,
"Song of the Violet," in *The Complete Works of William
Makepeace Thackeray* (Boston: Houghton Mifflin, 1889), 291.

172 *"For I've known"*   WFF to ESF, January 6, 1939, 23.
173 *stood at the top deck's rail*   Ibid.
     One night when it was hot   Ibid.
     "The ocean," he said   Ibid., 19.

CHAPTER 1: GRANDMOTHER DIED

177 *"Lights out 'cause I can see in the dark"*   Fugazi, "Caustic
     Acrostic," recorded March–September 1997 on End Hits,
     Dischord Records No. 110, compact disc.
     At 4 P.M. on August 31, 1939   Heinz Höhne, *The Order of the
     Death's Head: The Story of Hitler's SS,* trans. Richard Barry
     (New York: Penguin, 2000), 260–66.
178 *"Grossmutter gestorben"*   Ibid.
     an excuse to start a war   Ibid.
     250,000 by 1939   "The SS (Schutzstaffel): Background and
     Overview," Jewish Virtual Library, http://www.jewishvirtual
     library.org/background-and-overview-of-the-ss.
     "the guillotine used"   Höhne, *The Order of the Death's Head,* 3.
     "enveloped in the mysterious aura"   Ibid., 1.
179 *a forty-three-year-old Catholic farmer*   Bob Graham, "World
     War II's First Victim," *Telegraph* (London), August 29, 2009,
     http://www.telegraph.co.uk/history/world-war-two/6106566
     /World-War-IIs-first-victim.html.
     "Attention, this is Gliwice"   Höhne, *The Order of the Death's
     Head,* 260–66.
     At dawn   Steven M. Gillon, *FDR Leads the Nation into War*
     (New York: Basic Books, 2011), 8–9.
     woke to a ringing phone   Ibid.
     "Well, Bill"   Ibid.
     Later that morning   Franklin Delano Roosevelt press
     conference, September 1, 1939, http://www.presidency.ucsb.edu
     /ws/?pid=15798.
180 *the possibility of direct Nazi attacks*   Richard L. McGaha, "The
     Politics of Espionage: Nazi Diplomats and Spies in Argentina,
     1933–1945" (Ph.D. diss., Ohio University, 2009), 392–93.
     destabilize foreign governments   Ibid.
     Roosevelt was convinced   Ibid.
     a 1940 speech to Congress   Franklin D. Roosevelt, "Message
     to Congress on Appropriations for National Defense," speech,
     May 16, 1940, http://www.presidency.ucsb.edu/ws/?pid=15954.
181 *within reach of Nazi bombing raids*   Frank Knox, "Our
     Heavy Responsibilities to the Nation," speech to the St. Louis
     Conference of the United States Conference of Mayors,

February 20, 1941, http://www.ibiblio.org/pha/policy/1941
/1941-02-20a.html.

181 *140,000 arriving* Stefan Rinke, "German Migration to Latin
America (1918–1933)," in Thomas Adam, ed., *Germany and the
Americas: O-Z* (Santa Barbara, CA: ABC-CLIO, 2005), 27–31.
A "bewildering abundance" Stefan Zweig, *Brazil: Land of the
Future*, trans. Andrew St. James (New York: Viking Press, 1941), 82.
"The glare of the sun" Ibid.
"some gigantic building site" Ibid., 214.

182 *open windows of brothels* Waldo Frank, *South of Us: The
Characters of the Countries and the People of Central and South
America* (New York: Garden City Publishing, 1940), 113–14.
two hundred in Argentina alone "Interrogation of Edmund Von
Thermann, German Ambassador to the Argentine from 1934 to
1942," RG 59, General Records of the Department of State, Entry
188, box 26, NARA.
"a fair sale for German Bibles" U.S. Bureau of Foreign and
Domestic Commerce, *Commerce Reports, Part 1* (Washington,
D.C.: Government Printing Office, 1915), 1011, https://books
.google.com/books?id=1eA9AQAAMAAJ.
spoke only German "German Immigration to Brazil."
"The German spirit" Stephen Bonsal, "Greater Germany in
South America," *The North American Review* 176 (January
1903): 58–67.
"this part of the world" "German Political Designs with
Reference to Brazil," *The Hispanic American Historical Review*
2, no. 4 (November 1919): 586–610, http://www.jstor.org/stable
/2505875.
"complete subservience" Thermann interrogation.
uniforms of green Victoria González-Rivera and Karen
Kampwirth, eds., *Radical Women in Latin America: Left and
Right* (University Park: Pennsylvania State University Press,
2001), 241–42.

183 *"mustachioed like Hitler"* David Sheinin and Lois Baer Barr,
eds., *The Jewish Diaspora in Latin America: New Studies on
History and Literaure* (New York: Garland Publishers, 1996), 210.
"Discipline, Hierarchy, Order" McGaha, "The Politics of
Espionage," 271.
Adolfo Hirohito Ibid., 272.
found much to admire Uki Goñi, *The Real Odessa: Smuggling
the Nazis to Perón's Argentina* (London: Granta, 2003), 37–38.
found it "embarrassing" Thermann interrogation.
"Knowing that war was going" McGaha, "The Politics of
Espionage," 98–99.

183 *"without much effort"*   Thermann interrogation.
184 *course in espionage tradecraft*   George E. Sterling, "The History
    of the Radio Intelligence Division Before and During World
    War II," unpublished manuscript, PDF file, http://www.w3df
    .com, 78–79.
    actually looked like ink   Hedwig Elisabeth Weigelmayer
    Sommer interrogation by Boyd V. Sheets, RG 65, Classification
    64 (IWG), box 211, 57, NARA.
    shrunk documents   Ibid.
    All This and Heaven Too   Rachel Field, *All This and Heaven
    Too* (New York: Macmillan, 1939). For the ESF and Allied
    intelligence aspects of the book, see Sterling, "History of the
    Radio Intelligence Division," 60, and Rose Mary Sheldon, "The
    Friedman Collection: An Analytical Guide," rev. October 2013,
    Marshall Foundation, PDF file, 345–46.
185 *fit in a suitcase*   Sterling, "The History of the Radio Intelligence
    Division," 78–79.
    two kinds of suicide drugs   Sommer interrogation, 57.
    its most capable Funkmeister   "Gustav Utzinger" is an alias—his
    birth name was Wolf Emil Franczok—but he was known in South
    America primarily as Gustav Utzinger, so I am using that name
    in the text. The bulk of the information about Utzinger's career
    comes from a postwar interrogation at the Wannsee Internment
    Camp in Germany. See Robert Murphy, "Reporting the
    Interrogation of Wolf Emil Franczok, Alias Gustav Utzinger,"
    October 24, 1947, RG 65, Records of the Federal Bureau of
    Investigation, box 18, 64-27116, NARA. There are two sections
    of numbered pages in this document: an opening section of
    enclosures containing sworn statements by Utzinger, followed by
    a report of an interrogation.
186 *she freely shared anecdotes*   ESF speech to Mary Barteleme Club,
    Crystal Ballroom, Blackstone Hotel, Chicago, November 30,
    1951, box 17, folder 10, ESF Collection.
    "a vast dome of silence"   Ibid.
    her husband's biographer   ESF interview with Clark.
    "the world began to pop"   ESF interview with Pogue, 4.
    were still with her   ESF interview with R. Louis Benson,
    January 9, 1976, obtained under the Freedom of Information
    Act from NSA; received October 2015; originally requested by
    G. Stuart Smith.
187 *chomping on the pipe*   "Robert Gordon and Elizebeth S.
    Friedman at a Desk," photograph, 1940, ESF Collection.
    the cipher machines she kept   "History of USCG Unit #387."
    reported to the chief   ESF interview with Benson.

187 *dread phone calls*   ESF autobiography, 78.
the unit began to analyze   Ibid., 77.

188 *first gunfight*   Bennett Lessmann, "The Story of the SS Arauca: A Wartime Saga in Broward County," *Broward Legacy* 31, no. 1 (2011): 1–12.
"AM TRYING TO RUN"   "Arauca messages," box 6, folder 5, ESF Collection.
"THE CRUISER HAS TRAINED"   Ibid.
"CRUISER NAMED ORION"   Ibid.
"THREE AMERICAN ARMY PLANES"   Ibid.
"Exciting, round-the-clock adventures"   ESF, "foreword to uncompleted work," box 9, folder 11, ESF Collection. A six-page manuscript with handwritten corrections by ESF.
same kinds of call signs   L. T. Jones, "History of OP-20-GU (Coast Guard Unit of NCA)," October 16, 1943, RG 38, CNSG Library, box 115, 5750/193, NARA.
bearing fixes   Ibid.

189 *"the goose that lays the golden eggs"*   WFF, "Communications Intelligence and Security Presentation Given to Staff and Students" (lecture, Breckinridge Hall, Marine Corps School, April 26, 1960), 22.
"even more secret"   ESF interview with Benson.
adulterations of commercial codes   "History of USCG Unit #387," 5–7.
"the pie circuit"   Ibid., 11.
giving away   Ibid., 8.

190 *"an entering wedge"*   Ibid., 62.
common German words   Ibid., 62–67.
anagram the letters   Ibid.

191 *eleven letters of the alphabet*   Ibid., 5–7.
Durchwalken   Ibid.

192 *"the following was produced"*   Ibid.
messages that used   For *The Story of San Michele* as a book cipher, see Sterling, "History of the Radio Intelligence Division," 80. For *Soñar la vida* and *O Servo De Deus*, see "History of USCG Unit #387," 20.
One Nazi spy proposed   Hamburg to Rio, October 31, 1941, RG 457, SRIC, No. 3810.
a sophisticated process   Sterling, "History of the Radio Intelligence Division," 60–61. See also "History of USCG Unit #387," 67, and ESF's marginalia on newspaper clippings in Item 1006.1, WFF Collection.

193 *become the governess*   ESF's annotated, working copy of *All This and Heaven Too*, Item 1006, WFF Collection, 15.

195 *753,506 . . .*   Craig P. Bauer, *Secret History: The Story of Cryptology* (Boca Raton, FL: CRC Press, 2013), 255.

196 *first adopted them in 1926*   "The Man, the Machine, the Choice," in David Kahn, *Seizing the Enigma: The Race to Break the German U-boat Codes, 1939–1943* (Boston: Houghton Mifflin, 1991).
    Polish codebreakers were the first   Bauer, *Secret History,* 256–83.

197 *dramatically more powerful*   Ibid.
    "search engines for the keys"   Andrew Hodges, *Alan Turing: The Enigma* (London: Vintage, 2014), xviii.
    one to five messages per day   "History of USCG Unit #387," 216–30. All details from the coast guard's solution to this first Enigma machine are documented here.

200 *a linguist and scholar*   Mavis Batey, "Knox, (Alfred) Dillwyn (1884–1943)," 2004, rev. ed. 2006, Oxford Dictionary of National Biography, http://dx.doi.org/10.1093/ref:odnb/37641.

201 *"This recovery of wiring"*   "History of USCG Unit #387," 230.

202 *separate dining rooms*   Kent Boese, "Lost Washington: Harvey's Restaurant," *Greater Greater Washington,* June 23, 2009, https://ggwash.org/view/2073/lost-washington-harveys-restaurant.
    "Penalty of Leadership"   "Biography of John Edgar Hoover," John Edgar Hoover Foundation, http://www.jedgarhooverfoundation.org/hoover-bio.asp.
    a Caesar salad   Pamela Kessler, *Undercover Washington: Where Famous Spies Lived, Worked, and Loved* (Sterling, VA: Capital Books, 2005), 35–36.

203 *poured wine into the cryptologist's glass*   ESF interview with Benson.
    He got rid of them   Michael Newton, *The FBI Encyclopedia* (Jefferson, NC: McFarland, 2003), s.v. "women agents," 374.
    a secret dossier   Curt Gentry, *J. Edgar Hoover: The Man and His Secrets* (New York: W. W. Norton, 2001), e-book, location 5837.
    "dangerous than a man"   Henry M. Holden, *FBI 100 Years: An Unofficial History* (Minneapolis, MN: Zenith Press, 2008), 37.
    "don't tell 'em any secrets"   Ibid.

204 *a public-relations disaster*   Raymond J. Batvinis, *The Origins of FBI Counter-Intelligence* (Lawrence: University Press of Kansas, 2007), 10–26.
    a lot of bad blood   British Security Coordination, *The Secret History of British Intelligence in the Americas, 1940–1945* (New York: Fromm International, 1999), 468.
    "that Jew in the Treasury"   Gentry, *J. Edgar Hoover,* location 7845.
    "J. Edgar Hoover is a man"   British Security Coordination, *The Secret History,* 3.

204 *"It was once remarked"*   Ibid., 468.
205 *"no attack is so unlikely"*   Roosevelt, "Message to Congress on Appropriations."
"our teeming seaboard cities"   Knox, "Our Heavy Responsibilities."
"the common defense"   J. Edgar Hoover, "How the Nazi Spy Invasion Was Smashed," *The American Magazine* (September 1944): 20–21, 94–100.
a historic expansion   FBI, "History of the SIS Division," vol. 1, NARA, three PDF files provided by Richard McGaha.
The first five SIS agents   Ibid.
was actually Portuguese   Transcript of John J. Walsh (former SIS agent) interview with Stanley A. Pimentel, May 19, 2003, National Law Enforcement Museum, 25–26, http://www.nleomf .org/museum/the-collection/oral-histories/john-j-walsh.html.
206 *The FBI had neither*   British Security Coordination, *The Secret History*, 468.
She was to teach codebreaking   ESF interview with Benson.
She read pacifist poetry   All her life, Elizebeth wrote and typed copies of her favorite poems and kept them, including "Patterns" by Amy Lowell, about a young woman whose lover is killed in the First World War, and "Ultimatum for Man," a 1940 poem by Peggy Pond Church, a pacifist poet and schoolteacher in New Mexico whose land was taken by the government to build the Los Alamos facility for nuclear-weapons research. See box 11, folder 20, ESF Collection.
207 *"sleep is intermittent"*   ESF to WFF (addressed to Munitions Building), June 1940, box 2, folder 5, ESF Collection.
what kind of airline   Ibid., June 7, 1940.
fifty dollars   Ibid.
208 *"this morass of debt"*   WFF to ESF, June 4, 1940, box 3, folder 7, ESF Collection.
"not for some centuries"   Ibid., June 10, 1940.

CHAPTER 2: MAGIC

209 *One day in September 1940*   Frank Rowlett, *The Story of Magic: Memoirs of an American Cryptologic Pioneer* (Laguna Hills, CA: Aegean Park Press, 1998), 151–53.
called her Gene   Ibid.
a professor at George Mason   "Genevieve Grotjan Feinstein," NSA Cryptologic Hall of Honor, https://www.nsa.gov/about /cryptologic-heritage/historical-figures-publications/hall-of -honor/2010/gfeinstein.shtml.

209 *two patterns* Rowlett, *The Story of Magic*, 151–53. For a deft and clear technical explanation of Grotjan's insight and the larger process of breaking Purple, see the chapters on Purple in Craig P. Bauer, *Secret History: The Story of Cryptology* (Boca Raton, FL: CRC Press, 2013), 301–10.
her eyes were beaming Ibid.

210 *"Maybe I was just lucky"* Genevieve Grotjan Feinstein, NSA Oral History, May 12, 1991.
William agreed within seconds Ibid.
"Suddenly he looked tired" Ibid.
"The recovery of this machine" Ibid.
Someone went and got Cokes Ibid.
"sitting at his desk" Ibid.

211 *September 25, 1940* Jeffrey Kozak, "Marshall & Purple," Marshall Foundation, http://marshallfoundation.org/blog /marshall-purple/.
nothing at Purple's level WFF, "Contributions in the Fields of Communications Security and Communications Intelligence," undated, NSA.
a feat on par David Kahn, *The Codebreakers: The Comprehensive History of Secret Communication from Ancient Times to the Internet*, rev. ed. (New York: Scribner, 1997).
they demonstrated it Rowlett, *The Story of Magic*, 160–64.
"Last night your magicians" Ibid.
"By God" Ibid.

212 *"MAGIC Summaries"* Kahn, "One Day of MAGIC," in *The Codebreakers*, 1–67.
insisted on getting MAGIC raw David Stafford, "Churchill and Intelligence—Adventures in Shadowland, 1909–1953," *Finest Hour* 149 (Winter 2010–11), https://www.winstonchurchill.org /publications/finest-hour/finest-hour-149/churchill-and-intelli gence-adventures-in-shadowland-1909-1953.
"would be wiped out" George C. Marshall to Thomas E. Dewey, September 27, 1944, Papers of George Catlett Marshall, vol. 4: Aggressive and Determined Leadership, Marshall Foundation, http://marshallfoundation.org/library/digital -archive/to -thomas-e-dewey1/.

213 *He said hello* ESF interview with Pogue, 24; Rose Mary Sheldon, in discussion with the author, January 2015.
oily smoke rippling out Ulrich Steinhilper, *Spitfire on My Tail: A View from the Other Side* (Keston, UK: Independent Books, 2009), 306.
"The sky seemed full of them" Henry Steele Commager, *The*

Story of World War II, rev. Donald L. Miller (New York: Simon & Schuster, 2001), 38–41.

213 *telephones of Jewish families* "Historical Background: The Jews of Hungary During the Holocaust," Yad Vashem, http://www .yadvashem.org/yv/en/education/newsletter/31/jews_hungary .asp.
America had no standing Charles Lindbergh, "We Will Never Accept a Philosophy of Calamity," speech, Keep-America-Out -of-War rally, Chicago, August 4, 1940, http://www.ibiblio.org /pha/policy/1940/1940-08-04a.html.
"danger to this country" Charles Lindbergh, "Who Are the War Agitators?" speech, Des Moines, Iowa, September 11, 1941, http://www.charleslindbergh.com/americanfirst/speech.asp.

214 *British officers began to arrive* British Security Coordination, *The Secret History of British Intelligence in the Americas, 1940– 1945* (New York: Fromm International, 1999), Introduction by Nigel West.
thirty-fifth and thirty-sixth floors Ibid.
Ian Fleming Jennet Conant, *The Irregulars: Roald Dahl and the British Spy Ring in Wartime Washington* (New York: Simon & Schuster, 2008), 84–86.
twenty-three-year-old Roald Dahl Ibid., xiv, 10–11.
He seduced actresses and heiresses Ibid., 99–126.
"I would do my best to appear calm" Roald Dahl, "Lucky Break," in *The Wonderful Story of Henry Sugar* (New York: Puffin, 2000), 201.

215 *"All I can say"* Conant, 30.
planted anti-Nazi information British Security Coordination, *The Secret History,* 66–87.
staged protests at rallies Ibid.
sending gorgeous female spies Ibid., 193–96.
in the Western Hemisphere Ibid., "Part VII: Counter-Espionage," 345–403.
one day in June 1941 Mark Riebling, *Wedge: From Pearl Harbor to 9/11: How the Secret War Between the FBI and CIA Has Endangered National Security* (New York: Touchstone, 2002), 3–15; John Pearson, *The Life of Ian Fleming* (London: Bloomsbury, 2013).
"a chunky enigmatic man" Ibid.; see also John Bryden, *Best-Kept Secret: Canadian Secret Intelligence in the Second World War* (Toronto: Lester, 1993), 66.
"no conception of offensive" Ibid., 67.

216 *In July 1941, Roosevelt* Thomas F. Troy, "Donovan's Original Marching Orders," *Studies in Intelligence* 17, no. 2 (1973): 39–67,

https://www.cia.gov/library/center-for-the-study-of-intelli
gence/kent-csi/vol17no2/html/v17i2a05p_0001.htm.

216 *"incomparably better"* British Security Coordination, *The Secret History,* 471–72.

"The whole system" Ibid.

a husky, apple-cheeked colonel James Chadwick, "Frederick John Marrian Stratton, 1881–1960," *Biographical Memoirs of Fellows of the Royal Society* 7 (November 1961): 280–93.

217 *looked like Santa Claus* ESF interview with Benson.

The British operated Bob King, "The RSS from 1939 to 1946," November 22, 1944.

He wanted assistance "History of USCG Unit #387," Foreword.

others relied on "turning grilles" Ibid., 68–84.

made special devices and tools Jones, "History of OP-20-GU."

218 *"Fireside Chat" radio speech* Franklin Delano Roosevelt, "Fireside Chat 16: On the Arsenal of Democracy," December 29, 1940, University of Virginia Miller Center, https://millercenter .org/the-presidency/presidential-speeches/december-29-1940-fire side-chat-16-arsenal-democracy.

36 minutes and 56 seconds Ibid.

Hitler responded Associated Press, " 'God With Us Up to Now,' Hitler Says: Victory Sure in 1941 Army Men Are Told," *The Brownsville Herald,* December 31, 1940.

climbed into the blacked-out streets Associated Press, "Charred London Greets '41 With Cry 'To Hell With Hitler,' " *Washington Post,* January 1, 1941.

January 4, 1941 Ronald Clark, *The Man Who Broke Purple: Life of Colonel William F. Friedman, Who Deciphered the Japanese Code in World War II* (Boston: Little, Brown, 1977), 158.

the Neuropsychiatric Section Mary W. Standlee, *Borden's Dream: The Walter Reed Army Medical Center in Washington, D.C.* (Washington, D.C.: Borden Institute, 2007), 214, 299–304, 334–36.

219 *keeping the mentally ill* out *of the army* William C. Porter, "Psychiatry and the Selective Service," *War Medicine* 1 (May 1941): 364–71.

processing rather than healing Major M. R. Kaufman (physician assigned to Neuropsychiatric Section at Walter Reed), "The Problem of the Psychopath in the Army," in *Proceedings of the Annual Congress of Correction of the American Prison Association* 89 (1942): 128–38.

presumed to be less stressful William C. Porter, "The Military Psychiatrist at Work," *The American Journal of Psychiatry* 98, no. 3 (November 1941): 317–23.

220 *interfered with his ability to function*   Clark, *The Man Who Broke Purple*, 159.
"*the patient was isolated*"   Ibid.
"*mood swings*"   ESF to Ronald Clark, March 9, 1976, box 13, folder 30, ESF Collection.
221 *sailed across the Atlantic*   Robert L. Benson, "The Origin of U.S.-British Intelligence Cooperation (1940–1941)," *Cryptologic Spectrum* 7, no. 4 (1977): 5–8.
"*anxiety reaction*"   Ibid., 159.
discharged him on March 22   Ibid.
honorably discharged   Ibid.
temporary employee   " 'Court-martial' proceedings against William F. Friedman," box 13, file 14, ESF Collection.

CHAPTER 3: THE *HAUPTSTURMFÜHRER* AND THE *FUNKMEISTER*

223 *No code is ever completely solved*   ESF interview with Pogue, 79.
224 "*one of the most active*"   U.S. Department of Justice, Federal Bureau of Investigation, memorandum, *re: Siegfried Becker,* Francis E. Crosby to FBI Director, February 15, 1944, RG 65, box 18, 64-27116, NARA.
a "deft Teutonic hand"   Francis E. Crosby, "Memorandum for the Ambassador," February 4, 1944, RG 65, box 18, 64-27116, NARA.
He held the rank   Richard L. McGaha, "The Politics of Espionage: Nazi Diplomats and Spies in Argentina, 1933–1945" (Ph.D. diss., Ohio University, 2009), 22.
"*a sign of our loyalty*"   U.S. Department of Justice, Federal Bureau of Investigation, memorandum, *Subject: Johannes Siegfried Becker,* Francis E. Crosby to J. Edgar Hoover, November 22, 1944, including British translations of documents belonging to Becker, RG 65, box 18, 64-27116, NARA.
by direct or indirect steps   Crosby, "Memorandum for the Ambassador."
"*disliked his manner*"   "CSDIC Preliminary Interrogation Report on Heinrich VOLBERG," February 25, 1946, Records of the Security Service, KV2/89, TNA.
359,966   McGaha, "The Politics of Espionage," 211.
five foot ten   U.S. Department of Justice, Federal Bureau of Investigation, "Radio CEL, Albrecht Gustav Engels, Was., Et Al., Brazil—Espionage," RG 38, CNSG Library, box 77, NARA, 36.
225 *impregnating the wife*   McGaha, "The Politics of Espionage," 236.
grotesquely long fingernails   Charles F. Hemphill Jr., "Re:

Johannes Siegfried Becker," April 5, 1944, RG 65, box 18,
64-27116, NARA. "Subject has very unusual fingernails, in that
they curve straight down over the tips of his fingers. HANS
MUTH stated that this characteristic is so pronounced that they
appear to be deformed."

225 *a Jewish retirement home*   David Kahn, *Hitler's Spies: German
Military Intelligence in World War II* (New York: Macmillan,
1978), 266.
five hundred people   Ibid.
German for "informer"   McGaha, "The Politics of Espionage,"
185.
insufficiently ruthless   Katrin Paehler, "Espionage, Ideology, and
Personal Politics: The Making and Unmaking of a Nazi Foreign
Intelligence Service" (Ph.D. diss., American University, 2002), 46.
according to an SS handbook   Ibid., 215–16.
"no qualifications whatever"   Theodor Paeffgen interrogation
by Henry D. Hecksher, September 10, 1945, RG 65, box 183,
NARA.

226 *Gestapo thug named Kurt Gross*   Hedwig Elisabeth Weigelmayer
Sommer interrogation by Boyd V. Sheets, RG 65, Classification
64 (IWG), box 211, 11, NARA; see also W. Wendell Blanke,
"Interrogation Report of Karl Gustav Arnold," November 20,
1946, 13, NARA.
gladly told U.S. interrogators   Ibid.
a Jewish man from Holland   Ibid., 59–60.
"He was an intelligent person"   Ibid., 22.

227 *trunkful of explosives*   McGaha, "The Politics of Espionage,"
231–32.
he abandoned the sabotage   Sommer interrogation, 15.
carrying a revolver   McGaha, "The Politics of Espionage," 185.
a broad-shouldered German   U.S. Department of Justice,
"Radio CEL," 59–61.
jittery mechanical engineer   "The Starziczny Case," in Stanley E.
Hilton, *Hitler's Secret War in South America 1939–1945* (Baton
Rouge: Louisiana State University Press, 1999), e-book, location
1948; see also U.S. Department of Justice, "Radio CEL," 124.
lived with his Brazilian mistress   Ibid.
"the only real professional"   U.S. Department of Justice, "Radio
CEL," 58.

228 *the SS Windhuk*   Sommer interrogation, 17.
speaking to an American   Utzinger interrrogation.
"acted according to his own lights"   Ibid., 26.

229 *"in pursuit of his ends"*   Utzinger interrogation, enclosure no. 3.
Vladimir Bezdek   Vladimir Bezdek, Official Personnel Folder,

National Personnel Records Center, National Archives at
St. Louis, requested September 2016.
229 *read dictionaries in his free time*   Lekan Oguntoyinbo,
"Vladimir Bezdek: Retired WSU Professor, Linguist," *Detroit
Free Press,* May 19, 2000.
up and running in South America   U.S. Department of Justice,
"Radio CEL."
230 *an alphanumeric label*   "History of USCG Unit #387."
JOSE, or JUAN   Most of the Brazil decrypts from 1941 and
1942 are in RG 457, subseries SRIC, box 3, SRIC 1793–2591,
and box 5, SRIC 3723–3983, NARA; see also Hemphill Jr., "Re:
JOHANNES SIEGFRIED BECKER;" U.S. Department of
Justice, "Radio CEL."
He went by UTZ   Ibid.
231 *HUMBERTO was a piece of luck*   "History of USCG Unit
#387," 71.
"the greatest help to us"   "Final Report, British-Canadian-
American Radio Intelligence Discussions, Washington, D.C.,
April 6–17, 1942," RG 38, CNSG Library, Box 82, 5050/67,
NARA.
using book ciphers   "History of USCG Unit #387," 68.
switched to a grille-like cipher   Ibid.
232 *never sent useful information*   Ibid.
their own SIS filing system   FBI, "Subject: Frederick Duquesne,
Interesting Case Write-Up," March 12, 1985, eight PDF files.
The SIS used different serial numbers than the coast guard serial
numbers, but the texts of the messages are the same; for instance,
the handful of Mexico-to-Germany decrypts that Elizebeth
kept in box 6, folder 6 of her collection are identical to messages
included in the FBI/SIS Duquesne write-up.
"A considerable amount"   FBI, "History of the SIS Division,"
vol. 1, 288.
difficult for the coast guard   Jones, "History of OP-20-GU."
a massive spy investigation   Raymond J. Batvinis, "Ducase," in
The Origins of FBI Counter-Intelligence (Lawrence: University
Press of Kansas, 2007), 226–56.
233 *Sebold was secretly working*   Ibid.
The bureau reached out   "History of USCG Unit #387," 22–32;
Jones, "History of OP-20-GU."
to relay messages   Bativinis, "Ducase," in *The Origins of FBI
Counter-Intelligence.*
they were in an unknown code   Ibid.
Elizebeth broke it   "History of USCG Unit #387," 22–32.
Long Island to Hamburg   The coast guard called this radio link

Circuit 2-C and monitored it for the rest of the war. "History of USCG Unit #387," 22.

233 *"the greatest spy roundup"* Marc Wortman, "Fritz Duquesne: The Nazi Spy with 1,000 Faces," Daily Beast, February 26, 2017, http://www.thedailybeast.com/fritz-duquesne-the-nazi-spy -with-1000-faces.

234 *"gave birth to the popular cultural belief"* Batvinis, "Ducase," in *The Origins of FBI Counter-Intelligence,* 256.
"exposed the secret messages" Rose Mary Sheldon, "The Friedman Collection: An Analytical Guide," rev. October 2013, Marshall Foundation, PDF file, 345, text for Item 1006, WFF Collection.
too cavalier about publicity Jones, "History of OP-20-GU."

235 *disrupt their work* Diaries of Henry Morgenthau Jr., vol. 473, December 14–16, 1941, 37, Franklin D. Roosevelt Library and Museum website.
a Treasury staff meeting Diaries of Henry Morgenthau Jr., vol. 457, November 1–5, 1941, 237–64, Franklin D. Roosevelt Library and Museum website.
visible through a nearby window Peter Moreira, *The Jew Who Defeated Hitler: Henry Morgenthau Jr., FDR, and How We Won the War* (New York: Prometheus Books, 2014), 40.
at 10:45 A.M. Diaries of Morgenthau, November 1–5, 1941.
a birdlike man Moreira, *The Jew Who Defeated Hitler,* 85.
"She is very discontented" Diaries of Morgenthau, November 1–5, 1941.

236 *Harry Dexter White* James Nye, "Revealed: The Banker Who Shaped the Modern Financial World after WWII Was a Soviet Spy Who Wanted America to Become Communist," *Daily Mail* (London), March 5, 2013.
"I don't like to butt into this" Diaries of Morgenthau, November 1–5, 1941.
"But they knew" Ronald Clark, *The Man Who Broke Purple: Life of Colonel William F. Friedman, Who Deciphered the Japanese Code in World War II* (Boston: Little, Brown, 1977), 170.
Personnel started to stream in The most vivid recollection of the Munitions Building immediately after Pearl Harbor comes from John B. Hurt, the Japanese linguist on Friedman's team. Three pages, undated, 1944, NSA.

237 *1,177 crewmen* Wikipedia, s.v. "Attack on Pearl Harbor," last modified May 17, 2017, https://en.wikipedia.org/wiki/Attack _on_Pearl_Harbor.
wrote his will Hurt.
it would happen in Manila Ibid.

237 *a three-volume report* WFF, "Certain Aspects of 'MAGIC'
in the Cryptological Background of the Various Official
Investigations into the Attack on Pearl Harbor," March 1957,
NSA.
"cryptologic schizophrenia" WFF, "Second Period,
Communications Security" (lecture), NSA.

238 *a declassified NSA report* ESF interview with R. Louis Benson,
January 9, 1976, obtained under the Freedom of Information Act
from NSA; received October 2015.
Elizabeth would never understand Ibid.
"a date which will live in infamy" "FDR's Day of Infamy
Speech: Crafting a Call to Arms," *Prologue* 33, no. 4 (Winter
2001), https://www.archives.gov/publications/prologue/2001
/winter/crafting-day-of-infamy-speech.html.

239 *Elizabeth's ration book* ESF and WFF's Second World War
ration books are in a black folder of letters given to the Marshall
Foundation by John Ramsay Friedman.
appoint a new chief ESF interview with Benson.
This upset her Ibid.

240 *She got to know James Roosevelt* Colin Burke, "What OSS Black
Chamber? What Yardley? What 'Dr.' Friedman? Ah, Grombach?
Or Donovan's Folly," http://userpages.umbc.edu/~burke/whatoss
black.pdf.
a tall, irascible Evan Thomas, "Spymaster General," *Vanity Fair*
(March 2011), http://www.vanityfair.com/culture/2011/03/wild
-bill-donovan201103.
James Roosevelt approached Elizabeth Diaries of Morgenthau,
December 14–16, 1941, 37.
Donovan reinforced the demand William J. Donovan to
Morgenthau, December 14, 1941, in ibid., 53.
Morgenthau grumbled Ibid., 37.
three and a half weeks ESF to Colonel Donovan, via Chief
Liason Officer, Coordinator of Information, December 29, 1941,
box 15, folder 14, ESF Collection.
She built it from scratch Ibid.
a seethingly polite letter Ibid.
defined by recklessness Thomas, "Spymaster General."

241 *"My experience and observations"* ESF to Colonel Donovan.
an honorary L.L.D. ESF to Mrs. T. N. Alford, October 19,
1939, box 1, folder 9, ESF Collection.
Brazil had declared solidarity Boris Fausto, *A Concise History
of Brazil,* trans. Arthur Brakel (Cambridge, UK: Cambridge
University Press, 1999), 228.
firing torpedoes at Brazilian ships John Bryden, *Best-Kept*

Secret: Canadian Secret Intelligence in the Second World War (Toronto: Lester, 1993), 108–9.

241 *The positions of the ships* Ibid.
"Measures against members" Brazil to Germany, December 10, 1941, RG 457, SRIC, No. 2210.

242 *Operation Drumbeat* Bryden, *Best-Kept Secret,* 108–9.
the ruthless U-boats Ibid.
MARCH 14, 1942 South America to Germany, March 14, 1942, RG 457, SRIC, No. 2418.
an FCC listening station Rhode Island Radio, "Radio Intelligence Division," http://www.61thriftpower.com/riradio/rid.shtml.
8,398 American servicemen Eric Niderost, "Voyages to Victory: RMS Queen Mary's War Service," Warfare History Network, January 16, 2017, http://warfarehistorynetwork.com/daily/wwii/voyages-to-victory-rms-queen-marys-war-service/.
MARCH 7, 1942 South America to Germany, March 7, 1942, RG 457, SRIC, No. 2414.
MARCH 8 South America to Germany, March 8, 1942, RG 457, SRIC, No. 2413.

243 *MARCH 12* South America to Germany, March 12, 1942, RG 457, SRIC, No. 2418.
MARCH 13 Ibid.
MARCH 14 South America to Germany, March 14, 1942, RG 457, SRIC, No. 2419.
one million Reichsmarks Niderost, "Voyages to Victory."
able to take evasive maneuvers ESF wasn't the only Allied codebreaker who noticed that the *Queen Mary* was in peril; British and Canadian agencies solved similar messages. It was like multiple witnesses reporting the same crime to 911. See Bryden, *Best-Kept Secret,* 121.
"hiding place" Santiago to Hamburg, March 5, 1942, RG 457, SRIC, No. 3739.
31 degrees Celsius Brazil to Hamburg, March 7, 1942, RG 457, SRIC, No. 3799.
"Throughout country" Brazil to Germany, March 16, 1942, RG 457, SRIC, No. 3831.
the docked Swiss ship Brazil to Hamburg, March 17, 1942, RG 457, SRIC, No. 3821.
guessed that authorities Jones, "History of OP-20-GU."

244 *"You'll blow the house up!"* John Humphries, "The Man From Brazil," in *Spying for Hitler: The Welsh Double-Cross* (Cardiff: University of Wales Press, 2012), 199–211.
March 15, 1942 Leslie B. Rout Jr. and John F. Bratzel, "Climax

of the Espionage War in Brazil: 1942–55," in *The Shadow War: German Espionage and United States Counterespionage in Latin America during World War II* (Frederick, MD: University Publications of America, 1986), 172–222.

245 *Engels assumed* Ibid.
"MEYER CLASEN" Brazil to Germany, March 18, 1942, RG 457, SRIC, No. 3964.
That day Engels was arrested Rout and Bratzel, "Climax of the Espionage War in Brazil."
West believed Ibid.
Robert Linx drove around Rio George E. Sterling, "The History of the Radio Intelligence Division Before and During World War II," unpublished manuscript, PDF file, http://www.w3df.com, 85.

246 *verbatim copies* Ibid.; see also Jones, "History of OP-20-GU."
"delivered the complete information" J. Edgar Hoover, "How the Nazi Spy Invasion Was Smashed," *The American Magazine* (September 1944): 20–21, 94–100.
tougher on the prisoners Rout and Bratzel, "Climax of the Espionage War in Brazil."
fill in a handful of missing words Jones, "History of OP-20-GU."
went on a hunger strike Rout and Bratzel, "Climax of the Espionage War in Brazil."
looked the other way Ibid.

247 *one of the radio transmitters* Sommer interrogation, 23.
three long letters out of prison Rout and Bratzel, "Climax of the Espionage War in Brazil."
"Warning" Germany to Chile, March 23, 1942, RG 457, SRIC, No. 3809.
a pack of cigarettes arrived C. F. Hemphill, "Osmar Alberto Hellmuth," January 1, 1944, RG 65, box 18, 64-27116, NARA.

CHAPTER 4: CIRCUIT 3-N

249 *"Flight, fight, or neurosis"* Ronald Clark, *The Man Who Broke Purple: Life of Colonel William F. Friedman, Who Deciphered the Japanese Code in World War II* (Boston: Little, Brown, 1977), 258–59.

250 *"heebeegeebees . . . hbgbs* WFF, "Bletchley Park Diary," ed. Colin MacKinnon, http://www.colinmackinnon.com/files/The _Bletchley_Park_Diary_of_William_F._Friedman_E.pdf.
three consecutive days ESF to Barbara Friedman, May 22, 1942, box 3, folder 22, ESF Collection.
friends knocked on their door Ibid.
"Artiste de Boudoir" WFF to ESF, telegram beginning "YOUR

RENOWN," May 1942, box 1, General Correspondence, ESF Collection.

250 *"Doctor of Successful Marriage"* WFF to ESF, telegram beginning "BOARD OF OVERSEERS," May 1942, box 1, General Correspondence, ESF Collection.

251 *leftist political causes* Barbara Friedman to WFF, undated, box 4, folder 8, ESF Collection.
"I hope you will let nothing interfere" WFF to Barbara Friedman, October 11, 1944, box 3, folder 21, ESF Collection.

252 *Arlington Hall* Jennifer Wilcox, "Sharing the Burden: Women in Cryptology During World War II," Center for Cryptologic History, NSA, 2008.
Many were women Ibid.
secret cryptology courses Patricia Ryan Leopold, in discussion with the author, January 2015. See also Craig Bauer and John Ulrich, "The Cryptologic Contributions of Dr. Donald Menzel," *Cryptologia* 30, no. 4 (2006): 306–39. DOI: 10.1080/01611190600920951.
guarded by U.S. Marines ESF, "foreword to uncompleted work."
operated the bombes Wilcox, "Sharing the Burden."

253 *"rolling down one's legs"* Martha Waller, in discussion with the author, via e-mail, March 2015.
the Star of David "Star of David; Badges and Armbands," National Holocaust Centre and Museum, UK, https://www.nationalholocaustcentre.net/star-of-david.
Exactly as she had feared Jones, "History of OP-20-GU."
"thereafter completely changed" Ibid.
"the matter got out of hand" "R.I.P. No. 98, Appendix II, American Measures Against Communications Intelligence Publicity," April 5, 1943, RG 457, Friedman Collection, Entry UD-15D19, box 22, NARA, 400–401.

254 *also taken aback* F. H. Hinsley and C. A. G. Simkins, *British Intelligence in the Second World War, Vol. 4, Security and Counter-Intelligence* (London: Her Majesty's Stationery Office, 1990), 149.
"Rather shattering" [Redacted] to J. M. A. Gwyer, MI5, April 26, 1943, KV2/2845, TNA.
jurisdictional squabbles "R.I.P. No. 98, Appendix II," RG 457.
without their approval Ibid., 384.
in the Western Hemisphere Ibid., 394–97.
"more informal" BSC, 472.
"sotto voce" Ibid., 473.

255 *for tighter control* Jones, "History of OP-20-GU."
a weeklong conference "Final Report, British-Canadian-American Radio Intelligence Discussions."

255 *On the day she explained* "Brief of Minutes, Committee B, Method of Obtaining W/T Intelligence From Intercepted W/T Traffic, Including D/F Bearings," British-Canadian-American Radio Intelligence Discussions, Washington, D.C., April 8, 1942, RG 38, CNSG Library, Box 82, 5050/67, NARA. See also "Recordings of Final Report, British-Canadian-American Radio Intelligence Discussions, Washington, D.C.," April 6–17, 1942, envelope no. 2, list of April 8, 1942, speakers, RG 38, CNSG Library, Box 82, 5050/68, NARA.
fewer than twenty cryptanalysts ESF interview with Benson.
weirder, harder stuff "History of USCG Unit #387," 15.
ZUM NEUE JAHR Ibid., 95–96.

256 *a partial list* All of these circuits are described in "History of USCG Unit #387."
diplomats and even military officers L. T. Jones, "Memorandum to Op-20-G, Subj: Clandestine Radio Intelligence," September 7, 1944, obtained under the Freedom of Information Act from NSA; received October 2015; originally requested by G. Stuart Smith.

257 *just another communications channel* Ibid.
"when wireless is perfectly applied" John B. Kennedy, "When Woman Is Boss," interview with Nikola Tesla, *Collier's,* January 30, 1926, http://www.tfcbooks.com/tesla/1926-01-30.htm.
"a miscellany" Jones, "Memorandum to Op-20-G."
about the Nazi grasp Ibid.
the shapes were parallelograms "History of USCG Unit #387," 37–38.
others like labyrinths Ibid., 199.
"explosives from cacao" Argentina to Berlin, October 22, 1943, Serial CG3-2213, RG 38, CNSG Library, box 79, 3824/3, NARA.

258 *baby girl, Jutta* Hamburg to Iceland, June 1, 1944, RG 457, SRIC, No. 3687.
"My dear JOHNY" Berlin to Argentina, November 11, 1943, Serial CG3-2348, RG 38, CNSG Library, box 79, 3824/3, NARA.
Lieutenant Jones did that ESF interview with Benson; Jones, "History of OP-20-GU."
Jones did that, too ESF interview with Benson.
sometimes quarreling Ibid.
"one of the workers" Ibid.
$4,200 a year ESF payroll slip, July 1945, listing her previous salaries and government classification, Personnel Folder.
Her initials, ESF Germany to ?, 1942, RG 457, SRIC, No. 3648.
Her handwriting appeared See, for instance, Berlin to Argentina, April 6, 1944, Serial CG4-4142, RG 38, CNSG Library, box 79, 3824/3, NARA, with a handwritten stapled

note by ESF that begins "Comment," or see one of the messages
from the *Jolle* supply ship to Argentina, June 27, 1944, Serial
CG4-5077-A, RG 38, CNSG Library, box 79, 3824/2, NARA.

259 *a characteristic burst*   Berlin to Argentina, May 30, 1944, Serial
CG4-4847, RG 38, CNSG Library, box 79, 3824/3, NARA.
Berlin writes that a man named Curt "had his leg in a cast as a
result of a bombardment of BERLIN when he was going down
the stairs carrying a young lady on his back." ESF wrote in red
pencil, "neat trick."
described the meeting   Government Code and Cypher School,
memorandum, *CLANDESTINE*, Major G. G. Stevens to
D.D.(S), December 24, 1942, HW14/62, TNA.
at unexpected times   Ibid.
a device he called a "snifter"   George E. Sterling, "The History
of the Radio Intelligence Division Before and During World
War II," unpublished manuscript, PDF file, http://www.w3df
.com, 19–20.

260 *October 10, 1942*   David P. Mowry, "Cryptologic Aspects of
German Intelligence Activities in South America during World
War II," Series IV, vol. 11 (2011), Center for Cryptologic History,
National Security Agency, 85–86.
called it Circuit 3-N   "History of USCG Unit #387," 231.
chipped in with clues   Ibid.; Mowry, "Cryptologic Aspects."
twenty-eight encrypted messages   Ibid.

261 *1511 Calle Donado*   Utzinger interrogation, 4.
legitimate paying clients   Ibid.
girlfriend worked at AMT VI   In FBI memos exchanged after
Utzinger's arrest in August 1944, Bureau officials write that
before Utzinger left Germany for South America, he asked
friends to look after a woman named Hilde Burckhardt, who
told a roommate that she and Utzinger both worked for AMT
VI. See Federal Bureau of Investigation, memorandum, Subject:
"Gustav Utzinger, with aliases," James P. Joice Jr. to John Edgar
Hoover, October 5, 1945, RG 65, Classification 64 (IWG), box 14.
Also, in one of the Circuit 3-N decrypts, "Blue Eye" talks about
"participating in the construction of several directional shortwave transmitters," strongly suggesting that she was with AMT
VI. See Berlin to Argentina, October 26, 1943, Serial CG3-2236,
RG 38, CNSG Library, box 79, 3824/3, NARA.
signed the messages "blue eye"   Berlin to Argentina, October 26,
1943, Serial CG3-2236.
called herself "the Ahnfrau"   Berlin to Argentina, February 8,
1944, Serial CG4-3535, RG 38, CNSG Library, box 79, 3824/3,
NARA. The coast guard codebreakers weren't entirely sure how

"the Ahnfrau" was related to "Luna"—they sometimes wrote "grandmother?" or "mother?" or "wife?" next to her name on the decrypts—but I am fairly certain, from the context of the decrypts, that she was Utzinger's grandmother.

261 *"celebrating your birthday"*   Berlin to Argentina, October 28, 1943, Serial CG4-2837, RG 38, CNSG Library, box 79, 3824/3, NARA.
to brush his teeth   Berlin to Argentina, April 29, 1944, Serial CG4-4447, RG 38, CNSG Library, box 79, 3824/3, NARA.

262 *built transmitters*   Ibid., 3.
in the public square   Ibid., 4.
ancient Krupp cannon   Ibid., 3.
under extreme pressure   Ibid.
Waldo Frank toured Argentina   Waldo Frank, *South American Journey* (New York: Duell, Sloan and Pearce, 1943).
noticed "a spawn"   Ibid., 76.
a pro-Nazi newspaper   Ibid., 83–85, 128.
"a very uncertain grip"   Ibid., 213.
before he could escape, five cops   Ibid., 217.

263 *only listening posts*   Utzinger interrogation, 4–5.
a small farm   Arthur F. Carey, "Gustav Edward Utzinger, with Aliases, Espionage," August 15, 1945, RG 65, Classification 64 (IWG), box 14.
code name "Boss"   McGaha, "The Politics of Espionage," 189.
on a Spanish ship   Carey, "Gustav Edward Utzinger."
he had brought an Enigma   Utzinger interrogation, 5.
very sensitive information   Ibid., 4.

264 *a mythical radio organization*   Utzinger interrogation, enclosure no. 3.
Red, Green, and Blue   "Camp 020 Interim Report on the Case of General Friedrich Wolf," October 1945, RG 59, Entry 1088, box 26.
"seasoned collaborators"   Argentina to Berlin, October 14, 1943, Serial CG3-2179, RG 38, CNSG Library, box 79, 3824, NARA.
"I am teaching my boys"   Argentina to Berlin, January 20, 1943, Serial CG3-896, RG 38, CNSG Library, box 80, 3824/4, NARA.
another useful channel   Sommer interrogation, 38.
two Enigmas   Utzinger interrogation, Sommer interrogation. Dietrich Niebuhr gave Utzinger one Enigma and Becker gave him at least one other before the Red Enigma arrived via the wolf courier system.
a Liliput   "History of USCG Unit #387," 212.

265 *a cheerful progress update*   Argentina to Berlin, February 28, 1943, RG 38, Serial CG3-933, CNSG Library, box 80, 3824/4, NARA.
"Old boy, now we are off"   Berlin to Argentina, February 28,

1943, RG 38, Serial CG3-860, CNSG Library, box 80, 3824/4, NARA.

265 *Nazi censors* Earl R. Beck, *Under the Bombs: The German Home Front, 1942–1945* (Lexington: University Press of Kentucky, 1986), 35.
difficult to find toothbrushes Ibid., 24.
patrons stealing glasses Ibid.
the Battle of Stalingrad Joseph Goebbels, "Nation, Rise Up, and Let the Storm Break Loose," February 18, 1943, German Propaganda Archive, Calvin College, http://research.calvin.edu /german-propaganda-archive/goeb36.htm.
a fourteen-year-old girlfriend Goñi, *The Real Odessa*, xxiii.
a secret lodge Ibid., 20–22.

266 *a small notebook* "Summary of Traces, BECKER Siegfried," address book, July 11, 1945, KV2/89, TNA.
an alluring figure Utzinger interrogation, 8.
an informal deal Ibid., 2–5.
"Hitler's struggle" Goñi, *The Real Odessa*, 22.
a stack of text The coast guard documented its solution of the Green Enigma in "History of USCG Unit #387," 230–61. For an account of both the coast guard and British solutions for this machine, see Philip Marks, "Enigma Wiring Data: Interpreting Allied Conventions from World War II," *Cryptologia* 39, no. 1 (2015): 25–65. DOI: 10.1080/01611194.2014.915263.

267 *a unit called Intelligence Service Knox* Hinsley and Simkins, *British Intelligence, Vol. 4*, 182.
at about the same time Marks, "Enigma Wiring Data;" see also "History of USCG Unit #387," 262.
a G-model Enigma Ibid.
a "less superior Enigma" ESF interview with Clark.
"There was much celebration" ESF interview with Benson.

268 *"enciphered with LILY"* "History of USCG Unit #387," 212.
somewhat more complex Marks, "Enigma Wiring Data."
frequencies of certain juxtapositions "History of USCG Unit #387," 212–15.
never solve the Blue Ibid., 262.

269 *"We have antenna"* Argentina to Berlin, January 18, 1943, Serial CG3-921, RG 38, CNSG Library, box 80, 3824/4, NARA.
on the second floor ESF, "foreword to uncompleted work."
"Our work was compartmented" George Bishop to Vanessa Friedman, September 22, 1981, box 12, folder 14, ESF Collection.
a few of the SPARS ESF Personal History Statement.
her home for tea WFF to Barbara Friedman, January 16, 1944, box 3, folder 24, ESF Collection.

269 *"I think I was mesmerized"* Waller, in discussion with the
author, via e-mail, March 2015.
270 *a personal diary of the trip* WFF, "Bletchley Park Diary."
at least fourteen letters ESF to WFF, May 31, 1943, box 2, folder
7, ESF Collection.
"a momentary return for you" Ibid., May 16, 1943.
271 *a corsage of violets* Ibid., April 27, 1943.
cigarettes in the evenings Ibid., May 9, 1943.
110 degrees ESF, "foreword to uncompleted work."
rolled out extensive reforms "History of USCG Unit #387," 156.
Nazi cryptographer, Fritz Menzer David P. Mowry,
"Regierungs-Oberinspektor Fritz Menzer: Cryptographic
Inventor Extraordinaire," *Cryptologic Quarterly* 2, nos. 3 and 4
(1983–84): 21–36.
"Procedure 62" "History of USCG Unit #387," 195–202.
"Procedure 40" Ibid., 203–6.
5-by-5 square of letters Ibid.
272 *volume of traffic abruptly rose* Utzinger interrogation.
fifteen different messages Sommer interrogation, 27.
"Roseman, Morgenthau and Frankfurter?" Berlin to Argentina,
September 18, 20, 21, Serial CG3-1949, RG 38, CNSG Library,
box 79, 3824/3, NARA.
"The Fisher Co." Berlin to Argentina, November 21, 1943,
Serial CG3-2477, RG 38, CNSG Library, box 79, 3824/3, NARA.
273 *deposing the old regime* McGaha, "The Politics of Espionage," 269.
the Nazis called him "Godes" Argentina to Berlin, July 14,
1943, Serial CG3-1586, RG 38, CNSG Library, box 80, 3824/4,
NARA.
code name "Moreno" Argentina to Berlin, May 12, 1943, Serial
CG3-1788, RG 38, CNSG Library, box 80, 3824/4, NARA.
"the interests of the Axis powers" Argentina to Berlin, July 24,
1943, Serial CG3-1582, RG 38, CNSG Library, box 80, 3824/4,
NARA.
"ready in every respect" Argentina to Berlin, May 12, 1943,
Serial CG3-1788, RG 38, CNSG Library, box 80, 3824/4, NARA.
"USA is considered greatest enemy" Argentina to Berlin,
August 15, 1943, Serial CG3-1658, RG 38, CNSG Library, box
80, 3824/4, NARA.
274 *"V-men residing here"* Argentina to Berlin, July 24, 1943.
"completely in our camp" Argentina to Berlin, February 28,
1943, Serial CG3-858, RG 38, CNSG Library, box 80, 3824/4,
NARA.
Bolivian minister of mines Utzinger interrogation, enclosure no. 5.
Elias Belmonte Utzinger interrogation, 8.

274 "Final objective" Argentina to Berlin, August 28, 1943, Serial
      CG3-1893, RG 38, CNSG Library, box 80, 3824/4, NARA.
      a secret weapons deal McGaha, "The Politics of Espionage,"
      296–338.
275 *"An agent will depart"* Argentina to Berlin, July 14, 1943, Serial
      CG3-1608, RG 38, CNSG Library, box 80, 3824/4, NARA.
      anything quite this exciting "Interim Report on the Case of
      Osmar Alberto Hellmuth," RG 65, 64-27116, NARA.
      red mustache Ibid.
      was "easy prey" Ibid.
      an upscale neighborhood Goñi, *The Real Odessa*, xxiii.
      spoke often, in secret Utzinger interrogation, 8. In the decrypts,
      Becker and others refer to Perón by name as well as his group,
      which they called "The Colonels Lodge."
276 *he would board a ship* "Interim Report on the Case."
      "with good prospects" Ibid.
277 *"lack of imagination"* Argentina to Berlin, July 15, 1943, Serial
      CG3-1445, RG 38, CNSG Library, box 80, 3824/4, NARA.
      "such an easy time!" Argentina to Berlin, May 12, 1943, Serial
      CG3-1702, RG 38, CNSG Library, box 80, 3824/4, NARA.
      it was almost time Utzinger interrogation, enclosure no. 4.
      The Chilean's descriptions Argentina to Berlin, December 11,
      1943, Serial CG3-2746, RG 38, CNSG Library, box 81, NARA.
278 *sixty kilograms of gifts* Argentina to Berlin, October 8, 1943,
      Serial CG3-2103, RG 38, CNSG Library, box 80, 3824/4,
      NARA.
      "HELLMUTH enjoys" Argentina to Berlin, October 7, 1943,
      Serial CG3-2125, RG 38, CNSG Library, box 80, 3824/4, NARA.
      Osmar Hellmuth sailed "Interim Report on the Case."
      "Luna is probably Gustav Utzinger" Argentina to Berlin,
      June 13, 1944, Serial CG4-4991, RG 38, CNSG Library, box 79,
      3826/2, NARA.
      "bloc in South America" Argentina to Berlin, January 6,
      1944, Serial CG4-2907, RG 38, CNSG Library, box 81, 3824/4,
      NARA.
279 *in the middle of the night* "Interim Report on the Case."
280 *called Camp 020* Oliver Hoare, ed., *Camp 020: MI5 and the
      Nazi Spies* (Richmond, UK: Public Record Office, 2000).
      Colonel Robin Stephens Gilbert King, "The Monocled World
      War II Interrogator," Smithsonian.com, November 23, 2011,
      http://www.smithsonianmag.com/history/the-monocled-world
      -war-ii-interrogator-652794/.
      prisoners told credible stories Ian Cobain, "How Britain
      tortured Nazi PoWs," October 26, 2012, *Daily Mail* (UK), http://

www.dailymail.co.uk/news/article-2223831/How-Britain-tor
tured-Nazi-PoWs-The-horrifying-interrogation-methods-belie
-proud-boast-fought-clean-war.html.

280 *"I am speaking with authority"*   "Interrogation of Hellmuth at
Camp 020 by Lieut. Colonel Stephens," November 17, 1943, RG
65, box 19, 64-27116, NARA.

281 *"possessed, almost arrogant"*   Hoare, ed., *Camp 020*, 267.
"must necessarily deteriorate"   "Interim Report on the Case."
"Sargo," he told his captors   Ibid.

CHAPTER 5: THE DOLL LADY

283 *wearing out from overuse*   "History of USCG Unit #387," 215,
262.
"Enigma arrived via RED"   Ibid., 262.
"birthday surprise for LUNA"   Ibid.
a right-wing Bolivian general   Richard L. McGaha, "The
Politics of Espionage: Nazi Diplomats and Spies in Argentina,
1933–1945" (Ph.D. diss., Ohio University, 2009), 284–92.

284 *had set the coup into motion*   Ibid.; Becker's Bolivian friend,
Elias Belmonte, was the link between the Nazi/Argentine group
and the Bolivian coup plotters. See also the Berlin to Argentina
and Argentina to Berlin decrypts sent between January and April
1944 in RG 38, CNSG Library, Box 79.
the "first fruit"   Argentina to Berlin, January 17, 1944, Serial
CG4-3174, RG 38, CNSG Library, box 81, 3824/4, NARA.
"We extend hearty wishes"   Argentina to Berlin, December 28,
1943, Serial CG4-2758, RG 38, CNSG Library, box 81, 3824/4,
NARA.
had discussed its delivery   "History of USCG Unit #387," 262.
decided to double the crypto   Ibid., 263–66.

285 *to write a punch-card program*   Ibid., 270.
able to hammer Argentina   J. Lloyd Mecham, *The United States
and Inter-American Security, 1889–1960* (Austin: University of
Texas Press, 1965), 214–15.
sent a secret cable   Washington to ISK, February 19, 1944, cable
no. CXG204, HW 19, Records of the Government Code and
Cypher School: ISOS Section and ISK Section, subseries 361,
TNA.

286 *transmitted the wiring details*   Washington to ISK, February 24,
1944, cable no. CXG228, HW 19/361, TNA.
they had just solved Red themselves   ISK to Washington,
February 20, 1944, telegram no. CXG636, HW 19/361, TNA.
"We urgently need reports"   Argentina to Berlin, January 26,

1944, Serial CG4-3780, RG 38, CNSG Library, box 81, 3824/4, NARA.

286 *a new commander* Sommer interrogation, 5.
"a leak in your courier organization" Berlin to Argentina, February 21, 1944, Serial CG4-3831, RG 38, CNSG Library, box 81, 3824/3, NARA.
"go on in full revolutions" Berlin to Argentina, January 26, 1944, Serial CG4-3780, RG 38, CNSG Library, box 81, 3824/3, NARA.

287 *"Dear DARK EYE"* Berlin to Argentina, March 2, 1944, Serial CG4-3736, RG 38, CNSG Library, box 81, 3824/4, NARA.
"Here, life goes on" Berlin to Argentina, April 5, 1944, Serial CG4-4132, RG 38, CNSG Library, box 81, 3824/3, NARA.
vulnerability to chemical warfare Berlin to Argentina, February 4, 1944, Serial CG4-3785, RG 38, CNSG Library, box 81, 3824/3, NARA.
He wrote a story Sunday Express Correspondent, "Britain Smashes South America Spy Ring," *Sunday Express* (London), January 30, 1944.

288 *"Security Calypso"* C. H. Carson to Mr. Ladd, memorandum, *Subject: Osmar Alberto Hellmuth,* February 27, 1945, with attached "Security Calypso" lyrics by Young Ziegfield, RG 65, box 20, 64-27116, NARA.

289 *a peeved memo to Hoover* U.S. Department of Justice, Federal Bureau of Investigation, memorandum, *re: Osmar Alberto Helmuth* [sic], *Memorandum for the Ambassador,* D. M. Ladd to FBI Director, December 16, 1943, RG 65, box 18, 64-27116, NARA.
"The memorandum was written" Crosby to FBI Director, February 15, 1944; Crosby, "Memorandum to the Ambassador."

290 *to search their files* FBI, memorandum, *Memorandum No. 205, Series 1944, Memorandum for All Legal Attaches,* John Edgar Hoover, September 30, 1944, RG 65, box 19, 64-27116, NARA.

291 *direction-finding automobiles* George E. Sterling, "The History of the Radio Intelligence Division Before and During World War II," unpublished manuscript, PDF file, http://www.w3df.com, 91–92.
Becker assured him Utzinger interrogation, enclosure no. 4.
beneath a chicken coop Carey, "Gustav Edward Utzinger;" Rout and Bratzel, *The Shadow War.*
eardrum destroyed Argentina to Berlin, March 25, 1944, Serial CG4-3971, RG 38, CNSG Library, box 81, 3824/4, NARA.
"He fell on February 19" Argentina to Berlin, March 22, 1944, Serial CG4-3945, RG 38, CNSG Library, box 81, 3824/4, NARA.

292 *die for National Socialism* Argentina to Berlin, April 9, 1944, Serial CG4-4174, RG 38, CNSG Library, box 81, 3824/4, NARA.

292 *captured and hanged*   Sommer interrogation, 38.
sent him an ominous message   Berlin to Argentina, February 8, 1944, Serial CG4-3535.
an intriguing piece of news   Argentina to Berlin, March 22, 1944, Serial CG4-3890, RG 38, CNSG Library, box 81, 3824/4, NARA.
Velvalee Dickinson whirled around   John Jenkisson, "The FBI vs. New York Spies," *New York World Telegram*, June 22, 1945.
293 *five suspicious letters*   ESF to Edward C. Wallace (U.S. Attorney, Southern District of New York), April 1, 1944, box 7, folder 1, ESF Collection.
she had fallen into debt   Jenkisson, "The FBI vs. New York Spies."
he called the supervisor   R. A. Newby to D. M. Ladd, March 14, 1944.
294 *"According to Mr. Wallace"*   Ibid.
"performed in this connection"   Ibid.
"ADVISE AS TO SUBMISSION"   U.S. Department of Justice, Federal Bureau of Investigation, teletype, *Kin. Velvalee Dickinson*, New York to Director, March 18, 1944. Obtained under the Freedom of Information Act from FBI; received December 2015.
"Concerning the project"   U.S. Department of Justice, Federal Bureau of Investigation, memorandum, *Subject: Velvalee Dickinson*, J. Edgar Hoover to SAC, New York, March 23, 1944. Obtained under the Freedom of Information Act from FBI; received December 2015.
"My dear Mr. Wallace"   ESF to Wallace.
295 *an advantage, not a disadvantage*   Hoover to SAC, New York, March 23, 1944.
"the woman spy of this war"   George Kennedy, "The War's No. 1 Woman Spy," *West Sunday Star*, August 20, 1944, box 7, folder 2, ESF Collection.
296 *"Who are all these people?"*   "Doll Woman Enters Guilty Plea in Censor Case; Faces Ten Years," *New York Times*, July 29, 1944.
"except that it's larger"   Kennedy, "The War's No. 1 Woman Spy."
"the frustration of childlessness"   Ibid.
"one of the cleverest woman operators"   J. Edgar Hoover, "Hitler's Spying Sirens," *The American Magazine* (December 1944): 40–41, 92–94.
297 *"He is hidden"*   Argentina to Berlin, April 6, 1944, Serial CG4-4163, RG 38, CNSG Library, box 81, 3824/4, NARA.
"The enemy succeeded"   Argentina to Berlin, August 11, 1944, Serial CG4-5629, RG 38, CNSG Library, box 81, 3284/4, NARA.
arrested by the Argentine federal police   Utzinger interrogation, enclosure no. 4.
Juan Perón himself appeared   Ibid., 4.

297 *"the final chapter"*   Sommer interrogation, 32.

298 *"Technical advantages"*   Rout and Bratzel, *The Shadow War*, 454.

299 *a seven-page story*   J. Edgar Hoover, "How the Nazi Spy
Invasion Was Smashed," *The American Magazine* (September
1944): 20–21, 94–100.
   a fifteen-minute film   "Battle of the United States," *Army-Navy
Screen Magazine* 42, Steven Spielberg Film and Video Archive,
United States Holocaust Memorial Museum, https://collections
.ushmm.org/search/catalog/irn1003973; see also https://www
.youtube.com/watch?v=pdMTRjRvqGk for an uncut version.
   director Frank Capra   Mark Harris, *Five Came Back: A Story
of Hollywood and the Second World War* (New York: Penguin
Books, 2014), 233.

300 *the Friedman family Christmas card*   ESF and WFF, "B U L L E T I N
** 1944 ** F R I E D M A N," box 4, folder 6, ESF Collection.
   "Bill, Will, Billy"   Ibid.

301 *thirty thousand tons of bombs*   "1945, Summary of Air
Operations, January," in Royal Institute of International Affairs,
Chronology and Index of the Second World War, 1938–1945
(1947; repr., London: Meckler, 1975), 317.

302 *223 planes*   Randall Hansen, *Fire and Fury: The Allied Bombing
of Germany, 1942–1945* (New York: NAL Caliber, 2008), 260.
   found melted together   Ibid., 263.
   arrested in Buenos Aires   U.S. Department of Justice, Federal
Bureau of Investigation, personal and confidential memorandum
by special messenger, *Subject: Johannes Siegfried Becker, Buenos
Aires*, John Edgar Hoover to Frederick B. Lyon (chief, Division
of Foreign Activity Correlation, Department of State), April 21,
1945, RG 65, box 20, 64-27116, NARA.
   confiscated an address book   "Summary of Traces, BECKER
Siegfried."
   He gave statements   Utzinger interrogation, 14.
   "delicacies and champagne"   Ibid., 21.

303 *hardly enjoyed Becker's privileges*   Ibid., 12–14.
   He no longer feared   Ibid., enclosure no. 4.
   war criminals into Argentina   Uki Goñi, *The Real Odessa:
Smuggling the Nazis to Perón's Argentina* (London: Granta,
2003), xxiii, 107.
   an angry mob invaded   United Press, "Rebels Slay President,
Seize Power in Bolivia," *Washington Post*, July 22, 1946.
   William got bronchitis   ESF to Barbara Friedman, February 9,
1945, box 3, folder 26, ESF Collection.
   extra-long time getting dressed   ESF to Barbara Friedman,
February 22, 1945, box 3, folder 25, ESF Collection.

304 *"evil influences"*   ESF to Barbara Friedman, April 12, 1945, box 3, folder 26, ESF Collection.
a family friend died   WFF to Mrs. A. J. McGrail, November 29, 1945, attached to a "Harvard Honor Roll" sheet filled out by WFF and listing the accomplishments of Colonel A. John McGrail, NSA.
held her arm   ESF to Barbara Friedman, May 3, 1945, box 3, folder 27, ESF Collection.
"where I am, when I am"   Ibid.
"Remember, darling"   Ibid., May 11, 1945.
305 *"We have difficulty believing"*   Ibid., May 9, 1945.
"It's absolutely terrifying"   Ibid.
slept with the windows open   Ibid., June 18, 1945.
sent her a poem   Ibid., June 4, 1945.
a ninety-day assignment   Ibid., July 4, 1945.
"the last great secret"   Randy Rezabek, "TICOM: The Last Great Secret of World War II," *Intelligence and National Security* 27, no. 4 (2012): 513–30, http://dx.doi.org/10.1080/02684527.2012.688305.
306 *the military air terminal*   ESF to Barbara and John Ramsay Friedman, July 14, 1945, typed letter to her children, box 3, folder 27, ESF Collection.
She thought he looked handsome   Ibid.
waited until the C-54 took off   Ibid.

CHAPTER 6: HITLER'S LAIR

307 *rumbled up the twisting incline*   WFF diary of touring postwar Europe, dictated July 26, 1945, signed September 2, 1945, thirteen-page typescript in large blue binder, Marshall Foundation, 3–4.
a castle on a mountaintop   Ibid.
acoustics research   Army Security Agency, "European Axis Signal Intelligence in World War II as Revealed by TICOM Investigations and by Other Prisoner of War Interrogations and Captured Material, Principally German," May 1, 1946, NSA, 37–44.
The Allies wanted to know   Ibid.
308 *character in a murder mystery*   WFF diary of postwar Europe, 4.
Vierling's prototypes   "European Axis Signal Intelligence."
a brief supper of hot dogs   Ibid.
growing spookier   Ibid.
six TICOM teams   Randy Rezabek, "The Teams," TICOM Archive, http://www.ticomarchive.com/the-teams.
"a heavy feeling of sadness"   WFF diary of postwar Europe, 1.
309 *ate a C-ration*   Ibid., 4.
like LEPIDOPTERA   Ibid., 2.

309 *"The destruction to be seen"* WFF diary of postwar Europe, 8.
his own cryptologic publications TICOM discovered a copy
of his classic paper "The Index of Coincidence" that had been
translated into German from French. See Rose Mary Sheldon,
"The Friedman Collection: An Analytical Guide," rev. October
2013, Marshall Foundation, PDF file, 90.
for his own library Item 167.3, WFF Collection.
a grim duty WFF to ESF, October 6, 1917.

310 *"Zionism is only one"* WFF to Barbara Friedman, March 15,
1945, box 4, folder 8, ESF Collection.
did not sleep well WFF diary of postwar Europe, 5.
rode in the Army staff jeep Ibid., 5–7.
"He gave us a little speech" Ibid., 5.

311 *loved to entertain friends* Heike Görtemaker, *Eva Braun: Life
With Hitler* (New York: Alfred A. Knopf, 2011), 216.
"I think it is too bad" WFF diary of postwar Europe, 7.
"I shall have it made" Ibid., 6. The paperweight isn't part of
William's collection at the Marshall Foundation; no one seems to
know what happened to it.
this bowl of smoking ice ESF to WFF, July 26, 1945, box 2,
folder 8, ESF Collection.

312 *"At 1535 a visit with Dr. Turing"* WFF spiral-bound diary of
his 1945 England trip, box 13, folder 13, ESF Collection, 21.
stripped Turing's security clearance Andrew Hodges, *Alan
Turing: The Enigma* (London: Vintage, 2014), 574–664.
an apparent suicide Ibid., 614–15.
country village of Beaconsfield WFF diary of 1945 England trip, 16.
three high-value German POWs WFF to commanding general,
Army Security Agency, "Report on Temporary Duty, ETO,"
October 1, 1945, NSA.
unbreakable all the way Ibid.

313 *a burlesque show* WFF diary of 1945 England trip, 8.
asked him to tell stories Ibid., 36.
asleep and dreaming Ibid., 27–28.
"nearly as big as Dallas" Peter J. Kuznick, "Defending the
Indefensible: A Meditation on the Life of Enola Gay Pilot Paul
Tibbets Jr., *The Asia-Pacific Journal* 6, no. 1 (2008), http://apjjf
.org/-Peter-J.-Kuznick/2642/article.html.

314 *a sex dream about Enid* WFF diary of 1945 England trip, 28.
martinis in befuddled silence Ibid., 32–33.
"renewed call to surrender" Ibid.
listen to radio bulletins ESF to WFF, August 12, 1945, box 2,
folder 8, ESF Collection.

315 *"end the war P.D.Q."* Ibid., August 7, 1945.

315 *watch some tennis* WFF diary of 1945 England trip, 37–38.
"A day we will remember!" WFF to ESF, August 10, 1945, box 3, folder 9, ESF Collection.
"war will be a fact" Ibid., August 14, 1945.
praising his son's vocabulary WFF to John Ramsay Friedman, August 13, 1945, box 4, folder 8, ESF Collection.

316 *Elizebeth stayed in* ESF to WFF, August 15, 1945, box 2, folder 8, ESF Collection.
"Bobbie, darling" Ibid.
"The O.S.S is starting" Ibid., September 4, 1945.
Elizebeth heard a radio interview Ibid., August 16, 1945.

317 *a cold Sunday* Ibid., undated (late August 1945).
came over with his wife Ibid.
"I find it hard to tell you" WFF to ESF, August 19, 1945, box 3, folder 9, ESF Collection.
She realized how tricky ESF to WFF, August 26, 1945, box 2, folder 8, ESF Collection.
"Dearest" Ibid.
"very soon" WFF to ESF, August 29, 1945, box 3, folder 9, ESF Collection.
"I LOVE YOU!" Ibid.
General Douglas MacArthur Douglas MacArthur, "General MacArthur's Radio Address to the American People," September 2, 1945, https://ussmissouri.org/learn-the-history /surrender/general-macarthurs-radio-address.

318 *Elizebeth heard Truman say* ESF to WFF, September 4, 1945, box 2, folder 8, ESF Collection.
"It is our responsibility" Harry S. Truman, "Radio Address to the American People After the Signing of the Terms of Unconditional Surrender by Japan," September 1, 1945, http:// www.presidency.ucsb.edu/ws/?pid=12366.
no point in having fought a war ESF to Barbara Friedman, April 12, 1945, box 3, folder 26. She wrote that now Americans needed to fight "for a truly international post war world."
She wondered if William heard ESF to WFF, September 4, 1945, box 2, folder 8, ESF Collection.
"You are the dearest and best" Ibid.
Around September 12 WFF, "Report on Temporary Duty."
Elizebeth opened the door As best I can tell, neither Friedman ever wrote about this exact moment of reunion. I admit I'm inferring it from what WFF writes in his letters about the schedule of his trip home and the timing of his arrival.

319 *sorted through the voluminous files* ESF, "foreword to uncompleted work."

319 *a detailed technical account*   "History of USCG Unit #387."
    historians of codebreaking   Ibid., "Foreword."
    Five copies were printed   Ibid. Two copies went to the navy's
    OP-20-G, one to the Coast Guard brass, one to the Army, and
    one to British intelligence.
    destroy the rest   ESF, "foreword to uncompleted work."
    "government tombs"   Ibid.
    prepared to leave   Ibid.
    The navy forced her   Ibid.
    At the end of her final workday   ESF, "foreword to uncompleted
    work."

320 *"to that particular form"*   Ibid.
    "thrilling records"   Ibid.
    On August 14, 1946   H. L. Morgan (Acting Chief, Civilian
    Personnel Division, USCG) to ESF, August 14, 1946, box 6,
    folder 8, ESF Collection. This folder also contains the envelope in
    which the Reduction in Force letter arrived, and on the front and
    back of the envelope, ESF wrote a note explaining that it was her
    idea to eliminate her own job.

321 *show by the Amazing Dunninger*   ESF to Barbara Friedman,
    December 3, 1944, box 3, folder 25, ESF Collection.
    both debunker and illusionist   Wikipedia, s.v. "Joseph Dunninger,"
    last modified March 9, 2014, http://www.geniimagazine.com
    /magicpedia/Joseph_Dunninger.
    "came away with theories"   ESF to Barbara, December 3, 1944.

322 *alive and kicking*   WFF and ESF, *The Shakespearean Ciphers
    Examined* (London: Cambridge University Press, 1958), 9.
    asked the Friedmans   Ibid., 161–63.
    the following message   Ibid.

323 *"IN HER DAMP PUBES"*   Ibid., 258. They were renowned for
    their scientific insights and their serious feats of codebreaking,
    but this is one of those passages that shows how the Friedmans
    were also very funny, in a delicate and savage and wonderfully
    idiosyncratic way.
    "great natural gifts"   Ibid., 205.
    "a sincere and honourable woman"   Ibid., 264.
    "found in her texts"   Ibid.
    "was therefore at the mercy"   Ibid., 265.
    "whose work on the question"   Ibid., ix.
    they thanked Fabyan, too   Ibid.
    "give the devil his due"   ESF to Mrs. Percival White, March 28,
    1958, box 1, folder 23, ESF Collection.
    "Vile creature"   ESF interview with Valaki, transcribed
    February 16, 2012, 5.

323 *an accident*   ESF, "Pure Accident."

324 *She tries to imagine herself*   ESF must have done this when she wrote the book, because there's a passage by her describing "the gradual crystallization" of her opinions about the Bacon Cipher project in 1916 and 1917. The rest of the book is written in the first-person plural "we" but this passage uses the singular "I." WFF and ESF, *Shakespearean Ciphers*, 211.

EPILOGUE

327 *otherwise ordinary Tuesday*   Transcript of Donald F. Coffey oral history interview with NSA, November 4, 1982.
Scattered clouds   Weather Underground, "Weather History for KDCA, December 30, 1958," https://www.wunderground.com /history/airport/KDCA/1958/12/30/DailyHistory.html.
at least three men   Possibly four men. S. Wesley Reynolds, NSA director of security, writes that he visited the Friedman home with an NSA man named Cook and a third man from the attorney general's office. Coffey would make four. See S. Wesley Reynolds memo, January 2, 1959, RG457, Entry UD-15D19, "Reclassification of Friedman Articles," box 57.
a rented truck   Coffey oral history.

328 *a Defense Department order*   Garrison B. Coverdale to William G. Bryan (undated), RG457, Entry UD-15D19, box 57; see also Rose Mary Sheldon, "The Friedman Collection: An Analytical Guide," rev. October 2013, Marshall Foundation, PDF file, 5.
forty-eight items   Ronald Clark, *The Man Who Broke Purple: Life of Colonel William F. Friedman, Who Deciphered the Japanese Code in World War II* (Boston: Little, Brown, 1977), 252; see also "Inventory of the material taken from Friedman's house," RG457, Entry UD-15D19, box 57.
"went berserk"   Coffey oral history.
denied this   Ibid.
wrote in a memo   S. Wesley Reynolds memo.
The ciphers were obsolete   Sheldon, "Analytical Guide," 7.
"The NSA took away from me"   Clark, *The Man Who Broke Purple*, 252.

329 *silent rage*   Coffey oral history.
"The mad march of red fascism"   J. Edgar Hoover, "Speech Before the House Committee on Un-American Activities," speech, March 26, 1947, http://voicesofdemocracy.umd.edu /hoover-speech-before-the-house-committee-speech-text/.
"psychic giddiness"   Zigmond Lebensohn to Ronald Clark, May 10, 1976, box 1, folder 38, ESF Collection.

329 *unable to work or solve puzzles* Ibid.
a rope and a noose John Ramsay Friedman to Ronald Clark (undated), box 14, folder 14, ESF Collection.
"a tree to hang myself" Sheldon, "A Very Private Cryptographer," 15.
proponent of electroshock therapy Lebensohn, "Electroconvulsive Therapy . . . A Personal Memoir."
The first course of shocks Lebensohn to Clark, May 10, 1976.

330 *"was almost elated"* Ibid.
"Anxiety kept her figure slim" Murial Pollitt to ESF, October 6, 1981, box 12, folder 15, ESF Collection.
to get the pen moving Ibid.
"I found it an outlet" ESF to Anne [?], October 24, 1951, box 1, folder 17, ESF Collection.
stayed active Ibid.
"That part of my life is over" Ibid., October 13, 1951.
suitcase full of lantern slides Ibid.
at least fifteen mutilated sheets ESF speech to Mary Bartelme Club. Her draft of the speech begins with an unnumbered page called "Introduction" followed by fourteen more pages. The last page ends in midsentence; the concluding pages appear to have been lost or destroyed.
agonized about what to say ESF to Anne [?], November 8, 1951, box 1, folder 17, ESF Collection.

331 *pink ballroom at the Blackstone Hotel* Irene Powers, "Benefit Fetes Aglitter with Holiday Spirit," *Chicago Tribune*, November 18, 1951.
wasn't free to talk ESF speech to Mary Bartelme Club, "Introduction."
"Perhaps you may think" Ibid., 1.
showed slides of code messages Ibid., 1–7.
two and a half hours ESF to Anne [?], December 26, 1951, box 1, folder 17, ESF Collection.
a luncheon at Cambridge ESF, "foreword to uncompleted work."
"As befits a woman" Ibid.
sheet of lined yellow paper ESF, "Notes for 'Foreword, 1959," box 17, folder 20, ESF Collection. This is the seven-page handwritten draft of the "foreword to uncompleted work" typescript.
"FOREWORD" ESF, "foreword to uncompleted work."

332 *President Truman established* James Bamford, *The Puzzle Palace: A Report on NSA, America's Most Secret Agency* (New York: Houghton Mifflin, 1982).

332 *the most secret of agencies* Ibid.
"mostly nonsensical" WFF to Roberta Wohlstetter,
September 17, 1969, box 14, folder 12, ESF Collection.
"secrecy virus" Clark, *The Man Who Broke Purple,* 252.

333 *a nice ceremony* "Ceremony Honoring William F. Friedman,"
Arlington Hall Post Theatre, October 12, 1955, box 14, folder 12,
ESF Collection.
feared the NSA WFF undated letter, box 14, file 12, ESF
Collection.
"Frightening to be alone" Clark, *The Man Who Broke Purple,*
258–59.
"a great desire to live" WFF to Wohlstetter, September 17, 1969.
His feet swelled so much ESF to John Ramsay Friedman,
March 20, 1967, box 4, folder 2, ESF Collection.

334 *taking notes on his condition* ESF daybook, 1969, box 20, ESF
Collection.
The doctor stayed at the house Ibid.
"My beloved died at 12:15" Ibid.
"Dear heart be courageous" ESF Collection.
"the greatest brain of the century" Joe Mauborgne to ESF,
telegram, box 14, folder 1, ESF Collection.
"His effect on world history" Herman Wouk to ESF,
November 3, 1969, box 14, folder 1, ESF Collection.
"Our business now involves" Juanita Morris Moody to ESF,
November 7, 1969, box 14, folder 2, ESF Collection.

335 *"Woman's Privilege Card"* Cosmos Club "Woman's Privilege
Card," November 14, 1969, box 17, folder 24, ESF Collection.
She designed his tombstone ESF sketch of WFF's tombstone,
box 13, folder 31, ESF Collection.
specified that certain letters Ibid. The *a-* and *b-*forms are clear
on ESF's sketch, and she also included a more detailed tracing of
that specific line of text. Later, ESF explicitly told Ronald Clark
that "WFF" is the cipher message; see ESF to Clark, October 7,
1976, box 15, folder 4, ESF Collection. See also Elonka Dunin,
"Cipher on the Elizebeth and William Friedman tombstone at
Arlington National Cemetery Is Solved," http://elonka.com
/friedman/index.html.
an emotional thank-you letter John Ramsay Friedman to
Eugene McCarthy, November 12, 1969, box 293, Eugene J.
McCarthy Papers, Minnesota Historical Society.

336 *Immediately after his funeral* ESF letter to family and friends,
January 28, 1970, box 13, folder 31, ESF Collection.
a sense of duty Ibid.
entice a first-rate historian Ibid.

336 *paid for a typist*   Ibid.

six-hour days   ESF to Ronald Clark, June 12, 1974, box 15, folder 1, ESF Collection.

"entertained like a queen"   Ibid.

got her on tape   ESF interview with Marshall staff.

337 *had not indexed*   The only guide to the thousands of documents in the ESF Collection is an eighteen-page "Container List" that lists the names of the folders in the twenty-two boxes but does not describe their contents.

asked her to inspect   ESF to David Kahn, undated two-page letter on carbons, box 15, folder 2, ESF Collection.

The NSA men's chorus   "Dedication Ceremony for the William F. Friedman Memorial Auditorium," program, May 21, 1970, box 14, folder 12, ESF Collection.

338 *a competent account*   ESF to Marshall Foundation, July 14, 1977, box 15, folder 4, ESF Collection.

her savings dried up   ESF to Stuart and Mabel, April 30, 1974, box 15, folder 1, ESF Collection.

"There is just one thing"   ESF typewritten diary, February 7, 1967, box 3, folder 20, ESF Collection.

She gave an interview   Connie Lunnen, "She Has a Secret Side," *Houston Chronicle,* May 24, 1972.

"In a few years"   WFF and ESF burial wishes, box 16, folder 23, ESF Collection.

arteries failed   Maureen Joyce, "Elizabeth Friedman, U.S. Cryptanalyst, Pioneer in Science of Code-Breaking Dies," *Washington Post,* November 2, 1980.

Washington Post   Ibid.

New York Times   Alfred E. Clark, "E.S. Friedman, 88, Cryptanalyst Who Broke Enemy Codes, Dies," *New York Times,* November 3, 1980.

her ashes were scattered   John Ramsay Friedman eulogy at Arlington National Cemetery, November 1980, box 6, folder 26, ESF Collection.

339 *"beacon of hope"*   Barbara Osteika (ATF historian, Department of Justice), in discussion with the author, April 2015.

An FBI cryptanalyst   Jeanne Anderson (FBI cryptanalyst, Cryptanalysis and Racketeering Records Unit), in discussion with the author, via e-mail, September 2015.

briefed U.S. leaders   "Juanita Moody," NSA Center for Cryptologic Heritage, Hall of Honor, https://www.nsa.gov /about/cryptologic-heritage/historical-figures-publications /hall-of-honor/2003/jmoody.shtml.

Ann Caracristi   "Ann Caracristi," NSA Center for Cryptologic

Heritage, Hall of Honor, https://www.nsa.gov/about/crypto
logic-heritage/historical-figures-publications/women/honorees
/caracristi.shtml.

339 *brief, verifiably true comments*   Sheldon, "Analytical Guide."
See the entries for Items 658, 1006, and 1006.1.

340 *"There are plenty of mysteries"*   ESF interview with Valaki,
transcribed February 21, 2012, 8.
first joined the agency   Valaki obituary.
"Well, thanks again, Mrs. Friedman"   ESF interview with
Valaki, transcribed January 12, 2012, 5.

341 *"Girl cryptanalyst and all that"*   Ibid.
Valaki shut off the recorder   Ibid., 6.
"You mean to say"   ESF interview with Valaki, transcribed
February 21, 2012, 8.
"I'll bet no two women"   ESF interview with Valaki, transcribed
January 10, 2012, 8.
the women laughed   Ibid. The transcript reads, "((Both laugh.))"

INDEX